# Novel Perspectives in Forage Crops

# Novel Perspectives in Forage Crops

Edited by Octavia Barker

STATES
ACADEMIC PRESS
www.statesacademicpress.com

**States Academic Press,**
109 South 5th Street,
Brooklyn, NY 11249, USA

Visit us on the World Wide Web at:
www.statesacademicpress.com



ISBN: 978-1-63989-750-6



**Cataloging-in-Publication Data**

Novel perspectives in forage crops / edited by Octavia Barker.
p. cm.
Includes bibliographical references and index.
ISBN 978-1-63989-750-6
1. Forage plants. 2. Field crops. 3. Agronomy. 4. Crops. I. Barker, Octavia.
SB193 .N68 2023
633.2--dc23

# Table of Contents

# Preface

This book was inspired by the evolution of our times; to answer the curiosity of inquisitive minds. Many developments have occurred across the globe in the recent past which has transformed the progress in the field.

Forage refers to the edible parts of the plant such as stems and leaves. Grasses, herbaceous legumes, and tree legumes are examples of common forages. These are primarily consumed by grazing animals such as goats, horses, llamas, cows and sheep. Forage crops are crops that are cultivated for use as feeds for animals after their grain has been separated. Mostly, forage crops are grown in fragile habitats having poor fertility and various restrictions. Forage plants support both domesticated and wild herbivores. These plants comprise varied amounts of proteins, fiber, minerals, and lignin. This book explores all the novel perspectives related to forage crops in the present day scenario. It will also provide interesting topics for research, which interested readers can take up. The book will serve as a valuable source of reference for graduate and postgraduate students.

This book was developed from a mere concept to drafts to chapters and finally compiled together as a complete text to benefit the readers across all nations. To ensure the quality of the content we instilled two significant steps in our procedure. The first was to appoint an editorial team that would verify the data and statistics provided in the book and also select the most appropriate and valuable contributions from the plentiful contributions we received from authors worldwide. The next step was to appoint an expert of the topic as the Editor-in-Chief, who would head the project and finally make the necessary amendments and modifications to make the text reader-friendly. I was then commissioned to examine all the material to present the topics in the most comprehensible and productive format.

I would like to take this opportunity to thank all the contributing authors who were supportive enough to contribute their time and knowledge to this project. I also wish to convey my regards to my family who have been extremely supportive during the entire project.

**Editor**

# 1

# Water-Use Efficiency of Forage Crops in the Southeastern United States

**Christine H. Gelley [1], Amanda J. Ashworth [2], Patrick D. Keyser [3], Renata L. G. Nave [4,*] and Justin D. Rhinehart [5]**

1 Agriculture & Natural Resources Educator, Ohio State University Extension, Caldwell, OH 43724, USA; gelley.2@osu.edu
2 Poultry Production and Product Safety Research Unit, USDA-ARS, Fayetteville, AR 72701, USA; amanda.ashworth@usda.gov
3 Department of Forestry, Wildlife and Fisheries, University of Tennessee, Knoxville, TN 37996, USA; pkeyser@utk.edu
4 Department of Plant Sciences, University of Tennessee, Knoxville, TN 37996, USA
5 Department of Animal Science, University of Tennessee, Knoxville, TN 37996, USA; jrhinehart@utk.edu
* Correspondence: rnave@utk.edu

**Abstract:** Preparing agricultural producers to cope with volatile weather changes, specifically drought, requires a better understanding of forage water-use efficiency (WUE) potentials. Options to improve farm resiliency to drought may include the use of $C_4$ annual and perennial forages, which have greater production efficiency during drought than commonly used $C_3$ forages. Our objective was to measure WUE through real-time gas exchange measurements of photosynthesis and transpiration in (1) a greenhouse study and (2) under field-grazing conditions. Growth parameters, instantaneous water use efficiency (iWUE), and mass-based WUE (mWUE) data were collected under greenhouse conditions in Study 1 for the following species: crabgrass (*Digitaria sanguinalis* cv. 'Red River'), switchgrass (*Panicum virgatum* cv. 'Alamo'), big bluestem (*Andropogon gerardii* cv. 'OZ-70'), indiangrass (*Sorghastum nutans* cv. 'Rumsey'), eastern gamagrass (*Tripsacum dactyloides* cv. 'Pete'), bermudagrass (*Cynodon dactylon* cv. 'Vaughn's #1'), sorghum-sudangrass (*Sorghum bicolor* (L.) × *Sorghum sudanese* (P.) cv. 'Greengrazer'), and tall fescue (*Schedonorus arundinaceus (Schreb.) Dumort*). Study 2 occurred from 2014 to 2016, and evaluated iWUE of crabgrass, switchgrass, bermudagrass, eastern gamagrass, and a big bluestem/indiangrass mix under field conditions. Overall, in situ iWUE of crabgrass, switchgrass, eastern gamagrass, and bermudagrass did not differ, while iWUE of the big bluestem/indiangrass was less than switchgrass and crabgrass, an advantage for these species if the standardized precipitation index drops below zero. Bermudagrass, switchgrass, sorghum-sudangrass, pearl millet, and indiangrass had comparable mWUE values under greenhouse-simulated drought. These results will aid in the development of forage species recommendations for mitigating drought and improving resiliency.

**Keywords:** warm-season grasses; tall fescue; water-use efficiency; native grasses; drought tolerance

## 1. Introduction

The southeastern United States has been subjected to multiple severe droughts during the past decade and climatologists expect this trend to continue [1]. Drought conditions can drastically reduce availability of quality forage. In 2007, the southeast experienced an intense drought, forcing many beef producers to downsize or eliminate herds in response to forage deficits [2]. Increasing the use of warm-season forages in the southeastern region, which is primarily composed of cool-season forages, may minimize impacts of future droughts and assist in maintaining beef production, which is the United States' largest agricultural sector [2]. Improved warm-season forage production could better prepare producers for stochastic weather events, thereby minimizing profit loss under drought conditions. This is particularly true for the animal industry across the southeastern United States, where the predominant forage grass is a $C_3$ species—tall fescue. However, it remains unclear which warm-season forages have the greatest water-use efficiency (WUE) under water-limited and grazed conditions.

There are many grass species, especially $C_4$ grasses, that may provide compensation for forage loss in traditional grazing systems under drought [3–7]. Introduced annual species, such as sorghum-sudangrass, pearl millet (*Pennisetum glaucum*), and crabgrass, can maintain a high stocking density, and produce forage of equivalent nutritive value to many commonly used cool-season grasses in the southeast, such as tall fescue [8]. Longevity of the stand may be a limitation for utilization of warm-season annuals, as biomass accumulation is fast, but short-lived, and stands require reseeding and fertilizer on a regular basis [8]. When compared to native warm-season perennial grasses, root mass is less extensive and may limit access to water deeper in the soil profile during drought. Native warm-season perennial grasses, such as switchgrass, big bluestem, indiangrass, and eastern gamagrass, may also be suitable candidates for maintaining production during droughts [3–7]. These species are well-adapted to the region's climate, have few pest and disease problems, and have deep root systems [8]. In addition to large root masses, these species produce substantial aboveground herbage mass (HM), which, when managed for grazing or hay production, can provide enough energy to support animal growth and development, as they have yields greater than commonly used $C_3$ grasses in the region and require fewer inputs for production [3,7,9].

Drought tolerance is a complex combination of physiological responses during times of inadequate soil moisture that preserve plant growth. Mass-based water use efficiency (mWUE), a ratio of water input to dry matter output, has been successfully used as an indicator of plant tolerance to drought [3,10–15]. However, measurements of whole stand mWUE are challenging in field studies due to variations in soil type, topography, stand densities, and soil water content. Instantaneous measurements of WUE (iWUE) can be made based on gas exchange rates in real-time using photosynthesis and transpiration rates directly from the leaf, while creating a controlled micro-environment [16]. Utilizing measurements of iWUE under ambient conditions in the field may be additionally helpful for screening drought-tolerant forages.

Previous WUE studies have used iWUE to make inferences about the drought tolerance of forage crops, but few have examined the species included in this experiment and most have been conducted in a greenhouse setting. Beale et al. [17] compared iWUE of two perennial $C_4$ grasses (miscanthus (*Miscanthus* × *giganteus*) and big cordgrass (*Spartina cynosuroides*)) in a temperate climate, however, the results were inconclusive. A greenhouse experiment by Taylor et al. [13] studied many ecophysiological traits of $C_3$ and $C_4$ grasses including iWUE, which confirmed that $C_4$ grasses have greater WUEs, photosynthetic rates, and can use nitrogen more efficiently than $C_3$ grasses. Anyia and Herzog [12] conducted a greenhouse experiment that studied cowpea (*Vigna unguiculata*) under drought and showed that there were differences in WUE among varieties of cowpea, and that iWUE and mWUE are not necessarily correlated. Ghannoum et al. [11] conducted a greenhouse experiment using mWUE to measure drought resistance of Australian $C_4$ grasses and found that drought enhanced mWUE. Jongen et al. [15] used iWUE and mWUE to investigate drought responses of mixed grasslands, as did Hussain et al. [14] but using only mWUE values. Both observed that WUE was enhanced when evapotranspiration rates increased.

These studies document that iWUE is worthy of further investigation in regard to forage crop management. By monitoring and using transpiration and photosynthesis to quantify iWUE, physiological differences and optimum pasture species for climate resiliency may be discovered for the southeastern United States.

Drought conditions with high temperatures and low rainfall are increasingly prevalent in this region during summer, thus drastically influencing forage availability. Decreased forage reduces livestock-carrying capacities and requires supplemental feeding of purchased or conserved feedstuffs to maintain herd health. Therefore, the objective of this experiment was to quantify WUE (mWUE and iWUE) in order to evaluate forage productive potential of commonly recommended pasture crops for enhanced drought resilience.

## 2. Materials and Methods

In this experiment, we tested WUE responses to various precipitation, temperature, light, and soil conditions in both a greenhouse and a field study via mWUE and iWUE.

The greenhouse study served as a preliminary test of potential species for the field experiment, necessary parameters, and responses to imposed drought. Due to a limited number of paddocks, only one annual warm-season grass could be selected for the field experiment. The field study implemented knowledge gained from the greenhouse study into a pasture system that was actively grazed by cattle and measured under ambient conditions.

It was hypothesized that WUE of native warm-season perennial species would be greatest, as would iWUEs of these grasses when compared to annuals due to shallower rooting depths. It was also predicted that WUE fluctuations would coincide with changes in soil moisture.

### *2.1. Greenhouse Experiment*

#### 2.1.1. Site Description

This study was divided into two experiments, one under controlled conditions and the other in field settings. The greenhouse study took place in a University of Tennessee greenhouse in Knoxville, TN, in a randomized complete block design and four replications. Five perennial warm-season grasses (switchgrass, big bluestem, indiangrass, eastern gamagrass, and bermudagrass), three annual warm-season grasses (crabgrass, sorghum-sudangrass, and pearl millet), and one cool-season perennial grass (tall fescue) were evaluated from 29 October 2013 to 19 February 2014. Tall fescue was the only $C_3$ grass included to this study since it is the most widely grown forage in the region.

#### 2.1.2. Measurements

In preparation for the greenhouse experiment, seedlings of each species were grown in plug trays of potting soil and then transplanted on 29 October 2013 into 10 × 10 × 23 cm pots (1.6 L). Etowah silt loam (fine-loamy, siliceous, semiactive, thermic Typic Paleudults), a moderately fine, well-drained soil formed in alluvium or colluvium that is commonly underlain by limestone residuum [18], was collected at the East Tennessee AgResearch and Education Center's Plant Science Unit (35°54′ N 83°57′ W), autoclaved, and sifted for use. Pots were filled with soil, each to a consistent weight of 4.80 kg, with no supplementary nutrients applied. Upon transplanting, root material was washed and removed from trays. Subsequently, aboveground and belowground growth was trimmed to 14 cm (from the plant base to leaf tip and plant base to root tip) for all species, and total mass recorded thereafter. Seedlings of all species were in early vegetative stages with one tiller.

Immediately following transplanting, pots were watered until the soil was fully saturated and were left undisturbed over night to reach stable water-holding capacity (WHC), and then weighed. Weights served as a guide to determine appropriate saturation levels throughout the remainder of the experiment [19]. Four plants per species were randomly assigned to stressed (dry) and well-watered (wet) treatments. Stressed and well-watered treatments were dictated by pot WHC, which was equivalent to the maximum amount of water pots (averaged over all experimental units) could hold over 24 hours post soil saturation [19].

An additional set of plants (1 plant per species per treatment) was prepared in the same manner and observed separately under water stress conditions (defined as the WHC percentage at which plants were fully wilted, and when gas exchange measurements could no longer be taken from leaves). These observations were used to determine the appropriate percentage of WHC needed to induce drought stress and to determine photosynthetic cessation at the end of the experiment. Visual signs of drought stress included wilting, chlorosis, and reduced growth.

A random sample of each experimental unit (i.e., single pot with one plant) per species per treatment was weighed daily to determine water requirements for maintaining the prescribed percentage of WHC. Both wet and dry treatments were maintained at 90% WHC from 30 October through 22 November to monitor transplants and observe rate of tissue moisture loss (kg day$^{-1}$). From 23 through 30 November, dry treatments were maintained at 60% WHC and wet treatments were maintained at 90% WHC. The appropriate amount of water (mm) was applied to reach the desired capacity. Expecting that the annual forages would consume water more quickly than the perennial forages, weights from perennials were pooled and annuals were pooled together to determine the appropriate amount of water needed for each group at each watering.

To simulate the onset of mild drought conditions, a sub-lethal dry down was implemented from 1 through 7 December to induce drought stress. During this period, dry treatments dropped from 60% to 40% WHC and wet treatments from 90% to 70% WHC. Following the sub-lethal dry-down period, dry and wet regimes returned to 60% and 90% respectively, from 8 December to 16 December to mimic recovery from mild drought. A lethal dry-down period was implemented beginning on 17 December, at which time watering was terminated for the remainder of the experiment, thus simulating extreme drought conditions for the remainder of the study period. Observations of plant tissue health, leaf turgidity, and iWUE were taken every three days and the date of photosynthesis cessation was recorded for each pot, with the last observation taken on 19 February 2014. Date of photosynthesis cessation was determined based on the ratio of necrotic leaves to healthy green leaves, which was determined based on the point at which the gas exchange data could no longer be collected from the leaves. Once the ratio reached 4:1, it was assumed that the plants were no longer photosynthetically active. Following plant death, herbage and root dry matter were recorded.

Gas exchange measurements were taken on four dates during the greenhouse study with the LI-6400 XT Portable Photosynthesis System equipped with the 6400-02B LED Light Source (Lincoln, NE, USA) [16]. On each sampling date, one gas exchange data point was taken per plant on each of the plants in the experiment. The mean of four replicates per treatment was used to represent the iWUE of each species per sampling date.

Throughout data collection, $CO_2$ concentration in the leaf chamber was kept constant at 400 μmol $CO_2$ mol air$^{-1}$. Within the chamber, leaf temperature and photosynthetically active radiation (PAR) were adjusted to match ambient levels in the greenhouse on the day of sampling. Before each data point was recorded, a stabilization period of one to five minutes was implemented to reach consistently steady photosynthesis and iWUE rate values within one μmol CO2 mol H2O$^{-1}$. As dictated by the greenhouse winter management plan, greenhouse temperatures were set to heat when temperatures decreased below 18 °C and to cool when temperatures increased above 29 °C. Average daily temperature in the greenhouse was 21 °C. Leaf temperature ranged from 24.0 to 33.0 °C and PAR from the artificial light and filtered natural light in the greenhouse ranged from 190.0 to 260.0 μmol m$^{-2}$ s$^{-1}$. The newest emerged, fully expanded, photosynthetically active leaf of each plant were used to record photosynthesis rate, transpiration rate, and leaf area [20]. All data points were collected between 10:00 and 15:00 hours. Using these data points, an iWUE value was calculated on each experimental unit. Dates of data collections were as follows: 22 November 2013, during the acclimation period (post-transplant, pre-treatment), 26 November 2013, during the wet and dry regime implementation, 3 December 2013, during sub-lethal dry down, and 16 December 2013, following rewetting (post sub-lethal, pre-lethal).

After all watering was terminated, plants were monitored every three days with date of photosynthesis cessation recorded to determine the length of survival in days post-watering (DPW). At the end of the experiment, aboveground and belowground plant material was collected to determine mWUE in total dry matter produced per liter of water applied for each plant (g DM kg$^{-1}$ $H_2O$). The average of the four replications per treatment and sampling date was used to represent the iWUE of each species ($n$ = 32).

#### 2.1.3. Statistical Analysis

Differences among least squares means were evaluated using the PROC MIXED procedures [21]. Dependent variables were iWUE, mWUE, and survival in DPW. Each dependent variable was analyzed separately. Fixed effects were species, water treatment, and lifecycle (annual vs. perennial), and replication was a random effect. Pearson correlation coefficients (PROC CORR) were used in SAS [21] to test relationships between iWUE, temperature, humidity, and PAR during leaf gas exchange measurements. PROC MIXED procedures were used to run analysis of covariance in SAS [21] for mass as a covariate with iWUE. All results were evaluated for significance at $p < 0.05$ and checked for normality.

### *2.2. Field Experiment*

#### 2.2.1. Site Description

The second experiment was conducted at Highland Rim AgResearch and Education Center (HRREC) in Springfield, TN (36°28′ N, 86°49′ W, 203 m elevation), and at Ames Plantation AgResearch and Education Center (APREC) in Grand Junction, TN (35°8′ N, 89°12′ W, 174 m elevation), during growing seasons of 2014–2016. Treatments (species) were assigned to paddocks with an area of 1.2 ha (experimental units) under a completely randomized design with three replications. At HRREC, treatments consisted of four species: switchgrass, crabgrass, and a mixed pasture of big bluestem and indiangrass, totaling 9 paddocks. At APREC, treatments consisted of six species: switchgrass, crabgrass, bermudagrass, eastern gamagrass, and a mixed pasture of big bluestem and indiangrass, totaling 15 paddocks. Crabgrass was selected as the annual comparison.

At both locations, experimental units were continuously grazed by weaned heifers from 16 May to 6 August 2014, 15 May to 17 August 2015, and 12 May to 1 September 2016 at HRREC, and 13 May to 6 August 2014, 8 May to 17 August 2015, and 6 May to 16 August 2016 at APREC. Crabgrass at both sites was offset by 30 days to allow for establishment of the annual forage crop. Stocking rate was adjusted as needed to keep forage at a manageable height and allow adequate animal intake. Target heights were 60–76 cm (switchgrass), 40–46 cm (big bluestem and indiangrass blend), 45–60 cm (eastern gamagrass), and 7–20 cm (bermudagrass and crabgrass). These forage heights were selected to balance pasture and animal productivity with stand persistence [8,22,23]. Target heights differed between grass species, owing to differences in plant morphology and grow point heights.

Bermudagrass paddocks were established in 2013 at a seeding rate of 11 kg pure live seed (PLS) $ha^{-1}$ at both APREC and HRREC; however, following high seedling mortality resulting from an abnormally cold winter, bermudagrass was replanted at HRREC on 13 May 2014. Native perennial paddocks were established for a previous experiment in spring 2008 and 2009. Switchgrass and the indiangrass-big bluestem blend were seeded with a no-till drill at the following rates: indiangrass 4 kg PLS $ha^{-1}$, big bluestem 7 kg PLS $ha^{-1}$, and switchgrass 7 kg PLS $ha^{-1}$. Eastern gamagrass was planted with a corn planter at 13 kg PLS $ha^{-1}$. Crabgrass was seeded at 7 kg PLS $ha^{-1}$ on the following dates: 8 May (HRREC) and 21 April (APREC) in 2014, 8 May (HRREC) and 6 May (APREC) in 2015, and 20 April (HRREC) and 8 May (APREC) in 2016.

The most prominent soil type in paddocks at HRREC was a Sango silt loam (coarse-silty, siliceous, semiactive, thermic Glossic Fragiudults), a moderately well-drained soil formed in residuum weathered from cherty limestone, pH = 6.6 [18]. The most prominent soil type in paddocks at APREC was a Memphis silt loam (fine-silty, mixed, active, thermic Typic Hapludalfs), a well-drained soil formed in loess deposits, pH = 6.6 [18].

#### 2.2.2. Measurements

At both sites, iWUE was recorded approximately every 28 days. Measurements of iWUE were collected on four occasions per year at HRREC (28 May, 25–26 June, 23–24 July, and 29–30 August in 2014, 2–3 June, 29–30 June, 27–29 July, and 22–23 August in 2015, and 14–15 May, 11–12 June, 10–11 July, and 13–14 August in 2016) and at APREC (4–6 June, 8–10 July, 4–6 August, and 5–7 September in 2014, 25–27 May, 23–25 June, 22–23 July, and 13–16 August in 2015, and 16–18 May, 26–28 June, 30–31 July, and 27–28 August in 2016) (Figure 1).

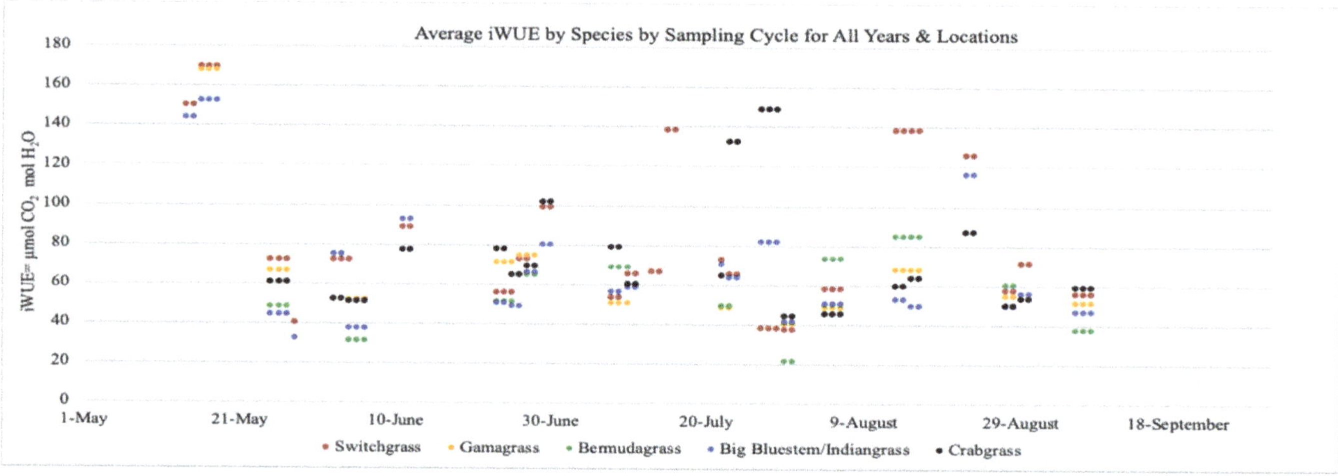


**Figure 1.** Mean instantaneous water-use (iWUE) per experimental unit (paddock) by species and sampling cycle at Highland Rim AgResearch and Education Center, Springfield, TN, and Ames Plantation AgResearch and Education Center, Grand Junction, TN, 2014–2016.

Three iWUE samples were taken from each of three randomly placed exclusion cages per paddock. Cages protected leaves from defoliation by cattle and were randomly moved to new locations within the paddock two weeks before each sampling date to ensure that the forage measured had been exposed to grazing but also had two weeks to regrow after grazing. Moving the cages also prevented sampling the same plant twice, as well as allowed for sampling of non-damaged, fully emerged sampling for iWUE.

In April 2015, Watermark 200SS soil moisture sensors (The Irrometer Company, INC. Riverside, CA, USA) were placed at both HRREC and APREC in each paddock to take observations of soil tension throughout the experiment. Within each paddock, one sensor was placed at each of two soil depths, 0–20 cm and 20–40 cm, and at mid-slope positions. Readings of soil water tension (ranging from 0 to 239 kPa, with zero being fully saturated) were collected from each sensor concurrent with iWUE sampling. Prior to installation, all sensors were calibrated to read 0 kPa at 100% saturation.

Gas exchange measurements were taken with the LI-6400 XT Portable Photosynthesis System (Li-Cor, INC. Lincoln, NE, USA). Leaf readings were taken from the most recently emerged, non-injured, fully expanded, photosynthetically active central leaf of the plant between 10:00 and 15:00 h. $CO_2$ (400 µmol $CO_2$ mol air$^{-1}$) and irradiance (1500 µmol m$^{-2}$ s$^{-1}$) were kept constant within the leaf chamber for all measurements. Appropriate PAR levels were established using data from light curves developed during the greenhouse experiment for each native species, with 1500 µmol m$^{-2}$ s$^{-1}$ being the average light saturation point and the light level used in the gas exchange chamber during sampling. Leaf temperature in the chamber was set based on the ambient temperatures at the start of each sampling day and kept constant until the end of daily sampling. Over the whole study, leaf temperature ranged between 17.6 and 39.1 °C. Leaf area was recorded for each data point.

Gas exchange measurements were used to calculate iWUE. Measurements per species were taken in random order (within replicates) and replicate order was also randomized on each sampling date to account for diurnal fluctuations.

#### 2.2.3. Statistical Analysis

Differences among least squares means were evaluated using the PROC MIXED procedures, adjusted for Tukey's method for least square means separation of SAS [21]. The dependent variables were mWUE and iWUE. Fixed effect was species. Random effects were year, location, date, and replicates. Residuals were checked for normality and interactions between year x species were not significant for all dependent variables, therefore data were analyzed across years. The PROC MIXED procedure was also used to run analysis of covariance in SAS [21], with soil moisture as a covariate for iWUE. All results were evaluated for significance at $p < 0.05$.

## 3. Results

### *3.1. Greenhouse Experiment*

#### 3.1.1. Water-Use Efficiency

iWUE differed ($p < 0.05$) by species, but there was no water treatment effect or species by water treatment interaction for iWUE (Table 1). Instantaneous WUE was significantly different when compared by species ($p < 0.05$). Therefore, iWUE by species did not change when subjected to the stressed and well-watered treatments. The three annual species did not differ, nor did they from switchgrass and eastern gamagrass ($p \geq 0.05$). Likewise, iWUE was not different among perennials. Sorghum-sudangrass (47.2 μmol $CO_2$ mol $H_2O^{-1}$) and pearl millet (46.0 μmol $CO_2$ mol $H_2O^{-1}$) had greater iWUE values than indiangrass (28.1 μmol $CO_2$ mol $H_2O^{-1}$), big bluestem (27.5 μmol $CO_2$ mol $H_2O^{-1}$), bermudagrass (20.3 μmol $CO_2$ mol $H_2O^{-1}$), and tall fescue (15.2 μmol $CO_2$ mol $H_2O^{-1}$). Crabgrass, switchgrass, and eastern gamagrass were not different than any of the species evaluated.

**Table 1.** Least square means of iWUE (μmol $CO_2$ mol $H_2O^{-1}$) by species in the greenhouse experiment across four sampling dates, for both dry and wet treatments, University of Tennessee, Knoxville, TN, 2013 ($n = 32$).

| Species | Mean | Minimum | Maximum | Standard Deviation |
|---|---|---|---|---|
| Sorghum-sudangrass | 47.2 a† | 31.9 | 64.9 | 11.5 |
| Pearl millet | 46.0 a | 25.0 | 54.2 | 9.1 |
| Crabgrass | 39.8 ab | 26.7 | 49.0 | 6.7 |
| Switchgrass | 35.2 ab | 21.0 | 52.1 | 8.2 |
| Eastern gamagrass | 34.8 ab | 31.2 | 39.2 | 2.4 |
| Indiangrass | 28.1 b | 21.0 | 36.0 | 4.9 |
| Big bluestem | 27.5 b | 21.5 | 37.4 | 4.6 |
| Bermudagrass | 20.3 b | 6.56 | 36.3 | 9.9 |
| Tall fescue | 15.2 b | 10.0 | 24.6 | 4.6 |

† Means within a column without a common superscript letter differ ($p < 0.05$).

Species and water treatment interacted to affect mWUE (Table 2). During the dry treatment, bermudagrass (25.9 g DM $kg^{-1}$ $H_2O$) accumulated greater mWUE than tall fescue (16.8 g DM $kg^{-1}$ $H_2O$), crabgrass (14.9 g DM $kg^{-1}$ $H_2O$), big bluestem (12.3 g DM $kg^{-1}$ $H_2O$), and eastern gamagrass (11.8 g DM $kg^{-1}$ $H_2O$), but was not greater than switchgrass, sorghum-sudangrass, pearl millet, or indiangrass ($p > 0.05$). Results of the wet treatment indicate that among perennials, only bermudagrass (14.9 g DM $kg^{-1}$ $H_2O$) and big bluestem (4.4 g DM $kg^{-1}$ $H_2O$) differed. Sorghum-sudangrass (23.9 g DM $kg^{-1}$ $H_2O$) had greater mWUE than bermudagrass, crabgrass (8.2 g DM $kg^{-1}$ $H_2O$), switchgrass (7.9 g DM $kg^{-1}$ $H_2O$), tall fescue (7.7 g DM $kg^{-1}$ $H_2O$), eastern gamagrass (7.5 g DM $kg^{-1}$ $H_2O$), indiangrass (6.7 g DM $kg^{-1}$ $H_2O$), and big bluestem. Pearl millet (17.0 g DM $kg^{-1}$ $H_2O$) had greater mWUE than crabgrass, switchgrass, tall fescue, eastern gamagrass, indiangrass, and big bluestem. Bermudagrass had greater mWUE than big bluestem.

**Table 2.** Least square means mWUE (g DM $kg^{-1}$ $H_2O$) of each species averaged across four sampling dates, University of Tennessee, Knoxville, TN greenhouse, 2013. Species differed between dry (60% water-holding capacity) and wet (90% water-holding capacity) treatments ($p < 0.0001$). Values are comparable for both the dry and wet treatments by column ($n = 16$ per species).

| Treatment | Species | Mean | Minimum | Maximum | Standard Deviation |
|---|---|---|---|---|---|
| Dry regime | | | | | |
| | Bermudagrass | 25.9 a† | 18.2 | 32.7 | 5.2 |
| | Switchgrass | 22.7 ab | 16.4 | 32.7 | 6.1 |
| | Sorghum-sudangrass | 21.5 ab | 17.5 | 26.7 | 4.1 |
| | Pearl millet | 18.4 abc | 16.0 | 20.9 | 1.9 |
| | Indiangrass | 16.8 abc | 12.7 | 23.6 | 4.1 |
| | Tall fescue | 16.8 bc | 9.1 | 21.8 | 4.9 |
| | Crabgrass | 14.9 bc | 13.1 | 16.0 | 1.1 |
| | Big bluestem | 12.3 bc | 9.1 | 14.5 | 2.4 |
| | Eastern gamagrass | 11.8 bc | 7.3 | 16.4 | 3.3 |
| Wet regime | | | | | |
| | Sorghum-sudangrass | 23.9 a | 11.4 | 36.8 | 12.4 |
| | Pearl millet | 17.0 ab | 15.1 | 20.8 | 2.3 |
| | Bermudagrass | 14.9 bc | 8.9 | 16.9 | 1.1 |
| | Crabgrass | 8.2 cd | 7.3 | 10.0 | 1.1 |
| | Switchgrass | 7.9 cd | 4.8 | 12.1 | 2.9 |
| | Tall fescue | 7.7 cd | 5.6 | 8.9 | 1.2 |
| | Eastern gamagrass | 7.5 cd | 3.2 | 16.9 | 5.5 |
| | Indiangrass | 6.7 cd | 3.2 | 10.5 | 2.8 |
| | Big bluestem | 4.4 d | 3.2 | 6.5 | 1.3 |

† Means without a common superscript letter within dry treatment and the wet treatment differ ($p < 0.05$).

### 3.1.2. Survival Length

Species affected ($p < 0.05$) survival length but water treatments did not (Table 3). Average survival length for all species ranged from 12 to 57 DPW. Eastern gamagrass (45.8 DPW), big bluestem (40.8 DPW), and indiangrass (39.6 DPW) survived longer than the remaining species, while switchgrass (29.8 DPW) survived longer than tall fescue (21.5 DPW), pearl millet (19.3 DPW), sorghum-sudangrass (18.1 DPW), bermudagrass (17.3 DPW), and crabgrass (15.8 DPW). There were no survival differences among annuals.

**Table 3.** Least square means for length of survival (days post-watering) by species in greenhouse experiment, University of Tennessee, Knoxville, TN, 2013.

| Species | Mean | Minimum | Maximum | Standard Deviation |
|---|---|---|---|---|
| Eastern gamagrass | 45.8 a† | 34 | 57 | 7.5 |
| Big bluestem | 40.8 a | 29 | 47 | 5.8 |
| Indiangrass | 39.6 a | 29 | 50 | 7.7 |
| Switchgrass | 29.8 b | 22 | 39 | 6.3 |
| Tall fescue | 21.5 c | 20 | 22 | 0.9 |
| Pearl millet | 19.3 c | 18 | 22 | 1.5 |
| Sorghum-sudangrass | 18.1 c | 12 | 22 | 2.9 |
| Bermudagrass | 17.3 c | 15 | 18 | 1.3 |
| Crabgrass | 15.8 c | 15 | 18 | 1.3 |

† Means within a column without a common superscript letter differ ($p < 0.05$).

3.1.3. Correlations

Correlations of iWUE, dry or wet treatment, total mass, and length of survival were evaluated. Results indicated that total mass was positively correlated to iWUE (0.61) and negatively correlated to length of survival for all species (−0.56). Mass was then used as a covariate to iWUE, but despite a moderately strong relationship ($R^2 = 0.66$), it was not significant for any species tested ($p = 0.49$).

*3.2. Field Experiment*

3.2.1. Weather

From May through September in 2014, precipitation at HRREC was 23% lower than average (525 mm) and precipitation at APREC was 46% greater than average (561 mm). In 2015, from May through September, precipitation at HRREC was 13% lower than average and at APREC, was 8% greater than average. In 2016, from May through September, precipitation at HRREC was 40% greater than average and at APREC, was 22% lower than average (Figure 2). Temperatures recorded at HRREC and APREC were near or somewhat above average during the study. In 2014, temperature from May through September at HRREC was 0.5 °C greater than average (22.9 °C), 0.9 °C greater than average in 2015, and 1.5 °C greater than average in 2016. In 2014, temperature from May through September at APREC was 0.2 °C lower than average (23.9 °C), 0.6 °C greater than average in 2015, and 1.42 °C greater than average in 2016.

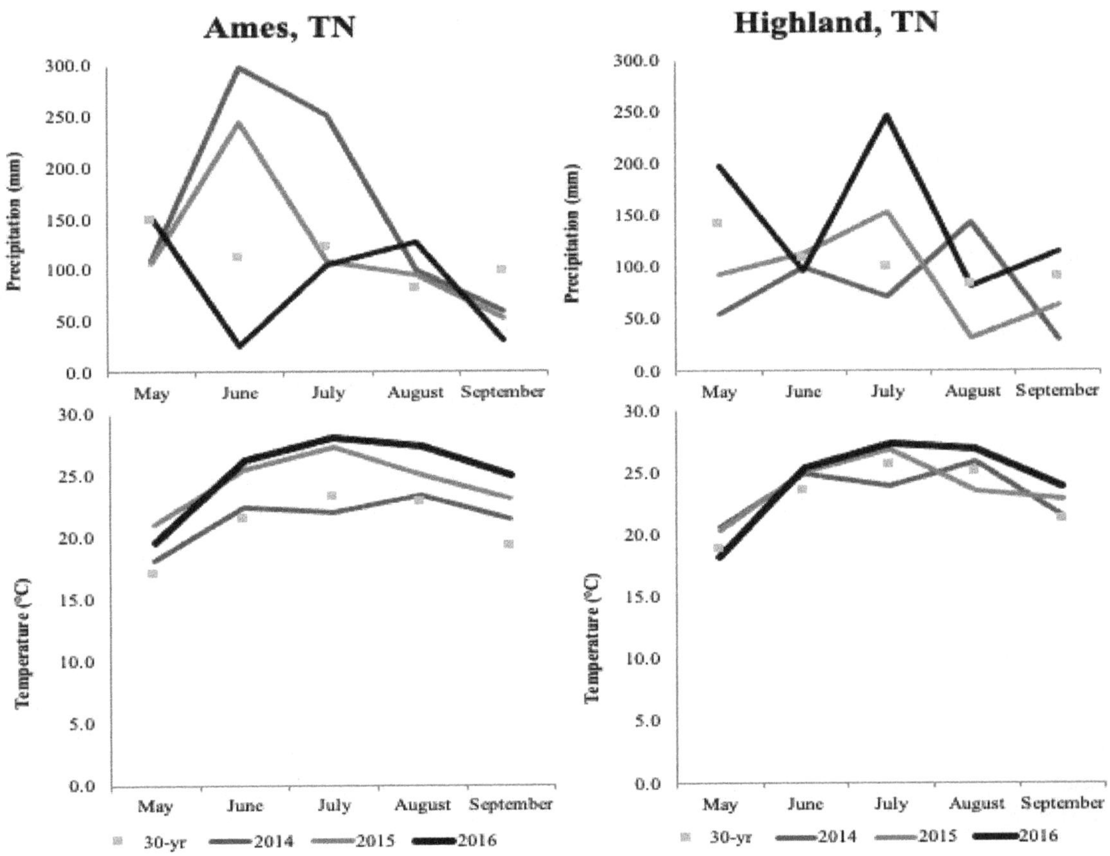


**Figure 2.** Monthly total precipitation and average monthly temperature at Highland Rim AgResearch and Education Center, Springfield, TN and Ames Plantation AgResearch and Education Center, Grand Junction, TN, 2014–2016 and 30 year means.

3.2.2. Water Use Efficiency

Across both locations (as locations did not differ), differences ($p < 0.05$) were observed among species (Table 4). Switchgrass, crabgrass, eastern gamagrass, and bermudagrass performed similarly ($\mu > 70.0$ μmol $CO_2$ mol $H_2O^{-1}$). Eastern gamagrass, bermudagrass, and indiangrass did not differ ($\mu > 65.0$ μmol $CO_2$ mol $H_2O^{-1}$), neither did bermudagrass, indiangrass, or big bluestem ($\mu > 63.0$ μmol $CO_2$ mol $H_2O^{-1}$) (Table 4).

**Table 4.** Least square means of iWUE (μmol $CO_2$ mol $H_2O^{-1}$) by species during field experiment conducted at Highland Rim AgResearch and Education Center, Springfield, TN and Ames Plantation AgResearch and Education Center, Grand Junction, TN, across years (2014–2016) and locations.

| Species | Mean | Minimum | Maximum | Standard Deviation |
|---|---|---|---|---|
| Switchgrass ($n$ = 24) | 74.5 [a†] | 37.3 | 169.1 | 4.4 |
| Crabgrass ($n$ = 21) | 74.1 [a] | 44.0 | 148.8 | 4.6 |
| Eastern gamagrass ($n$ = 12) | 74.4 [ab] | 40.1 | 167.5 | 5.3 |
| Bermudagrass ($n$ = 11) | 70.8 [abc] | 21.2 | 84.6 | 5.2 |
| Indiangrass ($n$ = 24) | 65.1 [bc] | 29.7 | 104.8 | 4.4 |
| Big bluestem ($n$ = 24) | 63.9 [c] | 32.1 | 152.1 | 4.4 |

† Means within a column without a common superscript letter differ ($p < 0.05$).

3.2.3. Soil Moisture

Throughout the experiment, soil moisture ranged from fully saturated to near the permanent wilting point (~200 kPa). However, plants did not exhibit drought symptoms, such as wilting or premature senescence (Table 5). On average, soil moisture was greater at the 20 cm depth ($\mu = 62.50$ kPa) than at the 40 cm depth ($\mu = 82.15$ kPa). Analysis of covariance did not indicate a consistent relationship between soil tension and iWUE among species.

**Table 5.** Soil tension (kPa) at two depths (20 and 40 cm) for experimental paddocks by species, 2015–2016 at Highland Rim AgResearch and Education Center, Springfield, TN, and Ames Plantation AgResearch and Education Center, Grand Junction, TN, across years and locations. Number of observations per species is represented by $n$.

| Depth | Crabgrass | Switchgrass | Bermudagrass | Eastern Gamagrass | Big Bluestem-Indiangrass |
|---|---|---|---|---|---|
| | ($n$ = 48) | ($n$ = 48) | ($n$ = 24) | ($n$ = 24) | ($n$ = 48) |
| 20 cm | 60.1 | 67.6 | 64.9 | 60.5 | 61.7 |
| 40 cm | 68.2 | 97.7 | 67.3 | 90.9 | 86.5 |

## 4. Discussion

### 4.1. Greenhouse Experiment

4.1.1. Water-Use Efficiency

Plants tested in the greenhouse experiment were seedlings as opposed to the fully mature plants in the field experiment. All plants were planted and transplanted on the same dates. However, the annual plants grew leaf and root tissue more quickly than perennials. This was expected due to the physiological differences between annuals and perennials and the completion of their lifecycles [24]. Therefore, we evaluated water treatment separately for annuals and perennials. Initial statistical analysis revealed that lifecycle was not significant for iWUE, mWUE, or survival length, and therefore, was omitted from the final statistical analysis.

During the greenhouse experiment, PAR did not reach a level that would allow any of the plant species to reach their light saturation points. This was because light from the source to the leaf canopy in the greenhouse during late fall–early winter dissipated rapidly. During greenhouse measurements, PAR ranged from 190.0 to 260.0 μmol $m^{-2}$ $s^{-1}$, while most species had a light saturation point around 1500 μmol $m^{-2}$ $s^{-1}$. Mitchell et al. [22] documented reduced photosynthetic rates for switchgrass under cool temperatures and low light intensities in a controlled greenhouse environment, such as ours. Light in the greenhouse was a limiting factor for iWUE across all species. However, matching PAR in the gas exchange readings to the ambient levels of the greenhouse was representative of regional growing conditions.

Species varied for each variable tested, while water treatment (dry vs. wet) was only significant for mWUE (Table 2). These results suggest that the dry treatment did not impact the carbon assimilation rates of the species tested or their lengths of survival after termination of watering (Table 3). Anyia and Herzog [12] conducted an imposed drought study under greenhouse conditions which had variable results for different varieties of cowpea, some of which increased iWUE while some showed no change in iWUE under severe drought stress. It is possible that species evaluated in the greenhouse truly do not differ in iWUE under drought stress, or it could be that the drought imposed was not severe enough to trigger a significant response. Byrd and May encountered similar results when imposing drought on switchgrass in a greenhouse setting and measuring transpiration efficiency [19].

Imposed drought did impact mWUE, with greater g DM $kg^{-1}$ $H_2O$ being obtained under drought conditions for all species, except for sorghum-sudangrass (21.5 g DM $kg^{-1}$ $H_2O$ in the dry treatment and 23.9 g DM $kg^{-1}$ $H_2O$ in the wet treatment) (Table 2). Similar responses were observed by Ghannoum et al. [11] on Australian $C_4$ grasses in an imposed drought greenhouse study.

Survival length was more favorable to the native warm-season perennial grasses, eastern gamagrass, big bluestem, indiangrass, and switchgrass than to the annuals or the introduced perennials. The more prevalent natural wax coating on the leaves of these species likely gave them an advantage compared to the annuals and bermudagrass. Tall fescue, which also has a waxy cuticle, was not as persistent as the native warm-season perennials but was similar to the warm-season annuals and bermudagrass.

Tall fescue performed fairly well compared to the other species and variables evaluated in the greenhouse. Tall fescue is more drought-tolerant than other $C_3$ grasses, which continues to lend it favor with livestock producers throughout the southeastern United States, despite the negative impacts of the fescue endophyte which can lead to vascular constriction in grazing animals that present symptoms including reproductive failure, difficulty regulating body heat, reduced appetite, and, in severe cases, the loss of hooves or tails [25]. The warm-season forages included in this study do not have endophyte issues and are often recommended to help mitigate the negative impacts of fescue endophyte along with drought conditions [25].

Correlations between variables suggest that as DM increases, plant iWUE increases. In addition, correlations from the greenhouse experiment suggest that as total DM accumulation increases, length of survival post-watering decreases. This corresponds with the theory that leaf expansion is highly sensitive to drought [26]. It has been repeatedly observed that when plants are under drought stress, leaves will have decreased surface areas compared to those produced when water is adequately available [12,14,26,27]. Producing leaves with smaller areas and less mass will reduce transpiration loss and thereby increase survival length during drought.

#### 4.1.2. Summary

Throughout the greenhouse study, annuals were evaluated for logical inclusion in the field study. While there were no differences between sorghum-sudangrass, pearl millet, and crabgrass for iWUE or DPW, crabgrass only differed from the native warm-season perennials for mWUE in the wet treatment. Due to the commonalities observed between crabgrass and the native warm-season perennials and a need for additional field tests of grazed crabgrass in the region, crabgrass was selected to be included in the field experiment. Although, all three warm-season annuals studied have potential for success mitigating drought on grazing lands in the southeastern United States.

### 4.2. Field Experiment

#### 4.2.1. Weather

Temperature and precipitation throughout the three years of field study were not conducive with drought conditions (Figure 2). Temperatures were similar to the 30-year average. Although precipitation varied between years and locations, soil moisture remained adequate for growth through all three summers.

#### 4.2.2. Water-Use Efficiency

Dates of sampling cycles varied between locations and among years (Figure 1). Given that these plants reach peak productivity mid-summer [28], the date at which sampling occurred in relationship to peak growth was not consistent among years, which would create variation in iWUE [29]. Also, total precipitation per growing season was not consistent, which would have created a wider spread of iWUE values observed [30]. However, this variation did not result in differences with respect to year, location, or sampling cycle for species or interactions, hence year was evaluated as a random effect throughout the experiment.

The iWUE values observed in this experiment were similar to those reported for warm-season grasses [13,17,29,31–33]. Switchgrass and crabgrass performed similarly and yielded greater iWUE values than indiangrass and big bluestem ($p \leq 0.05$). Bermudagrass iWUE was similar to all species tested (Table 4). These iWUE values for native species ordinated nearly the same in the field experiment as those observed in the greenhouse experiment, with the exception of bermudagrass (Table 1), suggesting that these may reflect real trends in iWUE.

Higher ratios of photosynthetic rate to transpiration rates were observed in the field at higher temperature and PAR, which is justified by the species light saturation points. The light saturation point is the point at which photosynthetic rate is maximized by PAR [34]. Although PAR can increase additionally, photosynthesis will not. Temperature is directly related to photosynthetic rate and therefore impacts WUE [35]. For $C_4$ plants, if moisture is not limiting, peak photosynthesis occurs near 34 °C [28], which leads us to believe that peak photosynthesis could have been inhibited early in the growing season and in the greenhouse by temperature and PAR. Beale et al. [17] observed similar patterns of limited growth due to light levels and temperatures when testing $C_4$ perennial grasses in the field in a temperate environment.

#### 4.2.3. Soil Moisture

Precipitation at HRREC was slightly below average in 2014 and 2015 (Figure 2). Average soil water tension was less than 100 kPa at both locations and both soil depths (Table 5), confirming that moisture conditions were adequate for plant growth as deep as 40 cm throughout the growing seasons [36]. Gas exchange is reduced by plant-available water when drought ensues, although leaf expansion is more sensitive to drought and is visually noticeable [26]. Signs of plant stress indicating drought were not observed during any of the sampling dates. Had drought occurred in the years of study, there could have been additional iWUE differences among species, such as was observed by Hussain et al. [14] in managed grasslands in Germany and by Jongen et al. [15] on a Mediterranean grassland in Portugal.

#### 4.2.4. Summary

The individual observations of iWUE from this study provided readings of what is physiologically occurring on one leaf of the plant at that moment, which can easily be influenced by small changes in the plant or environment, such as energy partitioning or soil moisture, that cannot be controlled within the leaf chamber of the infrared gas analyzer [30]. Therefore, additional investigations are needed in subsequent years during persistent drought conditions to gain an understanding of these interactions and how they influence iWUE. While iWUE is not a practical means of measuring drought tolerance or resilience by grazing managers in a typical farm or ranch setting, a dynamic dataset of iWUE values under various conditions could help farmers and ranchers make better decisions about what types of forages can improve grazing conditions during times of drought and heat stress. Though the moisture conditions during the years of this study were not considered droughty, there were still differences in WUE of the forages studied during 2014–2016 that could be helpful for future on-farm research.

## 5. Conclusions

Integration of perennial and annual warm-season grasses is of interest to beef cattle producers. Due to a lack of localized research on drought tolerance of many forages, this experiment tested multiple species in both a controlled greenhouse and a grazing field experiment. Data collected in the greenhouse confirmed that mWUE was enhanced by drought in all species tested. Overall results of the in-field grazing experiment were comparable to those of the greenhouse experiment. Switchgrass and crabgrass had the greatest iWUE compared to indiangrass and big bluestem. Eastern gamagrass and bermudagrass were intermediate among species tested. Results continue to support that native grasses have greater WUEs than cool-season grasses, which are commonly used in humid pastures of the southeastern United States. Based on these results, switchgrass and crabgrass are more efficient carbon assimilators than indiangrass and big bluestem. Further evaluation of these species should be conducted across a broader range of moisture conditions to detect differences between species during documented drought. In addition, dry matter production, leaf area, and whole canopy responses to drought are necessary for drawing field-landscape conclusions. Overall, results indicate that native warm-season grasses, bermudagrass, and crabgrass are complementary summer forages to help mitigate drought in pastures in the southeastern United States.

**Author Contributions:** Conceptualization, P.D.K. and J.D.R.; methodology, A.J.A.; software, C.H.G.; validation, C.H.G., A.J.A., and R.L.G.N.; formal analysis, C.H.G.; investigation, C.H.G., A.J.A., and R.L.G.N.; resources, P.D.K., R.L.G.N. and J.D.R.; data curation, C.H.G.; writing—original draft preparation, C.H.G.; writing—review and editing, C.H.G., A.J.A., R.L.G.N., and P.D.K.; visualization, C.H.G.; supervision, P.D.K. and R.L.G.N.; project administration, P.D.K. and R.L.G.N.; funding acquisition, P.D.K. and J.D.R. All authors have read and agreed to the published version of the manuscript.

**Acknowledgments:** The authors would like to thank the staff of Ames Plantation AgResearch and Education Center, and Highland Rim AgResearch and Education Center, Rondineli Barbero, Lilian Cordeiro, Thais Costa, David McIntosh, Ludmila Monteiro, Jessica Nelson, Leticia Tavares, and Aleksandra Wilson. We also thank Chuck West for providing helpful insights on data interpretation and project design.

## References

1. Fuhrman, C.; Konrad, C. A Trajectory Approach to Analyzing the Ingredients Associated with Heavy Winter Storms in Central North Carolina. *Weather Forecas* **2013**, *28*, 647–667. [CrossRef]
2. NASS—National Agricultural Statistics Service. *NASS National Agricultural Statistics Service*; USDA: Washington, DC, USA, 2015.
3. Koshi, P.T.; Stubbendieck, J.; Eck, H.V.; McCully, W.G. Switchgrasses: Forage yield, forage quality and water-use efficiency. *J. Range Manag.* **1982**, *35*, 623–627. [CrossRef]
4. Anderson, B.; Matches, A.G. Forage yield, quality, and persistence of switchgrass and caucasian bluestem. *Agron. J.* **1983**, *75*, 119–124. [CrossRef]
5. Sanderson, M.A.; Reed, R.L. Switchgrass growth and development: Water, nitrogen, and plant density effects. *J. Range Manag.* **2000**, *53*, 221–227. [CrossRef]
6. Stroup, J.A.; Sanderson, M.A.; Muir, J.P.; McFarland, M.J.; Reed, R.L. Comparison of growth and performance in upland and lowland switchgrass types to water and nitrogen stress. *Biores. Tech.* **2003**, *86*, 65–72. [CrossRef]
7. Temu, V.W.; Rude, B.J.; Baldwin, B.S. Nutritive value response of native warm-season forage grasses to harvest intervals and durations in mixed stands. *Plants* **2014**, *3*, 266–283. [CrossRef]

8. Keyser, P.D.; Holcomb, E.D.; Lituma, C.M.; Bates, G.E.; Waller, J.C.; Boyer, C.N.; Mulliniks, J.T. Forage attributes and animal performance from native grass inter-seeded with red clover. *Agron. J.* **2015**, *108*, 373–383. [CrossRef]
9. Griffen, J.L.; Jung, G.A. Leaf and stem forage quality of big bluestem and switchgrass. *Agron. J.* **1983**, *75*, 723–726. [CrossRef]
10. Kramer, P.J. Water use efficiency. *Water Relat. Plants* **1983**, *1*, 383–387.
11. Ghannoum, O.; Von Caemmerer, S.; Conroy, J.P. The effect of drought on plant water use efficiency of nine NAD-ME and nine NADP-ME Australian $C_4$ grasses. *Funct. Plant Biol.* **2002**, *29*, 1337–1348. [CrossRef]
12. Anyia, A.O.; Herzog, H. Water-use efficiency, leaf area and leaf gas exchange of cowpeas under mid-season drought. *European J. Agron.* **2004**, *20*, 327–339. [CrossRef]
13. Taylor, S.H.; Hulme, S.P.; Rees, M.; Ripley, B.S.; Woodward, F.I.; Osborne, C.P. Ecophysiological traits in $C_3$ and $C_4$ grasses: A phytogenetically controlled screening experiment. *New Phytolog.* **2010**, *185*, 780–791. [CrossRef] [PubMed]
14. Hussain, M.Z.; Grunwald, T.; Tenhunen, J.D.; Li, Y.L.; Mirzae, H.; Bernhofer, C.; Otieno, D.; Dinh, N.Q.; Schmidt, M.; Wartinger, M. Summer drought influence on $CO_2$ and water fluxes of extensively managed grassland in Germany. *Agricult. Ecosys. Environ.* **2011**, *141*, 67–76. [CrossRef]
15. Jongen, M.; Pereira, J.S.; Aires, L.M.; Pio, C.A. The effects of drought and timing of precipitation on the inter-annual variation in ecosystem-atmosphere exchange in a Mediterranean grassland. *Agricult. Forest Meteorol.* **2011**, *151*, 595–606. [CrossRef]
16. *LI-6400 XT Portable Photosynthesis Meter*, LI-COR Biosciences Inc.: Lincoln, NE, USA.
17. Beale, C.V.; Morison, J.I.L.; Long, S.P. Water use efficiency of $C_4$ perennial grasses in a temperate climate. *Agricult. Forest Meteorol.* **1999**, *96*, 103–115. [CrossRef]
18. NRCS. *Web Soil Survey*; NRCS—Natural Resource Conservation Service, USDA: Washington, DC, USA, 2014.
19. Byrd, G.T.; May, P.A. Physiological comparisons of switchgrass cultivars differing in transpiration efficiency. *Crop Sci.* **2000**, *40*, 1271–1277. [CrossRef]
20. Tucker, S.S.; Craine, J.M.; Nippert, J.B. Physiological drought tolerance and the structuring of tallgrass prairie assemblages. *Ecosphere* **2011**, *2*, 48. [CrossRef]
21. SAS. *Version 9.4.*; SAS Institute Inc.: Cary, NC, USA, 2009.
22. Mitchell, R.B.; Anderson, B.; Moser, L.E.; Waller, S.S. *Switchgrass and Big Bluestem for Grazing and Hay*; University of Nebraska: Lincoln, NE, USA, 1994.
23. Burns, J.C.; Fisher, D.S. Eastern gamagrass management for pasture in the Mid-Atlantic region: I. Animal performance and pasture productivity. *Agron. J.* **2010**, *102*, 171–178. [CrossRef]
24. Stern, K.R.; Bidlack, J.E.; Jansky, S. *Introductory Plant Biology*; McGraw-Hill Higher Education: Boston, MA, USA, 2008.
25. Elbersen, H.W.; West, C.P. Growth and water relations of field-grown tall fescue as influenced by drought and endophyte. *Grass Forage Sci.* **1996**, *51*, 333–342. [CrossRef]
26. Sadras, V.O.; Milroy, S.P. Soil-Water Thresholds for the Responses of Leaf Expansion and Gas Exchange: A Review. *Field Crops Res.* **1996**, *47*, 253–266. [CrossRef]
27. Feng, R.; Zhang, Y.; Yu, W.; Hu, W.; Wu, J.; Ji, R.; Wang, H. Zhao. Analysis of the relationship between spectral characteristics of maize canopy and leaf area index under drought stress. *Acta Ecol. Sinica* **2013**, *33*, 301–307. [CrossRef]
28. Yamori, W.; Hikosaka, K.; Way, D.A. Temperature response of photosynthesis in $C_3$, $C_4$, and CAM plants: Temperature acclimation and temperature adaptation. *Photosynth. Res.* **2013**, *119*, 101–117. [CrossRef] [PubMed]
29. Ribeiro, R.V.; Lyra, G.B.; Santiago, A.V.; Pereira, A.R.; Machado, E.C.; Oliveira, R.F. Diurnal and seasonal patterns of leaf gas exchange in bahiagrass (*Paspalum Notatum* Flugge) growing in a subtropical climate. *Grass Forage Sci.* **2006**, *61*, 293–305. [CrossRef]
30. Johnson, D.A.; Asay, K.A. Selection for improved drought response in cool-season grasses. *Soc. Range Manag.* **1993**, *16*, 194–202. [CrossRef]
31. Hovenden, M. Growth and photosynthetic responses to elevated [$CO_2$] in grasses from Tasmanian pasture. *Papers Proceed. Royal Soci. Tasmania* **2003**, *137*, 81–86. [CrossRef]
32. Xu, B.; Li, F.; Shan, L.; Ma, Y.; Ichizen, N.; Huang, J. Gas exchange, biomass partition, and water relationships of three grass seedlings under water stress. *Weed Biol. Manag.* **2006**, *6*, 79–88. [CrossRef]

33. Mantlana, K.B.; Veenendaal, E.M.; Arneth, A.; Grispen, V.; Bonyongo, C.M.; Heitkonig, I.G. Biomass and leaf-level gas exchange characteristics of three African Savanna $C_4$ grasses under optimum growth conditions. *Afr. J. Ecol.* **2009**, *47*, 482–489. [CrossRef]
34. Taiz, L.; Zeiger, E. Photosynthesis: Carbon reactions. *Plant Physiology.* **2002**, *3*, 70–145.
35. Long, S.P. $C_4$ photosynthesis at low temperature. *Plant Cell Environ.* **1983**, *6*, 345–363.
36. Enciso, J.; Jifon, J.; Wiedenfeld, B. Subsurface drip irrigation of onions: Effects of drip tape emitter spacing on yield and quality. *Agric. Water Manag.* **2007**, *92*, 126–130. [CrossRef]

2

# Socioeconomic Factors and Sustainable Forage Crops Production in Turkey Aegean Region: A Multivariate Modeling

**Celal Cevher [1,*] and Bulent Altunkaynak [2]**

1 Field Crops Central Research Institute, Şehit Cem Ersever 9-11, Yenimahalle, Ankara 06170, Turkey
2 Department of Statistics, Faculty of Science, Gazi University, Beşevler, Ankara 06500, Turkey; bulenta@gazi.edu.tr
* Correspondence: celal.cevher@tarimorman.gov.tr

**Abstract:** The aim of this study is to evaluate the effect of farmers' socioeconomic characteristics on sustainable forage crop production after forage crop subsidies. One of the innovative aspects of the study is the simultaneous modeling of alfalfa, silage corn and vetch production efficiency. For this, the multivariate linear regression model was used. In this way, the effect of socioeconomic characteristics on production is more clearly demonstrated by taking into account the dependency structure between forage crop production. For the study, 487 farmers in the Aegean region, where fodder crops can be produced throughout the year, were interviewed face-to-face and data were collected through a questionnaire. According to the results obtained, it was determined that the width of the land had a positive effect on the increase in alfalfa, silage corn, and vetch production. The number of animals was found to have a positive effect on alfalfa and silage maize production growth. The production of alfalfa was higher in rural farmers, and vetch production was higher in urban farmers. Farmers who do not have non-agricultural income focused on vetch production, and farmers with non-agricultural income focused on alfalfa production. It was seen that the majority of the farmers participating in the study were primary school graduates and lived in rural areas. In general, a significant increase was observed in the production of silage corn, vetch, and alfalfa, respectively. After the subsidies, it was concluded that this increase in the production of silage corn, vetch, and alfalfa was not at a level to meet the amount of forage crops needed by animal husbandry.

**Keywords:** forage crops; sustainability; subvention; socioeconomics; multivariate modelling; Aegean region

---

## 1. Introduction

Forage crop production is important for meeting the feed requirements of livestock and ensuring soil fertility [1]. One of the most important needs of animal husbandry is the production of forage. The feed requirements of a cattle of 500 kg body weight, to meet the nutrient requirement, are 4 kg/day of good quality dry matter and 10 kg/day of good quality green grass or silage corn [2]. In this context, as the number of animals increases, the need for feed increases. In animal production, feed inputs make up about 70% of the total production cost.

Turkey is conducive to small and large cattle-breeding because it has natural resources and favorable ecological conditions. Livestock, especially cattle and sheep, is an important component of farming systems in every region of Turkey. However, forage production in Turkey is not at a level that will meet the needs of the animals [2,3]. Alfalfa (*Medicago sativa* L.), silage corn (*Zea mays* L.), and vetch (*Vicia sativa* L.) are the most widely produced forages in Turkey. According to Turkish

Statistical Institute data (TurkStat, Ankara, Turkey, 2019), alfalfa, silage corn, and vetch, respectively, 641,212.8, 507,412.7 and 391,498.0 are cultivated in hectares (ha). The yields for these products were 28.0 tons/ha for alfalfa, 50.6 tons/ha for silage corn, and 11.0 tons/ha for vetch, respectively. On the other hand, according to TurkStat data, the number of bovine animals is about 14.2 million, the number of ovine animals is reported as 4.49 million [4]. While the feed requirement of these animals is about 95.6 million tons, total forage crop production is about 55.5 million tons. When 11.7 million tons of feed production from meadows and pastures are added, the forage deficit, which was reported as 28.4 million tons in 2019, could not be completed [5]. Farmers producing forage crops have been supported since 2002 in order to provide forage crops needed by livestock. Despite the agricultural subsidies provided to producers, the targeted feed crop production level could not be reached, indicating that these subsidies are not sufficient [5]. Accordingly, studies have been conducted to investigate the effect of subsidies on forage production [6–11]. These studies are generally aimed at examining the effects of macroeconomic variables on forage crop production. However, there is a limited number of studies on forage crop yield that have considered the socioeconomic structures of farmers [12,13]. It is known that socioeconomic characteristics of farmers affect agricultural information that has a high impact on production [14]. In addition, it has been observed that socioeconomic characteristics have an impact on farmers' adoption and use of new technologies [15,16].

The effects of farmers' socioeconomic characteristics on the variables, such as production, yield, technology adoption, and income, have been demonstrated by various studies [17–22]. No studies, however, on socioeconomic variables that affect the production and production patterns of forage crops after the subsidies were found in the time interval of this study. Forage crop supports have been given to farmers since 2002. Therefore, the effect of supports on forage crop production was examined between 2002 and 2015. Farmers can produce more than one forage crop within the production pattern. It is important to examine the effect of subsidies in more detail in order to determine which forage crops farmers choose, how much yield they have, and what socioeconomic characteristics affect this yield. On the other hand, farmers can change their production patterns in the following years, according to their forage support policies. For example, a farmer who sown alfalfa in his entire field, next year, could sow alfalfa in half of his field and silage corn in half. Therefore, there is a dependency, that is, a correlation between the support received by the farmer per hectare and the amount of forage crops. Regarding this relationship, if each of the variables that will affect the amount of forage crop production is examined separately for the forage crop, it may lead to the lack of real relationships or the emergence of non-causal relationships.

The objectives of this study were: (1) to determine the socioeconomic variables that affect farmers' alfalfa, silage corn, and vetch production patterns and (2) to compare the pre- and post- support forage crop production amounts.

## 2. Materials and Methods

### *2.1. Study Area and Sampling*

In Turkey, the Eastern Anatolia Region ranks first in terms of acreage for forage crops and intensity of their farming activities (35.15%), while the Aegean Region (our study area) ranks third with about 15%. However, when the planting area of forage crops and the amount of green grass production are examined, the Aegean Region is in the first place with 35,400 kg/ha in terms of green grass yield. Since the climatic conditions of the Aegean region are suitable for forage crop production all year round, more yields can be obtained in forage crop production [3]. Therefore, the Aegean region of Turkey, as an important forage crop producing region, has been identified as the study area (Figure 1). The list of names of all the farmers was obtained from the Farmer Registration System. Data were collected by face-to-face surveys of farmers selected with the cluster sampling method. Four hundred eighty-seven participant-farmers who supported the use of certified forage crops in the Aegean region were randomly selected out of 34,944 farmers. In terms of forages in Turkey, most grown forage crops are alfalfa, vetch, and silage corn. These three plants constitute the main sources of forage crops to meet the feed needs of livestock. Therefore, in our study, alfalfa, vetch, and silage corn were considered as feed sources.

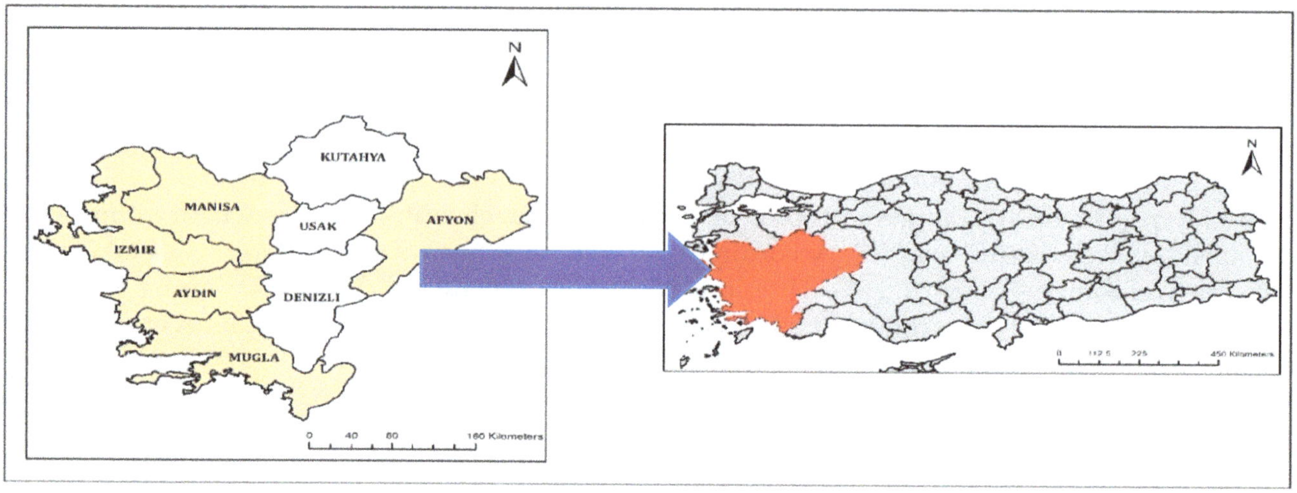


**Figure 1.** Map of the study area.

### 2.2. Farmer Inclusion Criteria

The study included farmers between the ages of 18–70 and with at least 10 bovine or ovine animals. In addition, since irrigated farming is practiced in a major part of the Aegean region, farmers who produced in irrigated land conditions were taken into consideration in the study. All the farmers included in the study sow forage crop and received state support according to the size of their cropland. At present, there is only a negligible number of woman-farmers in Turkey. Therefore, only male farmers were involved in the study. In this study, the enterprise characteristics of the farmers outside the socioeconomic variables were randomly selected to be homogeneous.

### 2.3. Statistical Analysis

For this purpose, alfalfa, silage maize, and vetch production amounts were considered as dependent variables and the multivariate regression model was used by considering the relationship structures between these variables. The independent variables of the study are the socioeconomic characteristics of the farmer. The production of sainfoin in the Aegean region is almost nonexistent. For this reason, alfalfa (A), silage corn (S), and vetch (V) are considered as roughage plants. The post-support (AS) and pre-support (BS) production differences of these forage crops constitute the dependent variables of the study. Socioeconomic variables, such as age (A), educational status (E), place of residence (PR), non-agricultural income status (I), total land (TL) and the number of animals (NA) belonging to farmers, were used as independent variables. Explanations of the variables are given in Table 1.

**Table 1.** Data description.

| Notation | Variable | Type | Description |
|---|---|---|---|
| A | Age | Independent | Year |
| E | Education | Independent | 1 = primary, 2 = middle school, 3 = high school, 4 = university |
| PR | Place of residence | Dummy | 0 = rural, 1 = city |
| I | Off-farm income | Dummy | 0 = no, 1 = yes |
| TL | Total land | Independent | da |
| NA | Number of animals | Independent | Animal Unit |
| $\Delta A$ | Difference in alfalfa production | Dependent | $\Delta A = C_{AS} - C_{BS}$ (kg/da) |
| $\Delta S$ | Difference in silage corn production | Dependent | $\Delta S = S_{AS} - S_{BS}$ (kg/da) |
| $\Delta V$ | Difference in vetch production | Dependent | $\Delta V = V_{AS} - V_{BS}$ (kg/da) |

Notes: $A_{AS}$: alfalfa production after subsidies; $A_{BS}$: alfalfa production before subsidies; $S_{AS}$: silage corn production after subsidies; $S_{BS}$: silage corn production before subsidies; $V_{AS}$: vetch production after subsidies; $V_{BS}$: vetch production before subsidies.

The conceptual structure of the multivariate regression model was created as in Figure 2, considering the possibility that the dependent variables may be related among themselves.

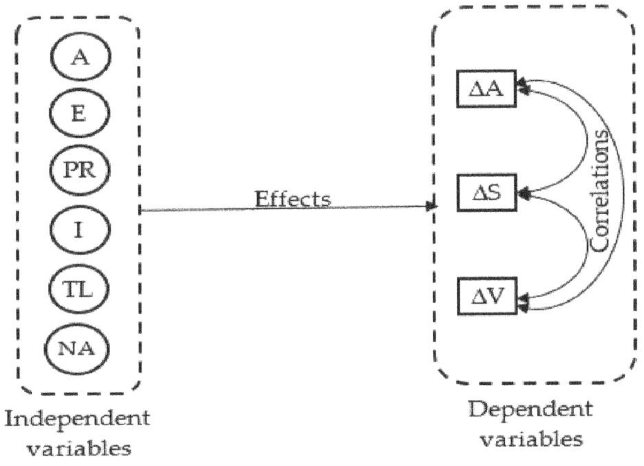


**Figure 2.** The conceptual model for determination of forage crops production pattern.

The multivariate regression model is used to examine the effect of independent variables on dependent variables, taking into account the covariance structure between dependent variables [23].

$$Y=X\beta+\varepsilon \tag{1}$$

where

$$Y=(\Delta A, \Delta S, \Delta V)^T \tag{2}$$

$$X=(1, A, E, PR, I, TL, NA) \tag{3}$$

$$\beta = \begin{bmatrix} \beta_{01} & \beta_{02} & \beta_{03} \\ \beta_{11} & \beta_{12} & \beta_{13} \\ \beta_{21} & \beta_{22} & \beta_{23} \\ \beta_{31} & \beta_{32} & \beta_{33} \\ \beta_{41} & \beta_{42} & \beta_{43} \\ \beta_{51} & \beta_{52} & \beta_{53} \\ \beta_{61} & \beta_{62} & \beta_{63} \end{bmatrix} \tag{4}$$

and

$$\varepsilon = (\varepsilon_1, \varepsilon_2, \varepsilon_3)^T \tag{5}$$

$\beta$ is a matrix showing the unknown parameters given in Equation (4). In Equation (5), there are assumptions $E(\varepsilon_i) = 0$ and $Cov(\varepsilon_i, \varepsilon_j) = \sigma_{ij}I$. Here, $\sigma_{ij}$ is the covariance between the ith dependent variable and the jth one. Thus, the error terms associated with different dependent variables may be correlated.

In this study, the dependent variables considered are ΔA, ΔS, and ΔV. Three different linear regression models can be established or a Structural Equation Modeling (SEM) can be created to examine the socioeconomic factors affecting each of these variables. However, in both cases, the covariance between dependent variables is ignored. In SEM models, you can create a covariance between independent, i.e., external variables, but the same is not true for dependent variables that are internal variables. Therefore, multivariate regression analysis was applied in our study. Multivariate regression analysis is a multivariate statistical method that takes into account the covariance between dependent variables.

## 3. Results

### 3.1. Socioeconomic Profile

The results of the socioeconomic profile are presented in Table 2.

**Table 2.** Socioeconomic profile.

| Variable | Level | M ± SD [¥] | *n* (%) |
|---|---|---|---|
| Age | - | 48.54 ± 9.98 | - |
| Education | primary | - | 336 (69.0) |
| | middle school | - | 59 (11.5) |
| | high school | - | 73 (15.0) |
| | university | - | 22 (4.5) |
| Place of residence | rural | - | 420 (86.2) |
| | city | - | 67 (13.8) |
| Off-farm income | no | - | 304 (62.4) |
| | yes | - | 183 (37.6) |
| Total land | - | 115.42 ± 119.9 | - |
| Number of animals | - | 23.24 ± 29.52 | - |

[¥] M: Mean; SD: Standard Deviation.

As Table 2 demonstrates, the average age of the farmers included in the study was 48.54 ± 9.98. Approximately seven out of every 10 farmers (69%) were primary school graduates. Only 4.5% of the surveyed farmers were university graduates. Farmers predominantly (86.2%) lived in rural areas. The proportion of farmers whose only income source is agriculture was 62.4%. The average land size owned by farmers is 115.42 ± 119.9 hectares. The average number of animals is 23.24 ± 29.52.

### 3.2. Descriptive Statistics

Descriptive statistics of dependent variables are given in Table 3.

**Table 3.** Descriptive statistics.

| Variable | ΔS | | ΔV | | Mean | Standard Deviation | Normal KS Test (*p* [¥]) |
|---|---|---|---|---|---|---|---|
| | r | *p* | r | *p* | | | |
| ΔA | 0.251 | 0.000 | 0.029 | 0.000 | 65.9 | 13.72 | 0.025 (0.200) |
| ΔS | | | −0.016 | 0.000 | 176.8 | 31.72 | 0.031 (0.200) |
| ΔV | | | 1.000 | 0.000 | 125.0 | 25.75 | 0.022 (0.200) |

[¥] Lilliefors significance correction, r: Pearson's correlation coefficient, KS: Kolmogorov–Smirnov.

As shown in Table 3, the correlation between the change in alfalfa production (ΔA) and the change in silage corn production (ΔS) after subsidies was significant ($p < 0.01$). The positive correlation coefficient means that after subsidies, alfalfa production and silage corn production increased together. This situation supports the necessity of using the multivariate linear regression model in the study. When average values were analyzed, there were increases in alfalfa, silage corn, and vetch productions after subsidies. The increase was most common in silage corn, followed by vetch and alfalfa, respectively. After subsidies, an average of 65.9, 176.8, and 125.0 kg increases per hectare were observed in alfalfa, silage corn, and vetch production, respectively. According to the test results in the last column of Table 3, the dependent variables provided the normal distribution assumption.

### 3.3. Multivariate Linear Regression Analysis

The analysis results of the conceptual model given in Figure 2 are presented in Table 4.

**Table 4.** Multivariate linear regression results.

| Variable | | Coefficient Results | | | | Model Results | | |
|---|---|---|---|---|---|---|---|---|
| Dependent | Independent | Coeff. (β) | SE | t | *p* | $R^2$ | F | *p* |
| ΔA | cons | 1.724 | 5.068 | 0.34 | 0.734 | 0.220 | 22.56 | 0.000 |
| | A | 0.030 | 0.059 | 0.51 | 0.609 | | | |
| | E | 1.740 | 0.681 | 2.56 | 0.011 | | | |
| | PR | −4.118 | 1.668 | −2.47 | 0.014 | | | |
| | I | 2.799 | 1.163 | −2.41 | 0.016 | | | |
| | TL | 0.032 | 0.005 | 6.58 | 0.000 | | | |
| | NA | 0.118 | 0.020 | 6.04 | 0.000 | | | |
| ΔS | cons | −1.431 | 11.390 | −0.13 | 0.900 | 0.263 | 28.56 | 0.000 |
| | A | −0.121 | 0.131 | −0.92 | 0.359 | | | |
| | E | 1.652 | 1.530 | 1.08 | 0.281 | | | |
| | PR | −2.431 | 3.748 | −0.65 | 0.517 | | | |
| | I | −2.742 | 2.613 | 1.05 | 0.294 | | | |
| | TL | 0.105 | 0.011 | 9.65 | 0.000 | | | |
| | NA | 0.224 | 0.044 | 5.07 | 0.000 | | | |
| ΔV | cons | −24.313 | 10.011 | −2.43 | 0.016 | 0.137 | 12.66 | 0.000 |
| | A | 0.165 | 0.115 | 1.43 | 0.153 | | | |
| | E | −0.646 | 1.344 | −0.48 | 0.631 | | | |
| | PR | 10.437 | 3.294 | 3.17 | 0.002 | | | |
| | I | −7.281 | 2.297 | 3.17 | 0.002 | | | |
| | TL | 0.069 | 0.010 | 7.12 | 0.000 | | | |
| | NA | −0.023 | 0.039 | −0.60 | 0.547 | | | |

As shown in Table 4, it was observed that farmer's educational status, place of residence, non-agricultural income status, total land, and the number of animals were effective on the difference (ΔA) in alfalfa production before and after subsidies. Educational status, total land size, and number of animals positively affected increasing alfalfa production ($p < 0.05$). After the support, rural farmers had a change in alfalfa production, producing about 42.0 kg per hectare higher compared to those living in the city ($p < 0.05$). Similarly, after subsidies, alfalfa production was increased by about 28.0 kg per hectare for farmers with non-agricultural income, as compared to farmers without non-agricultural income ($p < 0.05$). Silage corn production (ΔS) after subsidies was affected by total land and the number of animals with increased yield ($p < 0.05$). It was observed that the place of farmer's residence, non-agricultural income status, and total land were effective on the difference in vetch production (ΔV) before and after subsidies. In city farmers, after the subsidies, the change in vetch production was about 104.4 kg more per hectare compared to rural residents ($p < 0.05$). Similarly, in farmers without non-agricultural income, the change in vetch production after subsidies was about 73.0 kg more per hectare than farmers with non-agricultural income ($p < 0.05$).

## 4. Discussion

Despite the support of forage plants in Turkey, the production of fodder crops has not achieved the desired level. This situation made it important to determine the factors affecting the production increase. At this point, especially examining the socioeconomic characteristics of farmers can provide necessary measures to increase the level of production.

In this study, it was seen that most of the farmers are primary school graduates, that is, they have a low education level. Education is a productive value in agriculture because the farmer may be more skilled at managerial decisions on product selection and purchasing [20,24]. In addition, it was observed that a large part of the farmers lived in rural areas. This situation brings the farmers away from the current and technological information. For this reason, training activities for farmers should be emphasized and current information about the importance of forage crops and the techniques for growing them should be given.

A significant increase was observed in the production of alfalfa, silage corn, and vetch after subsidies. This increase occurred mostly in silage corn followed by vetch and alfalfa, respectively. In the Aegean region, almost all of the existing cattle are made up of culture race animals. Culture race animals prefer roughage with high protein and crude fiber content. The low cost of roughage increases the profitability of livestock enterprises. For this reason, the livestock farms in the Aegean region have gone on to increase the production of alfalfa, silage corn, and vetch in accordance with the ecology of the region. The ecological structure of the Aegean region is suitable for the cultivation of alfalfa, silage maize, and vetch, which are among the quality roughages. The fact that silage corn and vetch production was one-year roughage production and alfalfa production was a perennial forage plant affected the preference of farmers in roughage production. This situation causes the farmers to prefer the annual plants in order to benefit more from the subsidies. It has been observed that the width of land is important in the production of alfalfa, silage corn, and vetch.

It is known that as the amount of land increases, agricultural production also increases [20]. As the amount of land increases, farmers benefit more from the supports and cultivate more forage crops. The number of animals was effective on the increase in alfalfa and silage maize production. Dairy farming is widespread in the Aegean region and alfalfa is an important forage plant for every period of dairy cattle. From the nutritionally rich alfalfa plant, 8–10 harvests can be obtained annually in regions such as the Aegean and South Anatolia. In addition, in order to increase the meat and milk yield in the winter season, the green forage plants abundant in other seasons should be stored by silage. The silage feed used in animal husbandry is mostly obtained from corn plants in Turkey [25]. Therefore, as the number of animals increases, farmers are expected to increase their production of alfalfa and silage corn.

Considering the place where the farmer lives, it is determined that the increase in alfalfa production in rural farmers is 42.0 kg/ha more than the ones living in the city. On the contrary, in farmers living in the city, the increase in vetch production is 104.4 kg/ha more than those living in rural areas. As the farmer producing alfalfa spends more effort and time than the farmer producing vetch, it can be said that the farmers who live in rural areas prefer alfalfa production and the farmers who live in the city prefer to produce vetch, which requires less labor and time. In addition, since vetch production requires less cost than alfalfa production, it has been determined that farmers without non-agricultural income tend to produce vetch. On the other hand, it has been determined that farmers with non-agricultural income tend to produce alfalfa.

One of the important factors in increasing the amount of forage crop production is increasing the available forage crop supports. Farmers will also earn more income from other crop production by increasing the amount of forage crop support. This situation will naturally cause an increase in forage crop cultivation area in agricultural enterprises. In addition, providing farmers with up-to-date information on forage crop cultivation techniques may trigger an increase in production. One of the important results obtained in this study is the positive effect of enterprise scale on forage crop production area. According to the results of the model, a one-unit increase in the size of the farmland leads to an increase of 0.03 units in alfalfa, 0.10 units in silage corn and 0.07 units in vetch, respectively, in terms of production area. This result proves that enterprises with a large enterprise scale are advantageous in terms of the production of forage crops, which is the most important cost element of animal husbandry. This can be evaluated as an advantage of positive scale economy.

## 5. Conclusions, Limitations, and Future Research Directions

The production of forage crops and the amount of roughage obtained in pasture-pasture areas is not sufficient to meet the needs of sheep and cattle breeding in Turkey. In order to provide quality feed, which has become the most important problem of animal husbandry, it is important to take some precautions as soon as possible and to apply them carefully in order to improve the culture of forage crops in the field agriculture and meadow and pasture areas. There is a similar situation in the Aegean Region. When the climate conditions of the Aegean Region and the condition of the meadows and pastures are evaluated, the studies on forage crops and pasture presence should be given priority. With the training activities to be organized on a regional basis, forage crop cultivation should be encouraged, and updated information should be provided to producers on the importance of forage crops and growing techniques. In addition, support should be given to the production of alternative fodder crops that can support roughage production. In addition to the gradual increase of supports, the development of an appropriate support policy is also important for increasing forage crop production. Forage crop yield per hectare received in Turkey, which supports forage crops showed differences by region of fodder crops are given according to the size of arable land. This situation causes dissatisfaction among farmers. For this, supports can be applied in different ways depending on the regions or support can be made on the yield per hectare. We anticipate that improving this situation will significantly increase forage crop production. On the other hand, we suggest that it is important to provide frequent training to existing farmers.

No study is without limitations, and this study has some limitations and requires further scientific inquiry. First, as this study included only Aegean Region to collect data, these results cannot be generalized to all of Turkey, however, can be helpful to understand the behavior of sustainable forage crop production. Second, we think that working with more or different independent variables will increase the coefficient of determination. However, since this study focused only on the socioeconomic characteristics of the farmers, we had to keep the independent variables in a limited number, and we benefited from the literature while determining these variables. Third, while the research methodology adopted in this paper provided the collection of information with an adequate depth that met the paper's objective, the study would have been richer if a more extensive data collection methodology was used. We relied on a cross-sectional study approach that built on a farmer survey. A more extensive data gathering approach, such as a panel data study, would have provided more information. Finally, more detailed investigations can be made with techniques such as logistic regression analysis to examine the increase in forage crop production before and after support.

**Author Contributions:** conceptualization, C.C.; methodology, C.C., and B.A.; software, B.A.; validation, C.C., and B.A.; formal analysis, B.A.; writing—original-draft preparation, C.C., and B.A.; writing—review and editing, C.C., and B.A. All authors have read and agreed to the published version of the manuscript.

**Acknowledgments:** We are grateful for the unbelievably valuable comments of three anonymous reviewers and the Academic Editor. We thank the forage producers, agricultural institutions, and organizations in Aegean region from which we received the primary data of the study.

## References

1. Kosolapov, V.; Rud', V.; Korshunov, A.; Savchenko, I.; Switala, F.; Hogland, W. Scientific support of the fodder production: V.R. Williams All-Russian Fodder Research Institute (WFRI) activity. *IOP Conf. Ser. Earth Environ. Sci.* **2019**, *390*, 012010. [CrossRef]
2. Alçiçek, A.; Kılıç, A.; Ayhan, V.; Özdoğan, M. Türkiye'de kaba yem üretimi ve sorunlari. In Proceedings of the Türkiye Ziraat Mühendisliği VII Teknik Kongresi, Ankara, Turkey, 11–15 January 2010; pp. 1071–1080.
3. Topçu, G.D.; Özkan, Ş.S. General view to meadow-rangelands and forage crops cultivation of Aegean Region and Turkey. *COMU J. Agric. Fac.* **2017**, *5*, 21–28.
4. Ozkan, U. Comparative Overview and Evaluation of Turkey's Forage Crops Agriculture. *Turk. J. Agric. Eng. Res.* **2020**, *1*, 29–43.
5. Acar, Z.; Tan, M.; Ayan, İ.; Aşçı, Ö.Ö.; Mut, H.; Başaran, U.; Gülümser, E.; Can, M.; Kaymak, G. Türkiye'de Yem Bitkileri Tarımının Durumu ve Geliştirme Olanakları. In Proceedings of the Türkiye Ziraat Mühendisliği IX. Teknik Kongresi, Ankara, Turkey, 13–17 January 2020; pp. 529–553.
6. Yavuz, G.; Ceylan, İ.C. Polatlı ilçesinde üreticilerin yem bitkileri üretimine karar verme sürecinde etkili faktörlerin belirlenmesi üzerine bir araştırma. *Tarım Bilimleri Derg.* **2005**, *11*, 133–138.

7. Aşkan, E.; Dağdemir, V. Devlet desteklemelerinden faydalanan süt sığırcılığı yapan işletmelerin üretim değerini etkileyen faktörlerin analizi: Erzurum, Erzincan, Bayburt illeri örneği. *Turk. J. Agric. Econ.* **2015**, *21*, 69–76.
8. Saygı, Y.D.; Alarslan, Ö.F. Kaba yem destekleme uygulamalarının Yozgat yöresi süt sığırı yetiştiriciliğine etkileri. *Vet. Hekimler Dern. Derg.* **2012**, *83*, 25–35.
9. Ahn, B.; Han, S. Analysis on the effects of government's support for forage production. *Korean J. Agric. Econ.* **2016**, *57*, 55–78.
10. Chang, J.B. The effects of forage policy on feed costs in Korea. *Agriculture* **2018**, *8*, 72. [CrossRef]
11. Aksu, N.; Dellal, İ. Afyonkarahisar ilinde yem bitkileri desteğinin büyükbaş hayvancılık faaliyetleri ile ilişkisinin değerlendirmesi. *Yüzüncü Yıl Üniv. Tarım Bilimleri Derg.* **2016**, *26*, 52–60. [CrossRef]
12. Karadavut, U.; Genç, A.; Palta, Ç.; Çarkacı, A.; Kökten, K. Konya İli Yem Bitkileri Üreticilerinin Sosyo-Ekonomik Yapıları İle Başarılı Üretimi Etkileyen Faktörlerin Belirlenmesi. *Bingöl Üniv. Fen Bilimleri Derg.* **2011**, *1*, 1–12.
13. Aydoğdu, M.; Aydoğdu, İ.; Cevheri, C.; Sevinç, M.R.; Küçük, N. Şanlıurfa'daki yem bitkileri eken çiftçilerin sosyo-ekonomik profilinin analizi. *J. Ekon.* **2020**, *0*, 10–15.
14. Rehman, F.; Muhammad, S.; Ashraf, I.; Ruby, T. Effect of farmers' socioeconomic characteristics on access to agricultural information: Empirical evidence from Pakistan. *J. Anim. Plant Sci.* **2013**, *23*, 324–329.
15. Matata, P.; Ajay, O.; Oduol, P.; Agumya, A. Socio-economic factors influencing adoption of improved fallow practices among smallholder farmers in western Tanzania. *Afr. J. Agric. Res.* **2010**, *5*, 818–823.
16. Cruz, F. Adoption and diffusion of agricultural extensions. In *An Introduction to Extension Delivery Systems*; Valera, J.B., Martinez, V.A., Plopino, R.F., Eds.; Island Publishing House: Manila, Philippines, 1987; pp. 97–127.
17. Feder, G.; Just, R.E.; Zilberman, D. Adoption of agricultural innovations in developing countries: A survey. *Econ. Dev. Cult. Chang.* **1985**, *33*, 255–298. [CrossRef]
18. Baffoe-Asare, R.; Danquah, J.A.; Annor-Frempong, F. Socioeconomic factors influencing adoption of CODAPEC and cocoa high-tech technologies among small holder farmers in Central Region of Ghana. *J. Exp. Agric. Int.* **2013**, *3*, 277–292. [CrossRef]
19. Alam, M.; Siwar, C.; Talib, B.; Toriman, M. The relationships between the socioeconomic profile of farmers and paddy productivity in North-West Selangor, Malaysia. *Asia Pac. Dev. J.* **2011**, *18*, 161–173. [CrossRef]
20. Anigbogu, T.U.; Agbasi, O.E.; Okoli, I.M. Socioeconomic factors influencing agricultural production among cooperative farmers in Anambra State, Nigeria. *Int. J. Acad. Res. Econ. Manag. Sci.* **2015**, *4*, 43–58. [CrossRef]
21. Moloi, M.J. *A Comparison of Socioeconomic Characteristics that Determine the Farm Income of Emerging Lifestock and Horticultural Farmers in South Africa*; University of Limpopo: Polokwane, South Africa, 2010.
22. Ademiluyi, S.; Oladele, O.I.; Wakatsuki, T. Socioeconomic factors influencing power tiller use among Sawah farmers in Bida, Nigeria. *J. Food Agric. Environ. Finl.* **2008**, *6*, 387–390.
23. Johnson, R.A.; Wichern, D.W. *Applied Multivariate Statistical Analysis*, 3th ed.; Prentice Hall International: Englewood Cliffs, NJ, USA, 1992.
24. Fane, G. Education and the Managerial Efficiency of Farmers. *Rev. Econ. Stat.* **1975**, *57*, 452–461. [CrossRef]
25. Şahin, İ.F.; Zaman, M. An Important Cattle Feed: Silage. *Doğu Coğraf. Derg.* **2010**, *15*, 1–18.

# 3

# Morphological, Physiological and Genetic Responses to Salt Stress in Alfalfa

**Surendra Bhattarai [1], Dilip Biswas [1], Yong-Bi Fu [2] and Bill Biligetu [1,*]**

[1] Crop Development Center of Department of Plant Sciences, University of Saskatchewan, 51 Campus Drive, Saskatoon, SK S7N 5A8, Canada; surendra.bhattarai@usask.ca (S.B.); dkb283@mail.usask.ca (D.B.)

[2] Plant Gene Resources of Canada, Saskatoon Research and Development Centre, Agriculture and Agri-Food Canada, 107 Science Place, Saskatoon, SK S9H 3X2, Canada; yong-bi.fu@canada.ca

* Correspondence: bill.biligetu@usask.ca.

**Abstract:** Alfalfa (*Medicago sativa* L.) is an important legume forage crop. However, its genetic improvement for salt tolerance is challenging, as alfalfa's response to salt stress is genetically and physiologically complex. A review was made to update the knowledge of morphological, physiological, biochemical, and genetic responses of alfalfa plants to salt stress, and to discuss the potential of applying modern plant technologies to enhance alfalfa salt-resistant breeding, including genomic selection, RNA-Seq analysis, and cutting-edge Synchrotron beamlines. It is clear that alfalfa salt tolerance can be better characterized, genes conditioning salt tolerance be identified, and new marker-based tools be developed to accelerate alfalfa breeding for salt tolerance.

**Keywords:** alfalfa (*Medicago sativa* L.); salt stress; salt resistance

---

## 1. Introduction

Alfalfa (*Medicago sativa* L.) is a perennial forage legume that belongs to the sub-family of Papilionoideae. Though a diploid form exists, cultivated alfalfa is a predominantly cross-pollinated, tetraploid ($2n = 4x = 32$) species [1], which originated in southwestern Asia with Iran as the geographic center of origin [2,3]. Alfalfa is an important forage source for livestock industries around the world because of its wide adaptability, high yield, good quality, and resistance to frequent cuttings [3,4]. It can be used for pasture, hay, silage, dehydrated products, seed production, and soil improvement [3,4]. Globally, alfalfa is grown on about 30 million ha [5]. In Canada, more than four million ha of alfalfa is produced either in monoculture or in mixture with grasses [6], while approximately seven million ha is grown in the United States [7]. Alfalfa is a moderately saline-tolerant legume [8]. A number of alfalfa cultivars with improved salt tolerance have been developed using conventional breeding approaches [9]. However, the genetic improvement of salt tolerance in alfalfa is challenging, mainly as the response of alfalfa plants to salt stress is physiologically and genetically complex, because salt tolerance is controlled by multiple genes and involves various biochemical and physiological mechanisms [9].

Soil salinity is one of the most influential stressors that limits agricultural production. Saline soil is characterized by an excess concentration of soluble salts (chloride; sulfate; and carbonate of sodium, calcium, magnesium, potassium) in the root zone, making it difficult for plants to extract water and nutrients from the soil and causing plant injury [10]. Specifically, soil with an electrical conductivity (EC) of the saturation extract of more than 4 dS $m^{-1}$ in the root zone at 25 °C with 15% of exchangeable sodium is considered to be saline [11]. Salinization has long been known as a common environmental phenomenon worldwide [12,13] and is becoming a global issue of land degradation, with more prevalence in arid and semi-arid regions [14] as shown in Figure 1. More than 6% of the world's total land area is salt-affected, either by salinity (397 million hectares) or by the associated conditions of sodicity (434 million hectares) [15]. For example, salinization has affected the agricultural productivity of more than 10 million ha in the Great Plains of North America [16] with 6 million ha of agricultural land in the Canadian Prairies [17,18]. Due to salt build-up over time, the Food and Agricultural Organization (FAO) has estimated that 0.25–0.50 million ha of irrigated lands are becoming unsuitable for cultivation annually [19]. The development of crop cultivars resistant to soil salinity is one of the most effective strategies for maintaining sustainable crop production.

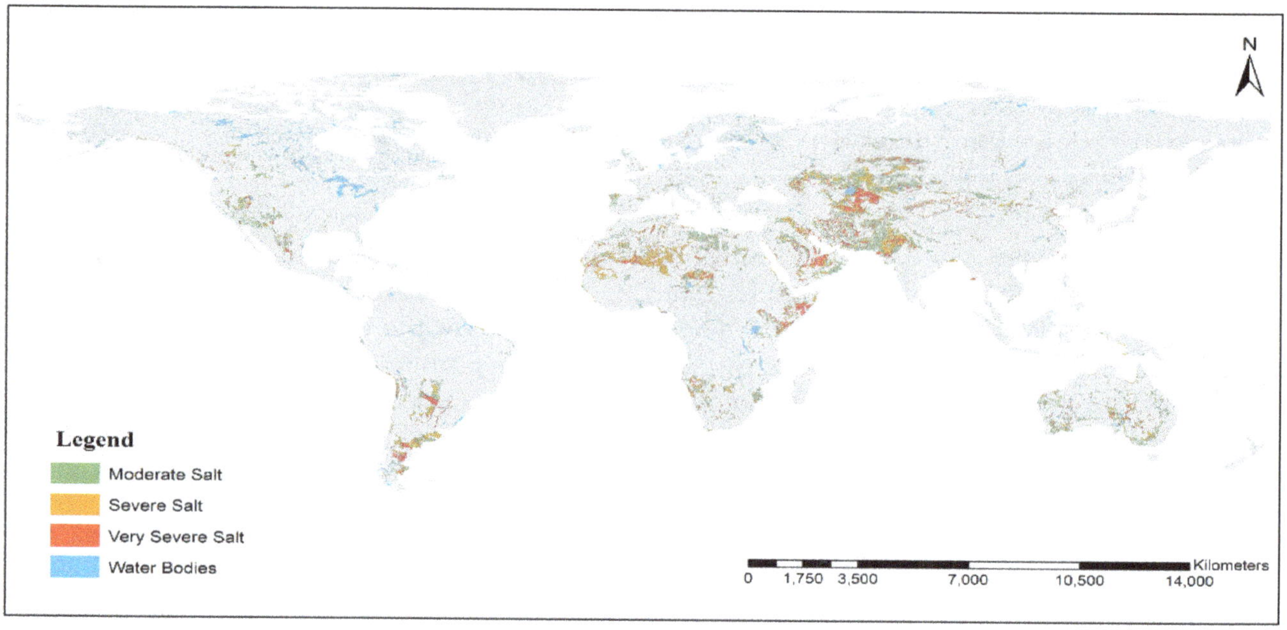


**Figure 1.** Global salt-affected regions and their severity levels (drawn based on Harmonized World Soil Database v. 1.2 [20]).

Understanding the mechanisms conferring salt tolerance and identifying heritable traits for screening for improved salt tolerance are crucial for alfalfa breeding. It is also important to identify and characterize the genes responsible for salt tolerance in alfalfa for the development of molecular markers for precise screening in breeding and genetic improvement. Thus, an attempt is made here to review research on the morphological, physiological, biochemical, and genetic responses of alfalfa plants to salt stress. We also discuss the potential of applying modern technologies to enhance alfalfa salt-resistant breeding, including genomic selection, RNA-Seq analysis, and cutting-edge Synchrotron beamlines.

## 2. Effect of Salt Stress on Morphology, Growth, Forage Yield and Nutritive Value

Salt stress reduces plant growth by influencing turgor, photosynthesis, and the activity of specific enzymes [21]. The influence can occur in two phases; the first phase is governed by the osmotic effect due to high salt concentration in root zones, whereas the second phase is governed by toxic effects due to high salt accumulation in leaf tissues [22] (Figure 2). The presence of high root zone salt concentration causes higher osmotic pressure in soil solution than in plant cells, reducing the ability of plants to uptake water and essential minerals like potassium and calcium [23]. In severe salt stress, the soil solution becomes hyper-osmotic, causing the root cells to lose water, which results in severe wilting or plant senescence [24]. Osmotic stress initially reduces leaf growth and eventually causes a reduction in shoot development and reproductive growth [25], primarily due to the water deficit in plant tissues [24]. Salt stress induces decreased photosynthetic rate due to osmotic stress-induced partial stomatal closure [25]. Sodium ions absorbed by plant roots, if present at a high concentration in cytosol, can be harmful to the plant [26]. Since sodium and potassium ions are both monovalent cations, they compete for uptake by the plant under fully hydrated saline conditions [27] resulting in a deficiency in potassium, an essential macronutrient required for normal metabolic function [23,28]. Increased concentrations of sodium and chloride ions in the cytoplasm can disrupt cellular processes, causing damage to photosynthetic apparatus as well as the dehydration of cells [25,29]. Therefore, limiting excess amounts of sodium in cytosol and increasing the cellular potassium to sodium ratio are crucial for salt tolerance [30,31]. This means the maintenance of a regular photosynthetic rate and stable K+/Na+ ratio are important traits for salt-tolerant alfalfa cultivars. The development of molecular markers should target the specific traits mentioned above to provide informative markers.

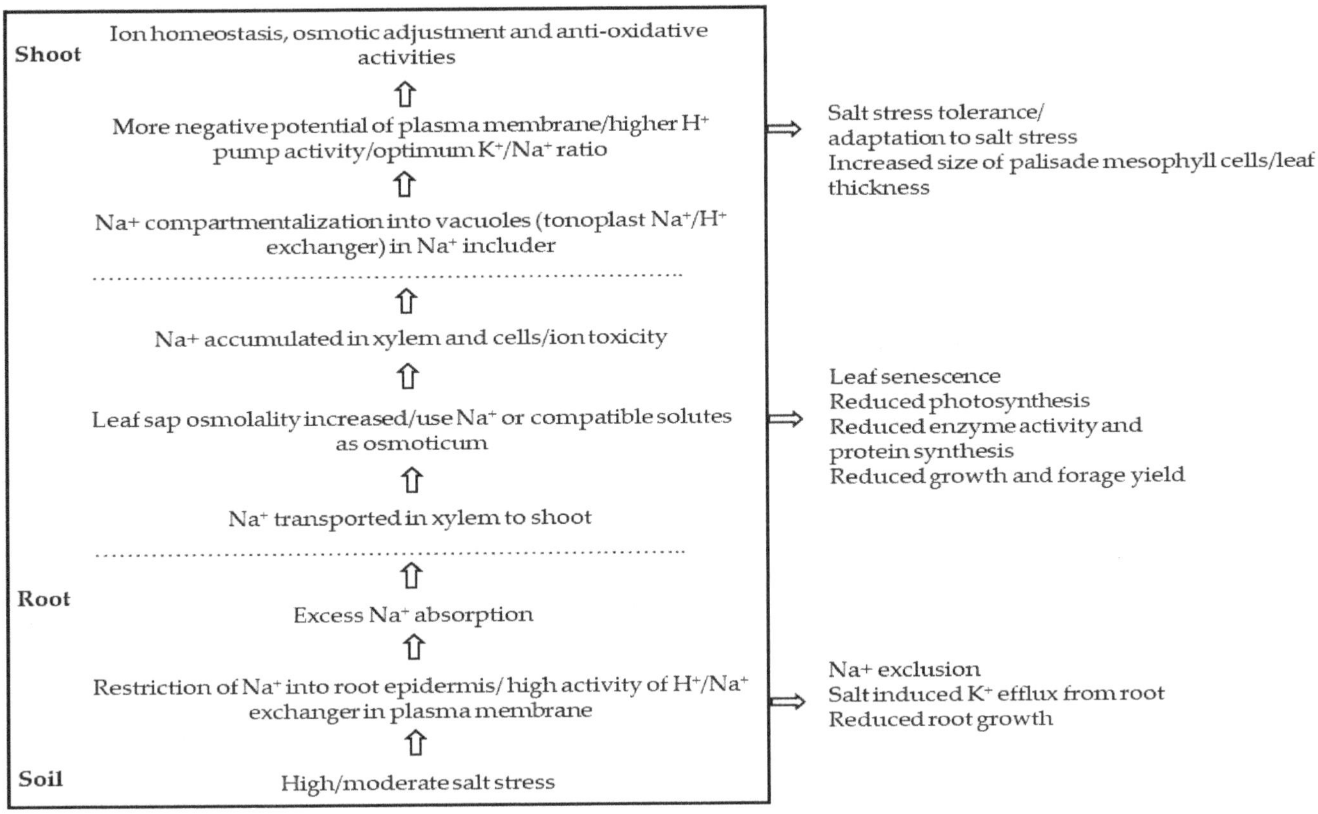


**Figure 2.** Summary of salt stress tolerance in alfalfa.

Smith [32] described three different growth stages at which alfalfa plants may be affected by salinity: germination, seedling growth, and mature plant growth. The seed germination and seedling stages of alfalfa are highly sensitive to salt stress [33]. Though many soluble salts exist in the natural environment, the majority of research studies on alfalfa salt tolerance used a NaCl solution, with few studies evaluating the response of alfalfa biomass production under a mixture of different salts [34,35]. Alfalfa seed germination is more susceptible to $CaCl_2$ and NaCl salts and more resistant to KCl [36]. Salt stress can significantly reduce the germination and seed vigor of alfalfa [36,37]. The effect of $Na_2SO_4$ solution on alfalfa plants from emergence to maturity was examined and it was found that the relative emergence (%) decreased dramatically at 12.7 dS $m^{-1}$, with no surviving plants at 30 dS $m^{-1}$ [38]. In another study, a mixture of $CaCl_2$, NaCl, $MgSO_4$, and $Na_2SO_4$ was used to obtain the required EC levels to evaluate alfalfa seedling emergence [16]. In this study, alfalfa seedlings emerged equally under the 1.5 dS $m^{-1}$ and 8.0 dS $m^{-1}$ treatments, but the emergence decreased by 3–30% at 15.6 dS $m^{-1}$, except for a few tolerant cultivars [16]. In the past, many breeding efforts have been targeted to improve the seed germination of alfalfa under salt stress, but the selection for salt tolerance should focus on the whole life cycle of alfalfa rather than focusing on a particular growth stage, as there is no clear correlation between seed germination to post germination performance [39,40].

The shoot growth of alfalfa had been found to be more adversely affected by salinity than root growth [41]. A study on 15 alfalfa populations under salt stress imposed by a mixture of $MgSO_4$, $Na_2SO_4$, $CaCl_2$, NaCl, and KCl revealed that root mass per plant at 18.3 dS $m^{-1}$ and 24.5 dS $m^{-1}$ EC was reduced by 18% and 49%, while the shoot mass was reduced by 50% and 73%, respectively [35]. The average shoot mass of alfalfa during the first, first + second, and first + second + third cuts decreased by 50%, 44%, and 38% at 8.0 dS $m^{-1}$ (chloride and sulfate salt solution) and by 80%, 73%, and 67% at 15.6 dS $m^{-1}$, respectively [16]. Salt stress in alfalfa caused a reduction in the relative growth rate [42,43]. Moderate NaCl salt stress (9 ± 0.2 dS $m^{-1}$) reduced the plant height, leaf, and stem masses of alfalfa by 32%, 34%, and 35%, respectively [44]. Sibole et al. [45] reported that stem and petiole growth was sensitive to NaCl salt stress in two different salt-tolerant Mediterranean *Medicago* species. As compared to the salt stress imposed by NaCl, other saline solutions containing $Ca^{2+}$ and $K^+$ ions not only have less toxic effects on plants, but also they can mitigate the negative effects on a plant under salt stress [46,47].

Though a reduction in growth rate and shoot mass were common, high genetic diversity existed among alfalfa populations under salt stress [16,35,48], suggesting an adequate genetic variation for further selection for improved salt tolerance. Lei et al. [49] compared the performance of seven alfalfa cultivars under NaCl stress and found that the growth rate of salt-tolerant cultivars was less affected by a high salt stress of ~50 dS $m^{-1}$. However, in another study, no significant variation in both shoot and root biomasses was observed between salt-tolerant and intolerant alfalfa cultivars under NaCl stress [50]. Differences among studies might be because of the genetic backgrounds of the alfalfa, as the studies used different salt-tolerant cultivars.

The forage nutritive value was affected by the salinity and varied among the cultivars [51]. Salt stress up to 20 dS $m^{-1}$ increased the leaf-to-stem ratio of alfalfa and its crude protein content [52] and reduced acid detergent fiber (ADF) and neutral detergent fiber (NDF) [51,53]. The reduction in stem height may be the main factor causing changes in alfalfa nutritive value, as it increases the leaf-to-stem ratio.

## 3. Effect of Salt Stress on Physiological Responses in Alfalfa

The mechanisms underlying the ability of alfalfa seedlings to tolerate salt stress are complex processes involving photosynthesis, detoxifying, antioxidant, secondary metabolism, and ion transport [54]. Salinity alters photosynthetic pigments (i.e., chlorophyll content), therefore it reduces the maximum photochemical efficiency of alfalfa genotypes [55,56] and increases the dark respiration rate in alfalfa [43]. Salinity reduces membrane stability, relative water content, and growth rates and increases lipid peroxidation, proline, and hydrogen peroxide ($H_2O_2$) contents in the leaf tissue of alfalfa [57]. In fact, studies have shown that an increased level of proline content is associated with improved salt tolerance [41,58]. In addition, effective osmoregulation in salt-tolerant alfalfa cultivars is also associated with increased levels of sucrose and pinitol in leaves and a high accumulation of starch in roots [50]. The pinitol accumulation is a characteristic of halophytic plants growing in a saline environment. Anower et al. [59] characterized the physiological traits of two alfalfa half-sib (HS) families, HS-A and HS-B, selected for improved salt tolerance. Under salt treatment, HS-B showed a greater leaf number (72%) and stem length (44%) and HS-A displayed better leaf production (84%) compared to the unselected initial population. This improved growth was associated with a 208% and 78% greater accumulation of chlorophyll content in HS-B and HS-A, respectively. An increase in protein concentration in salt-tolerant alfalfa cultivars had been reported in previous studies [51,53]; it may be associated with an increase in chlorophyll and enzyme activities in the leaves. It is common for plant breeders to measure the plant vigor and shoot mass to determine the performance of alfalfa under salt stress. However, this analysis could be enhanced by the determination of leaf-to-stem ratio and certain physiological traits such as chlorophyll content and protein concentration.

## 4. Effect of Salt Stress on Oxidative Stress and Anti-Oxidative Activities

Salinity induces oxidative stress in plants at the sub-cellular level [60]. Salt stress increases the accumulation of superoxide radicals ($O_2^-$) and hydrogen peroxide ($H_2O_2$) in cell compartments including chloroplast and mitochondria [60]. Therefore, lipid peroxidation and protein oxidation occur in the apoplastic space. An increase in peroxidase (POX) reduces $H_2O_2$ to water using various substrates as electron donors in salt-tolerant alfalfa [61,62], suggesting that the analysis of antioxidant enzymes could be useful in understanding the salt tolerance of alfalfa. During seed germination and seedling growth, increased salt stress also increases the activities of superoxide dismutase (SOD), catalase (CAT), and ascorbate peroxidase activity (APX)—and the increase is higher in salt-tolerant alfalfa cultivars [62,63]. Under salt stress, the unsaturated fatty acids of plant membranes are decomposed to malondialdehyde (MDA), and the rate of lipid peroxidation in terms of MDA can be used as a biochemical indicator to evaluate salt-tolerant cultivars [30,64]. The salt tolerance of alfalfa is also improved by melatonin application, which acts as an antioxidant in scavenging $H_2O_2$ and enhances antioxidant enzymes' activities [65]. Under salt stress, tolerant alfalfa shows less severe cell membrane damage and lower accumulation of reactive oxygen species (ROS) than salt sensitive types [66].

## 5. Effect of Salt Stress on Ion Uptake in Alfalfa Plants

Under salt stress, about 98% of the ions in soil solutions are excluded from the root in most plants, while the rest are transported to shoot tissues [22]. To prevent tissue ion toxicity, the ion exclusion mechanism can restrict excessive ion transport from root to shoot. The ion exclusion mechanism includes: (1) the minimal uptake of particular toxic ions by the root and maximization of ion efflux to the soil, (2) restricting excessive ion loading into the xylem, (3) increasing the ion retrieval from xylem to other tissues like the root and stem, and (4) increasing the ion transport from shoot to root through the phloem [67]. High concentrations of sodium and chloride ions in leaves generally lead to leaf senescence. However, the tissue tolerance to such a stress is genotype-specific [24]. Tolerance to high concentrations of sodium and chloride ions in leaves can be achieved by the intercellular partitioning of ions, thus avoiding their accumulation in the photosynthetic organelles [25]. For example, salt-tolerant species can sequestrate ions in the cell vacuoles in leaves [25,68]. Thus, the comparison of leaf injury scores under high salt stress can be used as a morphological marker for identifying salt-tolerant genotypes in early phases with similar $LD_{50}$ values (median lethal dose to kill 50% of population) [33].

Salt stress can lead to nutritional imbalances due to competitive absorption and translocation of elements, which may result in reduced physiological activity [69]. This might in part explain the plant growth reduction and low shoot mass of alfalfa at high salt stress, as discussed in the previous section. The high concentrations of sodium and chloride ions in soil solutions may decrease the calcium, potassium, and magnesium activities in alfalfa [70] and also decrease nitrogen accumulation [71,72]. High salt stress increases the concentrations of sodium, total sulfur, chloride, magnesium, and phosphorus, and decreases the concentrations of potassium and calcium in the shoots of alfalfa genotypes [38]. The maintenance of the $K^+/Na^+$ ratio is necessary through $Na^+$ exclusion from the root epidermis, but NaCl-induced $K^+$ efflux from the root causes the deficiency of macronutrients (nitrogen and potassium), reducing plant production and productivity.

Knowledge of the biochemical composition of plant tissues is crucial for salt tolerance studies, as cutting-edge synchrotron-based approaches can help plant scientists to quantify the biochemical compounds and image their structures with minimal sample modifications [73]. The synchrotron is a powerful facility that accelerates charged particles such as electrons in a large ring-like trajectory at relativistic (near-light) speed [73], where the energy of the electron generates light ranging from infrared to soft and hard X-rays at high intensities [74]. In recent years, synchrotron techniques have been employed to study a number of biotic and abiotic stresses, such as the identification of heat-tolerant field pea (*Pisum sativum* L.) genotypes [75,76], drought tolerance traits in spring wheat (*Triticum aestivum* L.) [77], and *Fusarium* head blight tolerance in wheat [78,79]. Synchrotron beamlines were also applied to study the accumulation and distribution of organic compounds and salt ions in leaf, stem, and root tissues of alfalfa cultivars with different salt tolerances [80]. A comprehensive study on ion localization in sub-cellular levels using synchrotron beamlines would provide further insights into salt tolerance in alfalfa.

## 6. Proteome and Transcriptomic Analyses

Several transcriptome-based and proteome studies have been conducted to understand salinity stress in alfalfa [49,54,66,81–87]. After a three-day salt treatment, a proteomic study on the roots of two-week-old seedlings found 83 differentially expressed proteins in alfalfa cultivars with contrasting tolerances to salinity (Table 1) [81]. These proteins are involved in ion homeostasis, protein turnover and signaling, protein folding, cell wall components, carbohydrate and energy metabolism, reactive oxygen species (ROS) regulation and detoxification, and purine and fatty acid metabolisms. Proteins such as peroxidase, protein disulfide-isomerase, nicotinamide adenine dinucleotide (NAD) synthetase, and isoflavone reductase were significantly up-regulated in the salt-tolerant alfalfa cultivar "nonomu" [81]. The proteomic analysis of 30-day-old alfalfa seedling roots treated with ~30 dS $m^{-1}$ NaCl for 8 h found 93 and 30 differentially expressed proteins in salt-tolerant alfalfa ("Zhongmu-1") and salt-sensitive *Medicago truncatula* ("Jemalong A17"), respectively (Table 1) [82].

These proteins primarily play roles in molecule binding and catalytic activities. Xiong et al. [54] identified 26 (shoot) and 35 (root) differentially abundant proteins in salt-stressed alfalfa compared to alfalfa that had experienced no salt treatment. Similarly, the proteomic analysis of osmo-primed alfalfa seeds that germinated under salinity stress found they contained 94 proteins with different responses to salt treatments [86]. Based on their functions, the proteins were classified as protein destination and storage (seed storage proteins and small heat-shock proteins), cell growth/division (late embryogenesis abundant proteins and seed maturation proteins), metabolism (methionine synthase, cysteine synthase and haem oxygenase), and disease and defense (glutathione S-transferase).

**Table 1.** Number of differently expressed genes and proteins and their functions in salt-tolerant and intolerant alfalfa cultivars.

| Alfalfa Genotype | Tissue | Salt Stress | Total Number of Differentially Expressed Genes/Proteins | Major Pathway/Function | Reference |
|---|---|---|---|---|---|
| NM-801 (tolerant), Vernal (intolerant) | root | 2-week-old seedlings treated with ~5, ~10 dS $m^{-1}$ NaCl for 3 days | 83 | Ion homeostasis, protein turnover and signaling, protein folding, cell wall components, carbohydrate and energy metabolism, reactive oxygen species regulation and detoxification, and purine and fatty acid metabolism. | [81] |
| Zhongmu-1 (*M. sativa*, tolerant), Jemalong A17 (*M. truncatula*, intolerant) | root | 1-month-old seedlings treated with ~30 dS $m^{-1}$ NaCl for 8 h | 93 (tolerant)<br>30 (intolerant) | Molecule binding and catalytic activity. Defense against oxidative stress, metabolism, photosynthesis, protein synthesis and processing, and signal transduction. | [82] |
| AZ-88NDC (intolerant), AZ-GERM SALT-II (tolerant) | root | 1-week-old seedlings treated with ~15 dS $m^{-1}$ NaCl for 7 days | 288/273 and 468/337 up/down-regulated in intolerant and tolerant, respectively | Response to stress, kinase activity, hydrolase activity, oxidoreductase activity, and extracellular region. | [83] |
| Zhongmu No. 1 (tolerant) | root | 12-day-old seedlings treated with ~25 dS $m^{-1}$ NaCl for 1, 3, 6, 12, 24 h | 8861 at one or more time points | Iron ion transport, ion homeostasis, antiporter, signal perception, signal transduction, transcriptional regulation, and antioxidative defense. | [84] |
| Zhongmu No. 1 (tolerant) | root, shoot | 1-week-old seedlings treated with ~10, ~20 dS $m^{-1}$ NaCl for 7 days | 26 (shoot)<br>35 (root) | Photosynthesis (31%) and stress and defense (20%) in the shoot. Defense (26%); metabolism (17%); and protein translation, processing, and degradation (17%) in the root. | [54] |
| CW064027, Bridgeview (tolerant), Rangelander (intolerant) | shoot | 4th-cut treated with 1.53, 8, 15.6 dS $m^{-1}$ maintained by sulphate- based sodium, calcium, and magnesium salts | 685/527, 368/139 up/down-regulated in CW064027 and Bridgeview at control, 537/949, 375/1045 up/down-regulated in CW064027 and Bridgeview at 8 dS $m^{-1}$, 1129/1196, 843/1516 up/down-regulated in CW064027 and Bridgeview at 15.6 dS $m^{-1}$ | Redox-related genes, B-ZIP transcripts, cell wall structural components, lipids, secondary metabolism, auxin and ethylene hormones, development, transport, signaling, heat shock, proteolysis, pathogenesis-response, abiotic stress, RNA processing, and protein metabolism. | [85] |
| Zhongmu-1 (tolerant), Xingjiang Daye (intolerant) | leaf | 30-day-old plants treated with ~50 dS $m^{-1}$ NaCl for 7 days | 1125 and 2237 between cultivars at control and stress, respectively | Response to stimulus, reactive oxygen species, responding to stress, and response to hormone and other stress-responsive processes. | [49] |

Transcriptomic approaches have been employed for alfalfa to understand the gene expression associated with salt stress (Table 1) [49,66,83–85,87]. Eighty-two unique transcripts were found from the salt-stressed seedlings of alfalfa by sampling at different time intervals from 10 min to 24 h, including 24% that were proteins related to plant metabolism and 9% that were related to abiotic stress [87]. Arshad et al. [88] found that the over-expression of microRNA156 in alfalfa resulted in increased biomass production, stem number, concentration of crude protein, and reduced uptake of $Na^+$ under salt stress. In another transcriptome study, there were 876 and 1303 differentially expressed genes (DEGs) under salt stress in the root tissues of one-week-old seedlings in salt-intolerant and tolerant alfalfa genotypes respectively, with 604 DEGs specific to the salt-tolerant type [83]. Similarly, an RNA-Seq analysis displayed 2237 and 1125 DEGs between "Zhongmu-1" (salt-tolerant) and "Xingjiang Daye" (salt-intolerant) in the presence and absence of salt stress, among which were many genes that are involved in stress-related pathways (Table 1) [49]. After a salt treatment, the number of DEGs in "Xingjiang Daye" (19,373 DEGs) compared with the control treatment was about four times that of "Zhongmu-1" (4833 DEGs). Compared with "Xingjiang Daye", "Zhongmu-1" maintained a more stable expression of genes related to ROS, calcium pathways, phytohormone biosynthesis, and Na+/K+ transport [49]. The transcriptome responses of salt-tolerant ("211609") and salt-intolerant ("Xinjiang Daye") alfalfa cultivars revealed significantly higher expression levels of *NHX1*, *ZFG*, *CBF4*, and *HSP23* genes in "211609" than in "Xinjiang Daye" [66]. In addition, a transcriptomic analysis of alfalfa roots under ~25 dS $m^{-1}$ NaCl identified 8861 NaCl-regulated DEGs in alfalfa [84]. These DEGs were categorized into 13 gene ontology categories: oxidoreductase activity, oxidation-reduction process, structural constituent of cytoskeleton, hydrolase activity, carbohydrate metabolic process, negative regulation of catalytic activity, polysaccharide catabolic process, iron ion binding, transmembrane transporter activity, cytoskeleton, trehalose biosynthetic process, protein polymerization, and ion homeostasis [84]. In previous transcriptomic studies, an individual alfalfa genotype was sampled as a replicate for RNA-Seq analysis. As alfalfa is generally seeded as a synthetic population in a field, it may be more suitable to use a group of genotypes as a replicate for a transcriptomic study. Therefore, 100 alfalfa genotypes were divided into four independent replicates to represent an alfalfa population and researchers found that 50% of DEGs were down-regulated in a salt-intolerant population [85]. The functions of the genes down-regulated in the salt-intolerant alfalfa cultivar were grouped into cell wall structural components, lipids, secondary metabolism, auxin and ethylene hormones, transport, signaling, pathogenesis response, and abiotic stress [85]. The transcriptome analysis identified many annotated sequences homologous to genes involved in osmolyte synthesis (beta-amylase, fructose-1, 6-bisphosphate, aldolase, and sucrose synthase) and ion homeostasis (H+-PPase, cation/H+ exchanger3, Ca-related channel, sodium symporter, nitrate and K+ channels, *p* translocator, and metal transporters) [49,85,87]. In summary, a large number of candidate genes responsible for salt tolerance in alfalfa have been identified in various studies, and these candidate genes can be used as a baseline for further genetic analyses of salt tolerance in alfalfa. However, it is essential to identify key genes associated with important morphological and physiological traits (i.e., chlorophyll content, sucrose synthase or plant height), and validate them in different alfalfa breeding populations to apply them in salt-tolerant breeding.

## 7. Breeding for Salt Tolerance

Breeding efforts for salt tolerance in alfalfa have focused on high seed germination under stress to improve stand establishment. There are more than 60 registered alfalfa cultivars with improved salinity tolerance in the USA, representing different fall dormancy categories (Table 2) [89]. The majority of the registered cultivars were selected for salinity tolerance at the germination stage, while only 13 of them were selected for salinity tolerance at mature growth stages. Although certain progresses have been made through conventional plant breeding, the genetic improvement of salt tolerance in alfalfa is low due to several factors. Firstly, alfalfa is polyploidy and outcrossing in nature [90] and genetic studies of its salt tolerance can be complicated and less informative. Secondly, the perennial growth form and low heritability of salt-tolerant traits have further complicated the breeding effort [91]. Thirdly, as previously stated, there is no clear relationship for salt tolerance between the germination stage and post-germination performance [39], indicating a need for a whole life cycle selection from germination to the flowering stage.

**Table 2.** Salt-tolerant alfalfa cultivars and their fall dormancy (FD) and growth stages (G, germination and/or F, forage production) for salt tolerance (ST) (Source: National Alfalfa & Forage Alliance, [89]).

| Variety | FD | ST | Variety | FD | ST |
|---|---|---|---|---|---|
| Foothold | 2 | G | GUNNER | 5 | G |
| Spredor 5 | 2 | G | MPIII Max Q | 5 | G |
| Hi-Gest 360 | 3 | G | RR NemaStar | 5 | G |
| LegenDairy XHD | 3 | G | RR Tonnica | 5 | G |
| HVX Tundra II | 3 | G | WL 365HQ | 5 | G |
| LegenDairy AA | 3 | G | 6610N | 6 | G |
| RR Presteez | 3 | G | Cisco II | 6 | G/F |
| Rugged | 3 | G | Hi-Gest 660 | 6 | G |
| WL 336HQ.RR | 3 | G | Revolt | 6 | G |
| 6401N | 4 | G | RRALF 6R200 | 6 | G |
| 6472A | 4 | G | WL 454HQ.RR | 6 | G/F |
| 6497R | 4 | G | 6829R | 7 | G |
| AFX 457 | 4 | G | AFX 779 | 7 | G |
| AFX 469 | 4 | G | AmeriStand 715NT RR | 7 | G/F |
| AmeriStand 415NT RR | 4 | G | Sun Titan | 8 | G |
| AmeriStand 427TQ | 4 | G | SW8421S | 8 | F |
| AmeriStand 455TQ RR | 4 | G | WL 535HQ | 8 | G |
| AmeriStand 457TQ RR | 4 | G | WL 552HQ.RR | 8 | G |
| AmeriStand 480 HVXRR | 4 | G | 6906N | 9 | G |
| Barricade SLT | 4 | G/F | AFX 960 | 9 | G |
| DKA40-16 | 4 | G | AmeriStand 901TS | 9 | G |
| DKA44-16RR | 4 | G | LG 9C300 | 9 | G |
| Integra 8444R | 4 | G/F | PGI 908-S | 9 | G/F |
| Magnum Salt | 4 | G/F | RRALF 9R100 | 9 | G |
| Rebound AA | 4 | G | Sun Quest | 9 | G |
| RR Stratica | 4 | G | SW 9215 | 9 | F |
| RR VaMoose | 4 | G | SW 9720 | 9 | F |
| WL 356HQ.RR | 4 | G | SW9215RRS | 9 | G/F |
| 6516R | 5 | G | WL 656HQ | 9 | G |
| 6547R | 5 | G/F | WL 668HQ.RR | 9 | G |
| AFX 579 | 5 | G | 6015R | 10 | G |
| Nimbus | 5 | F | AFX 1060 | 10 | G |

A number of studies have evaluated the performance of alfalfa germplasm under salt stress [16,92–94]. The genetic variability among nine alfalfa populations in response to sulfate salt revealed a high relative shoot mass of 8 dS $m^{-1}$ in "Halo" alfalfa, which had a lower seed emergence than that of "Rugged" [16]. Benabderrahim et al. [92] studied the genetic diversity of 36 alfalfa populations from Tunisia using 12 agronomic and physiological traits and identified three salt-tolerant types. Jiang et al. [93] have applied random amplified polymorphic DNA (RAPD) markers and clustered 25 salt-tolerant alfalfa populations into nine clusters, suggesting that salt-tolerant alfalfa germplasm possess a certain genetic diversity which can be utilized in salt-tolerant breeding. Azzam et al. [94] identified a highly salt-tolerant alfalfa population by screening 16 alfalfa populations using RAPD and inter simple sequence repeat markers. Therefore, the available genotypic diversity for salt tolerance in alfalfa provides an opportunity for plant breeders to select and develop superior salt-tolerant cultivars. However, there is still a lack of well-characterized genetic materials and common phenotyping protocols for parental selection. Due to the spatial heterogeneity of soil properties and seasonal variation in rainfall, it is difficult to perform plant screening for salinity tolerance in the field [95]. Thus, early phase screening in controlled environments is a feasible option before testing the advanced lines in the saline field with the use of proper salinity induction.

The success of a breeding program lies in the precise phenotyping of the traits of interest (seed germination, ion transport, osmolyte synthesis, signaling, photosynthesis, and protein synthesis) and correlating them with genes or quantitative trait loci for marker-assisted selection [96]. A comprehensive breeding platform, including high-throughput phenotyping and genotyping, is needed for the efficient improvement of salt tolerance in alfalfa. High-throughput phenotyping contributes directly to the genetic gain by evaluating genetic variation more efficiently. [97]. A hyperspectral imaging platform has the potential for the detection of stress responses as it can capture wavelengths within and beyond the visible spectrum [98].

Hyperspectral cameras detect both spectral and spatial information; each spatially located pixel contains full wavelength (~350–2500 nm) information. Imaging within the visible spectrum (~400–700 nm) can measure the morphological and color properties of plants, while hyperspectral imaging can also study the radiative properties of plant leaves to facilitate the early detection of abiotic stress [98,99]. Hyperspectral imaging has been successfully applied to study the vegetation indices of crops under salt stress [100–102]. The physiological reflectance index and normalized difference vegetation index are useful for the early detection of salinity stress in plants [100,101]. Sytar et al. [102] reviewed the application of hyperspectral imaging as a fast and reliable technique for the detection of quantitative and qualitative changes in plants to evaluate plant variation under salt stress.

Several studies have focused on the development of molecular markers associated with salt tolerance in alfalfa at different growth stages [93,94,103–105]. The most significant salt tolerance markers during the germination stage were identified on chromosomes 1, 2, and 4, while the marker located on chromosome 6 overlapped with drought resistance [104,106]. During alfalfa seed germination, a broad-sense heritability of the germination rate was observed at 0.60 at 0.5% (~8 dS $m^{-1}$ NaCl) salt concentration, which decreased to 0.24 at 0.75% (~12 dS $m^{-1}$ NaCl) and 0.27 at 1.0% (~17 dS $m^{-1}$ NaCl) [104]. The broad-sense heritability of leaf chlorophyll content increased from 0.22 at 0 dS $m^{-1}$ to 0.34 at 8 dS $m^{-1}$ NaCl [105]. As salt-tolerant traits are genetically complex and multi-gene controlled, the development of genome-wide markers might be useful for conducting a genomic selection based on the predictive breeding values of genotypes. Genomic selection can enhance the genetic gain and reduce the breeding cycle when it increases the selection accuracy [107,108].

The trait prediction accuracy remains generally low in current genomic selection models, even with the aid of dense, genome-wide single nucleotide polymorphism (SNP) markers [108]. As demonstrated by Fu et al. [108], salt tolerance-associated SNP markers in alfalfa could be explored to achieve a more accurate trait-specific prediction in genomic selection. This is feasible, as the application of the RNA-Seq technique in alfalfa salinity tolerance studies is relatively common and a large amount of genomic information is already available. With the development of modern breeding techniques, such as high-throughput phenotyping and genotyping-by-sequencing platforms, genomic selection can enhance the salt-tolerant breeding of alfalfa at a reasonable cost.

**Author Contributions:** B.B. and S.B. conceived this review paper; S.B. wrote the original manuscript; D.B. reviewed and contributed to the writing the physiological section; B.B. and Y.-B.F. reviewed and revised the manuscript. All authors have read and agreed to the published version of the manuscript.

**Acknowledgments:** We would like to thank Anuja Thapa of the School of Environment and Sustainability at the University of Saskatchewan for her technical help during 'World Salt Map' development using ArcMap software.

## References

1. Lesins, K.A.; Lesins, I. *Genus Medicago (Leguminosae): A Taxogenetic Study*; Dr. W. Junk bv Publishers: The Hague, The Netherlands, 1979; p. 228.
2. Bolton, J.L. *Alfalfa Botany, Cultivation and Utilization*; Leonard Hill Ltd.: London, UK; Interscience Publishers, Inc.: New York, NY, USA, 1962.
3. Goplen, B.P.; Baenziger, H.; Bailey, L.D.; Gross, A.T.H.; Hanna, M.R.; Michaud, R.; Richards, K.W.; Waddington, J. *Agriculture Canada: Growing and Managing Alfalfa in Canada*; Publication 1705/E; Agriculture Canada: Ottawa, ON, Canada, 1982.
4. Coburn, F.D. *The Book of Alfalfa: History, Cultivation and Merits. Its Uses as a Forage and Fertilizer*; Orange Judd Co: New York, NY, USA, 1907.
5. Yuegao, H.; Cash, D. Global status and development trends of alfalfa. In *Alfalfa Management. Guide for Ningxia*; Cash, D., Ed.; United Nations Food and Agriculture Organization: Beijing, China, 2009; pp. 1–14.
6. Statistics Canada. Census of Agriculture, Hay and Field Crops. Table 32-10-0416-01. 2016. Available online: http://www.statcan.gc.ca/eng (accessed on 5 February 2020).
7. USDA–NASS. Crop Production 2018 Summary. USDA–NASS. 2018. Available online: http://www.nass.usda.gov (accessed on 9 February 2020).
8. Maas, E.V.; Hoffman, G.J. Crop salt tolerance-current assessment. *J. Irrig. Drain. Div.* **1977**, *103*, 115–134.
9. Flowers, T.J. Improving crop salt tolerance. *J. Exp. Bot.* **2004**, *55*, 307–319. [CrossRef] [PubMed]
10. Szabolcs, I. *Salt-Affected Soils*; CRC Press: Boca Raton, FL, USA, 1989.
11. Shrivastava, P.; Kumar, R. Soil salinity: A serious environmental issue and plant growth promoting bacteria as one of the tools for its alleviation. *Saudi J. Biol. Sci.* **2015**, *22*, 123–131. [CrossRef] [PubMed]
12. Kassas, M. Seven paths to desertification. *Desert. Control Bull.* **1987**, *15*, 24–26.
13. Thomas, D.S.G.; Middleton, N.J. Salinization: New perspectives on a major desertification issue. *J. Arid Environ.* **1993**, *24*, 95–105. [CrossRef]
14. Tanji, K.K. Nature and extent of agricultural salinity. In *Agricultural Salinity Assessment and Management*; Tanji, K.K., Ed.; American Society Civil Engineers: New York, NY, USA, 1990; pp. 1–13.
15. FAO. *Global Network on Integrated Soil Management for Sustainable Use of Salt-Affected Soils*; FAO Land and Plant Nutrition Management Service: Rome, Italy, 2005; Available online: http://www.fao.org/ag/agl/agll/spush (accessed on 11 February 2020).
16. Steppuhn, H.; Acharya, S.N.; Iwaasa, A.D.; Gruber, M.; Miller, D.R. Inherent responses to root-zone salinity in nine alfalfa populations. *Can. J. Plant Sci.* **2012**, *92*, 235–248. [CrossRef]
17. Steppuhn, H. What is soil salinity? In *Proceedings Soil Salinity Assessment Workshop*; Alberta Agriculture: Lethbridge, AB, Canada, 1996; pp. 1–5.
18. Wiebe, B.H.; Eilers, R.G.; Eilers, W.D.; Brierley, J.A. Application of a risk indicator for assessing trends in dryland salinization risk on the Canadian Prairies. *Can. J. Soil Sci.* **2007**, *87*, 213–224. [CrossRef]
19. Martinez, J.; Manzur, C.L. Overview of salinity problems in the world and FAO strategies to address the problem. In Proceedings of the International Salinity Forum, Riverside, CA, USA, 25–27 April 2005; pp. 311–313.
20. FAO; IIASA; ISRIC; ISS-CSA; JRC. *Harmonized World Soil Database (Version 1.2)*; FAO: Rome, Italy; IIASA: Laxenburg, Austria, 2008.
21. Munns, R. A leaf elongation assay detects an unknown growth inhibitor in xylem sap from wheat and barley. *Aust. J. Plant Physiol.* **1992**, *19*, 127–135. [CrossRef]
22. Munns, R. Genes and salt tolerance: Bringing them together. *New Phytol.* **2005**, *167*, 654–663. [CrossRef]
23. Munns, R.; James, R.A.; Lauchli, A. Approaches to increasing the salt tolerance of wheat and other cereals. *J. Exp. Bot.* **2006**, *57*, 1025–1043. [CrossRef]
24. Munns, R. Comparative physiology of salt and water stress. *Plant Cell Environ.* **2002**, *25*, 239–250. [CrossRef] [PubMed]
25. Munns, R.; Tester, M. Mechanisms of salinity tolerance. *Annu. Rev. Plant Biol.* **2008**, *59*, 651–681. [CrossRef] [PubMed]
26. Tuteja, N. Mechanisms of high salinity tolerance in plants. *Methods Enzymol.* **2007**, *428*, 419–438.
27. Schachtman, D.; Liu, W.H. Molecular pieces to the puzzle of the interaction between potassium and sodium uptake in plants. *Trends Plant Sci.* **1999**, *4*, 282–287. [CrossRef]

28. Bhandal, I.S.; Malik, C.P. Potassium estimation, uptake, and its role in the physiology and metabolism of flowering plant. *Int. Rev. Cytol.* **1998**, *110*, 205–254.
29. Ashraf, M.; Harris, P.J.C. Photosynthesis under stressful environment: An overview. *Photosynthetica* **2013**, *51*, 163–190. [CrossRef]
30. Annunziata, M.G.; Ciarmiello, L.F.; Woodrow, P.; Maximova, E.; Fuggi, A.; Carillo, P. Durum wheat roots adapt to salinity remodeling the cellular content of nitrogen metabolites and sucrose. *Front. Plant Sci.* **2017**, *7*, 2035. [CrossRef]
31. Carillo, P.; Cirillo, C.; De Micco, V.; Arena, C.; De Pascale, S.; Rouphaelb, Y. Morpho-anatomical, physiological and biochemical adaptive responses to saline water of *Bougainvillea spectabilis* Willd. trained to different canopy shapes. *Agric. Water Manag.* **2019**, *212*, 12–22. [CrossRef]
32. Smith, S.E. Salinity and the production of alfalfa (*Medicago sativa* L.). In *Handbook of Crop Stress*; Pessarakli, M., Ed.; Marcel Dekker, Inc.: New York, NY, USA, 1993; pp. 431–448.
33. Peel, M.D.; Waldron, B.L.; Jensen, K.B.; Chatterton, N.J.; Horton, H.; Dudley, L.M. Screening for salinity tolerance in alfalfa. *Crop Sci.* **2004**, *44*, 2049–2053. [CrossRef]
34. Soltanpour, P.N.; Ippolito, J.A.; Rodriguez, J.B.; Self, J.; Gillaume, M.; Al-Wardy, M.M.; Mathews, D. Chloride versus sulfate salinity effects on alfalfa shoot growth and ionic balance. *Soil Sci. Soc. Am. J.* **1999**, *63*, 111–116. [CrossRef]
35. Cornacchione, M.V.; Suarez, D.L. Evaluation of alfalfa (*Medicago sativa* L.) populations' response to salinity stress. *Crop Sci.* **2017**, *57*, 137–150. [CrossRef]
36. Azhdari, G.; Tavili, A.; Zare, M.A. Effects of various salts on the germination of two cultivars of *Medicago sativa*. *Front. Agric. China* **2010**, *4*, 63–68. [CrossRef]
37. Soltani, A.; Khodarahmpour, Z.; Jafari, A.A.; Nakhjavan, S. Selection of alfalfa (*Medicago sativa* L.) cultivars for salt stress tolerance using germination indices. *Afr. J. Biotechnol.* **2012**, *11*, 7899–7905.
38. Cornacchione, M.V.; Suarez, D.L. Emergence, forage production, and ion relations of alfalfa in response to saline waters. *Crop Sci.* **2015**, *55*, 444–457. [CrossRef]
39. Al-Niemi, T.S.; Campbell, W.F.; Rumbaugh, M.D. Response of alfalfa cultivars to salinity during germination and post-germination growth. *Crop Sci.* **1992**, *32*, 976–980. [CrossRef]
40. Johnson, D.W.; Smith, S.E.; Dobrenz, A.K. Selection for increased forage yield in alfalfa at different NaCl levels. *Euphytica* **1992**, *60*, 27–35.
41. Torabi, M.; Halim, M.R.A. Variation of root and shoot growth and free proline accumulation in Iranian alfalfa ecotypes under salt stress. *J. Food Agric. Environ.* **2010**, *8*, 323–327.
42. Shannon, M.C.; Grieve, C.M.; Francois, L.E. Whole-plant response to salinity. In *Handbook of Plant-Environment Interactions*; Wilkinson, R.E., Ed.; Marcel Dekker Inc.: New York, NY, USA, 1994; pp. 199–244.
43. Khavarinejad, R.A.; Chaparzadeh, N. The effects of NaCl and $CaCl_2$ on photosynthesis and growth of alfalfa plants. *Photosynthetica* **1998**, *35*, 461–466. [CrossRef]
44. Valizadeh, M.; Moharamnejad, S.; Ahmadi, M.; Jalaly, H.M. Changes in activity profile of some antioxidant enzymes in alfalfa half-sib families under salt stress. *J. Agric. Sci. Technol.* **2013**, *15*, 801–809.
45. Sibole, J.V.; Cabot, C.; Poschenrieder, C.; Barcelo, J. Ion allocation in two different salt-tolerant Mediterranean *Medicago* species. *J. Plant Physiol.* **2003**, *160*, 1361–1365. [CrossRef]
46. Hanley, M.E.; Sanders, S.K.D.; Stanton, H.M.; Billington, R.A.; Boden, R. A pinch of salt: Response of coastal grassland plants to simulated seawater inundation treatments. *Ann. Bot.* **2020**, *125*, 265–275. [CrossRef]
47. Tootoonchi, M.; Gettys, L.A. Testing salt stress on aquatic plants: Effect of salt source and substrate. *Aquat. Ecol.* **2019**, *53*, 325–334. [CrossRef]
48. Sandhu, D.; Cornacchione, M.V.; Ferreira, J.F.S.; Suarez, D.L. Variable salinity responses of 12 alfalfa genotypes and comparative expression analyses of salt-response genes. *Sci. Rep.* **2017**, *7*, 42958. [CrossRef]
49. Lei, Y.; Xu, Y.; Hettenhausen, C.; Lu, C.; Shen, G.; Zhang, C.; Li, J.; Song, J.; Lin, H.; Wu, J. Comparative analysis of alfalfa (*Medicago sativa* L.) leaf transcriptomes reveals genotype-specific salt tolerance mechanisms. *BMC Plant Biol.* **2018**, *18*, 35. [CrossRef]
50. Bertrand, A.; Dhont, C.; Bipfubusa, M.; Chalifour, F.P.; Drouin, P.; Beauchamp, C.J. Improving salt stress responses of the symbiosis in alfalfa using salt-tolerant cultivar and rhizobial strain. *Appl. Soil Ecol.* **2015**, *87*, 108–117. [CrossRef]

51. Robinson, P.H.; Grattan, S.R.; Getachew, G.; Grieve, C.M.; Poss, J.A.; Suarez, D.L.; Benes, S.E. Biomass accumulation and potential nutritive value of some forages irrigated with saline-sodic drainage water. *Anim. Feed Sci. Technol.* **2004**, *111*, 175–189. [CrossRef]
52. Al-Khatib, M.; McNeilly, T.; Collins, J. The potential of selection and breeding for improved salt tolerance in lucerne (*Medicago sativa* L.). *Euphytica* **1992**, *65*, 43–51. [CrossRef]
53. Suyama, H.; Benes, S.E.; Robinson, P.H.; Grattan, S.R.; Grieve, C.M.; Getachew, G. Forage yield and quality under irrigation with saline-sodic drainage water: Greenhouse evaluation. *Agric. Water. Manag.* **2007**, *88*, 159–172. [CrossRef]
54. Xiong, J.; Sun, Y.; Yang, Q.; Tian, H.; Zhang, H.; Liu, Y.; Chen, M. Proteomic analysis of early salt stress responsive proteins in alfalfa roots and shoots. *Proteome Sci.* **2017**, *15*, 19. [CrossRef]
55. Smethurst, C.F.; Rix, K.; Garnett, T.; Auricht, G.; Bayart, A.; Lane, P.; Wilson, S.J.; Shabala, S. Multiple traits associated with salt tolerance in lucerne: Revealing the underlying cellular mechanisms. *Funct. Plant Biol.* **2008**, *35*, 640–650. [CrossRef]
56. Shone, M.G.T.; Gale, J. Effects of sodium chloride stress and nitrogen source on respiration, growth and photosynthesis in lucerne (*Medicaqo sativa* L.). *J. Exp. Bot.* **1983**, *34*, 1117–1125. [CrossRef]
57. Chaparzadeh, N.; Mehrnejad, F. Oxidative markers in five Iranian alfalfa (*Medicago sativa* L.) cultivars under salinity stress. *Iran. J. Plant Physiol.* **2013**, *3*, 793–799.
58. Campanelli, A.; Ruta, C.; Morone-Fortunato, I.; De Mastro, G. Alfalfa (*Medicago sativa* L.) clones tolerant to salt stress: In vitro selection. *Cent. Eur. J. Biol.* **2013**, *8*, 765–776. [CrossRef]
59. Anower, R.M.; Mott, I.W.; Peel, M.D.; Wu, Y. Characterization of physiological responses of two alfalfa half-sib families with improved salt tolerance. *Plant Physiol. Biochem.* **2013**, *71*, 103–111. [CrossRef]
60. Acosta-Motos, J.R.; Ortuño, M.F.; Bernal-Vicente, A.; Diaz-Vivancos, P.; Sanchez-Blanco, M.J.; Hernandez, J.A. Plant responses to salt stress: Adaptive mechanisms. *Agronomy* **2017**, *7*, 18. [CrossRef]
61. Babakhani, B.; Khavari-Nejad, R.A.; Hassan Sajedi, R.; Fahimi, H.; Saadatmand, S. Biochemical responses of Alfalfa (*Medicago sativa* L.) cultivars subjected to NaCl salinity stress. *Afr. J. Biotechnol.* **2011**, *10*, 11433–11441.
62. Ashrafi, E.; Razmjoo, J.; Zahedi, M.; Pessarakli, M. Screening alfalfa for salt tolerance based on lipid peroxidation and antioxidant enzymes. *Agron. J.* **2015**, *107*, 167–173. [CrossRef]
63. Wang, X.S.; Han, J.G. Changes of proline content, activity, and active isoforms of antioxidative enzymes in two alfalfa cultivars under salt stress. *Agric. Sci. China* **2009**, *8*, 431–440. [CrossRef]
64. Jain, M.; Tiwary, S.; Gadre, R. Sorbitol-induced changes in various growth and biochemical parameters in maize. *Plant Soil Environ.* **2010**, *56*, 263–267. [CrossRef]
65. Cen, H.; Wang, T.; Liu, H.; Tian, D.; Zhang, Y. Melatonin application improves salt tolerance of alfalfa (*Medicago sativa* L.) by enhancing antioxidant capacity. *Plants* **2020**, *9*, 220. [CrossRef]
66. Quan, W.L.; Liu, X.; Wang, H.Q.; Chan, Z.L. Physiological and transcriptional responses of contrasting alfalfa (*Medicago sativa* L.) varieties to salt stress. *Plant Cell Tissue Organ Cult.* **2016**, *126*, 105–115. [CrossRef]
67. Tilbrook, J.; Roy, S.J. Salinity tolerance. In *Plant Abiotic Stress*, 2nd ed.; Jenks, M.A., Hasewaga, P.M., Eds.; Wiley-Blackwell: New York, NY, USA, 2014; pp. 134–178.
68. Roy, S.J.; Negrão, S.; Tester, M. Salt resistant crop plants. *Curr. Opin. Biotechnol.* **2014**, *26*, 115–124. [CrossRef]
69. Marschner, H. *Mineral Nutrition of Higher Plants*; Academic Press: London, UK, 1995.
70. Ashrafi, E.; Razmjoo, J.; Zahedi, M. Effect of salt stress on Growth and Ion accumulation of alfalfa (*Medicago sativa* L.) cultivars. *J. Plant Nutr.* **2018**, *41*, 818–831. [CrossRef]
71. Pessarakli, M.; Huber, T.C.; Nakabayashi, K. Growth response of barley and wheat to salt stress. *J. Plant Nutr.* **1991**, *14*, 331–340. [CrossRef]
72. Khan, M.G.; Silberbush, M.; Lips, S.H. Physiological study on salinity and nitrogen interaction in alfalfa II photosystem and transpiration. *J. Plant Nutr.* **1994**, *17*, 669–684. [CrossRef]
73. Vijayan, P.; Willick, I.R.; Lahlali, R.; Karunakaran, C.; Tanino, K.K. Synchrotron radiation sheds fresh light on plant research: The use of powerful techniques to probe structure and composition of plants. *Plant Cell Physiol.* **2015**, *56*, 1252–1263. [CrossRef]
74. Duncan, W.; Williams, G. Infrared synchrotron radiation from electron storage rings. *Appl. Opt.* **1983**, *22*, 2914–2923. [CrossRef]
75. Jiang, Y.; Lahlali, R.; Karunakaran, C.; Kumar, S.; Davis, A.R.; Bueckert, R.A. Seed set, pollen morphology and pollen surface composition response to heat stress in field pea. *Plant Cell Environ.* **2015**, *38*, 2387–2397. [CrossRef]

76. Lahlali, R.; Jiang, Y.; Kumar, S.; Karunakaran, C.; Liu, X.; Borondics, F.; Hallin, E.; Bueckert, R. ATR-FTIR spectroscopy reveals involvement of lipids and proteins of intact pea pollen grains to heat stress tolerance. *Front. Plant Sci.* **2014**, *5*, 747. [CrossRef]
77. Willick, I.R.; Lahlali, R.; Vijayan, P.; Muir, D.; Karunakaran, C.; Tanino, K.K. Wheat flag leaf epicuticular wax morphology and composition in response to moderate drought stress are revealed by SEM, FTIR-ATR and synchrotron X-ray spectroscopy. *Physiol. Plant.* **2017**, *162*, 316–332. [CrossRef]
78. Lahlali, R.; Karunakaran, C.; Wang, L.; Willick, I.; Schmidt, M.; Liu, X.; Borondics, F.; Forseille, L.; Fobert, P.R.; Tanino, K.; et al. Synchrotron based phase contrast X-ray imaging combined with FTIR spectroscopy reveals structural and biomolecular differences in spikelets play a significant role in resistance to Fusarium in wheat. *BMC Plant Biol.* **2015**, *15*, 24. [CrossRef]
79. Lahlali, R.; Kumar, S.; Wang, L.; Forseille, L.; Sylvain, N.; Korbas, M.; Muir, D.; Swerhone, G.; Lawrence, J.R.; Fobert, P.R.; et al. Cell wall biomolecular composition plays a potential role in the host type II resistance to Fusarium head blight in wheat. *Front. Microbiol.* **2016**, *7*, 910. [CrossRef]
80. Bhattarai, S.; Karunakaran, C.; Tanino, K.K.; Fu, Y.B.; Coulman, B.; Biligetu, B. Physiological and biochemical responses of alfalfa (*Medicago sativa* L.) to salt stress. In Proceedings of the Communicating Innovation in Plant Science, Plant Canada 2019, Guelph, ON, Canada, 7–10 July 2019; p. 158.
81. Rahman, M.A.; Alam, I.; Kim, Y.G.; Ahn, N.Y.; Heo, S.H.; Lee, D.G.; Liu, G.; Lee, B.H. Screening for salt-responsive proteins in two contrasting alfalfa cultivars using a comparative proteome approach. *Plant Physiol. Biochem.* **2015**, *89*, 112–122. [CrossRef]
82. Long, R.; Li, M.; Zhang, T.; Kang, J.; Sun, Y.; Cong, L.; Gao, Y.; Liu, F.; Yang, Q. Comparative proteomic analysis reveals differential root proteins in *Medicago sativa* and *Medicago truncatula* in response to salt stress. *Front. Plant Sci.* **2016**, *7*, 424. [CrossRef]
83. Postnikova, O.A.; Shao, J.; Nemchinov, L.G. Analysis of the alfalfa root transcriptome in response to salinity stress. *Plant Cell Physiol.* **2013**, *54*, 1041–1055. [CrossRef]
84. Luo, D.; Zhou, Q.; Wu, Y.G.; Chai, X.T.; Liu, W.X.; Wang, Y.R.; Yang, Q.C.; Wang, Z.Y.; Liu, Z.P. Full length transcript sequencing and comparative transcriptomic analysis to evaluate the contribution of osmotic and ionic stress components towards salinity tolerance in the roots of cultivated alfalfa (*Medicago sativa* L.). *BMC Plant Biol.* **2019**, *19*, 32. [CrossRef]
85. Gruber, M.; Xia, J.; Yu, M.; Steppuhn, H.; Wall, K.; Messer, D.; Sharpe, A.; Acharya, S.; Wishart, D.; Johnson, D.; et al. Transcript analysis in two alfalfa salt tolerance selected breeding populations relative to a non-tolerant population. *Genome* **2017**, *60*, 104–127. [CrossRef]
86. Yacoubi, R.; Job, C.; Belghazi, M.; Chaibi, W.; Job, D. Proteomic analysis of the enhancement of seed vigour in osmoprimed alfalfa seeds germinated under salinity stress. *Seed Sci. Res.* **2013**, *23*, 99–110. [CrossRef]
87. Jin, H.; Sun, Y.; Yang, Q.; Chao, Y.; Kang, J.; Jin, H.; Li, Y.; Margaret, G. Screening of genes induced by salt stress from alfalfa. *Mol. Biol. Rep.* **2010**, *37*, 745–753. [CrossRef]
88. Arshad, M.; Gruber, M.Y.; Wall, K.; Hannoufa, A. An insight into microRNA156 role in salinity stress responses of alfalfa. *Front. Plant Sci.* **2017**, *8*, 356. [CrossRef]
89. National Alfalfa and Forage Alliance. Alfalfa Variety Ratings. 2020. Available online: https://www.alfalfa.org/pdf/2020_Alfalfa_Variety_Leaflet.pdf (accessed on 15 January 2020).
90. Annicchiarico, P.; Barrett, B.; Brummer, E.C.; Julier, B.; Marshall, A.H. Achievements and challenges in improving temperate perennial forage legumes. *Crit. Rev. Plant Sci.* **2015**, *34*, 327–380. [CrossRef]
91. Allen, S.G.; Dobrenz, A.K.; Schonhorst, M.H.; Stoner, J.E. Heritability of NaCl tolerance in germinating alfalfa seeds. *Agron. J.* **1985**, *77*, 99–101. [CrossRef]
92. Benabderrahim, M.A.; Guiza, M.; Haddad, M. Genetic diversity of salt tolerance in tetraploid alfalfa (*Medicago sativa* L.). *Acta Physiol. Plant.* **2020**, *42*, 5. [CrossRef]
93. Jiang, J.; Yang, B.L.; Xia, T.; Yu, S.M.; Wu, Y.N.; Jin, H.; Li, J.R. Analysis of genetic diversity of salt-tolerant alfalfa germplasms. *Genet. Mol. Res.* **2015**, *14*, 4438–4447. [CrossRef] [PubMed]
94. Azzam, C.R.; Naby, Z.M.A.E.; Mohamed, N.A. Salt tolerance associated with molecular markers in alfalfa. *J. Biosci. Appl. Res.* **2019**, *5*, 416–428.
95. Munns, R.; James, R.A. Screening methods for salinity tolerance: A case study with tetraploid wheat. *Plant Soil.* **2003**, *253*, 201–218. [CrossRef]

96. Tiwari, R.; Mamrutha, H.M. Precision phenotyping for mapping of traits for abiotic stress tolerance in crops. In *Biotechnology: Prospects and Applications*; Salar, R., Gahlawat, S., Siwach, P., Duhan, J., Eds.; Springer: New Delhi, India, 2013.
97. Araus, J.L.; Kefauver, S.C.; Zaman-Allah, M.; Olsen, M.S.; Cairns, J.E. Translating high-throughput phenotyping into genetic gain. *Trends Plant Sci.* **2018**, *23*, 451–466. [CrossRef]
98. Fahlgren, N.; Gehan, M.A.; Baxter, I. Lights, camera, action: High-throughput plant phenotyping is ready for a close-up. *Curr. Opin. Plant Biol.* **2015**, *24*, 93–99. [CrossRef]
99. Römer, C.; Wahabzada, M.; Ballvora, A.; Pinto, F.; Rossini, M.; Panigada, C.; Behmann, J.; Léon, J.; Thurau, C.; Bauckhage, C.; et al. Early drought stress detection in cereals: Simplex volume maximisation for hyperspectral image analysis. *Funct. Plant Biol.* **2012**, *39*, 878–890. [CrossRef]
100. Naumann, J.C.; Young, D.R.; Anderson, J.E. Spatial variations in salinity stress across a coastal landscape using vegetation indices derived from hyperspectral imagery. *Plant Ecol.* **2009**, *202*, 285–297. [CrossRef]
101. Behmann, J.; Steinrucken, J.; Plumer, L. Detection of early plant stress responses in hyperspectral images. *ISPRS J. Photogramm. Remote Sens.* **2014**, *93*, 98–111. [CrossRef]
102. Sytar, O.; Brestic, M.; Zivcak, M.; Olsovska, K.; Kovar, M.; Shao, H.B.; He, X.L. Applying hyperspectral imaging to explore natural plant diversity towards improving salt stress tolerance. *Sci. Total Environ.* **2017**, *578*, 90–99. [CrossRef] [PubMed]
103. Liu, X.P.; Hawkins, C.; Peel, M.D.; Yu, L.X. Genetic loci associated with salt tolerance in advanced breeding populations of tetraploid alfalfa using genome wide association studies. *Plant Genome* **2019**, *12*, 180026. [CrossRef] [PubMed]
104. Yu, L.X.; Liu, X.; Boge, W.; Liu, X.P. Genome-wide association study identifies loci for salt tolerance during germination in autotetraploid alfalfa (*Medicago sativa* L.) using genotyping-by-sequencing. *Front. Plant Sci.* **2016**, *7*, 956. [CrossRef] [PubMed]
105. Liu, X.P.; Yu, L.X. Genome-wide association mapping of loci associated with plant growth and forage production under salt stress in alfalfa (*Medicago sativa* L.). *Front. Plant Sci.* **2017**, *8*, 853. [CrossRef]
106. Zhang, T.; Yu, L.-X.; Zheng, P.; Li, Y.; Rivera, M.; Main, D.; Greene, S.L. Identification of loci associated with drought resistance traits in heterozygous autotetraploid alfalfa (*Medicago sativa* L.) using genome-wide association studies with genotyping by sequencing. *PLoS ONE* **2015**, *10*, e0138931. [CrossRef]
107. Meuwissen, T.H.; Hayes, B.J.; Goddard, M.E. Prediction of total genetic value using genome-wide dense marker maps. *Genetics* **2001**, *157*, 1819–1829.
108. Fu, Y.B.; Yang, M.H.; Zeng, F.; Biligetu, B. Searching for an accurate marker-based prediction of an individual quantitative trait in molecular plant breeding. *Front. Plant Sci.* **2017**, *8*, 1182. [CrossRef]

# 4

# Factors Underlying Seed Yield in Red Clover: Review of Current Knowledge and Perspectives

**Tim Vleugels [1,*], Helga Amdahl [2], Isabel Roldán-Ruiz [1,3] and Gerda Cnops [1]**

1 Plant Sciences Unit, Flanders Research Institute for Agriculture, Fisheries and Food (ILVO), Caritasstraat 39, 9090 Melle, Belgium; isabel.roldan-ruiz@ilvo.vlaanderen.be (I.R.-R.); gerda.cnops@ilvo.vlaanderen.be (G.C.)
2 Graminor Breeding Ltd., Hommelstadvegen 60, NO-2322 Ridabu, Norway; helga.amdahl@graminor.no
3 Department of Plant Biotechnology and Bioinformatics, Ghent University, Technologiepark 71, 9052 Zwijnaarde, Belgium
* Correspondence: tim.vleugels@ilvo.vlaanderen.be.

**Abstract:** Red clover is a valuable forage crop, but often copes with unsatisfactory seed yield. Management practices to increase seed yield include fertilization, adequate weed and pest control, the synchronization of flowering through pre-cutting in spring, and the application of plant growth regulators to prevent lodging. Seed yield problems may have variable underlying reasons, such as inadequate pollination, fertility or genetic problems, or a combination of such traits. In this review paper, we summarize the results of recent publications that shed new light on the traits explaining seed yield differences between red clover genotypes. The main seed yield components are the number of flower heads per plant and the seed yield per flower head. Seed yield differences between diploid and tetraploid red clover are largely explained by the lower seed number per head. Recent research showed that, although inadequate pollination can lead to poor seed yield in certain areas, pollination cannot explain differences in seed yield between genotypes, cultivars, or ploidy levels. Correspondingly, corolla tube dimensions are not associated with seed yield, in spite of what is often believed by seed producers. On the other hand, fertility problems such as aberrations during male meiosis tend to occur more frequently in tetraploid genotypes and/or genotypes with low seed yield. A recent genetic study revealed 34 candidate genes for seed development, which opens perspectives for marker-assisted breeding. A final and remarkable finding is the occurrence of self-fertility in tetraploid red clover and its association with high seed yield. Breeders should be aware that selection for seed yield in tetraploid red clover may lead to unintentional selection for self-fertility, with possible consequences for inbreeding. The implications of recent findings for seed yield breeding and for the creation of novel tetraploids are discussed. Future research opportunities are considered.

**Keywords:** *Trifolium pretense*; pollination; flowering traits; fertility; seed development; breeding

## 1. Introduction

Red clover (*Trifolium pratense* L.) is a perennial forage legume belonging to the Fabaceae family. This crop is the second largest forage legume crop in the world in terms of production area, after alfalfa (*Medicago sativa* L.) [1]. It is grown in temperate areas of the world, commonly in combination with grasses such as timothy (*Phleum pratense* L.), meadow fescue (*Festuca pratensis* Huds.), tall fescue (*Festuca arundinacea* Schreb.), and/or perennial ryegrass (*Lolium perenne* L.) [2]. Red clover has the ability to fix atmospheric nitrogen through symbiosis with the bacteria *Rhizobium leguminosarum* biovar *trifolii*, which allows it to attain high forage yields (up to 15 tons DM/ha in monoculture)—higher than most forage grasses—even without N fertilization. As a consequence, red clover is a valuable rotational crop, especially in organic agriculture where no synthetic N fertilizers are used [3]. Red clover has a crude protein content of around 18% (similar to alfalfa and white clover), which remains stable throughout the growing season, in contrast to forage grasses [2]. Moreover, red clover increases palatability and digestibility of the forage mixture, resulting in higher weight gains and reproduction rates of cattle [4,5]. When red clover is included in the ruminant's diet, it provides higher crude protein and polyunsaturated fatty acids content in the final product (meat and milk) [4].

Red clover is, by nature, a diploid species (2n = 2x = 14), but tetraploid (2n = 4x = 28) varieties also exist in commercial production. Tetraploid red clover was first produced in 1939 by treating germinating seeds with colchicine [6,7]. Other chemicals that can be used to generate tetraploids through chromosome-doubling are trifluralin, oryzalin, and nitrous oxide gas. Alternatively, tetraploids can be obtained by gametophytic non-reduction, i.e., by crossing one diploid plant and one tetraploid plant [8]. The first generations of newly created tetraploids usually suffer from inbreeding, through the partial loss of heterozygosity during the chromosome-doubling event. Intercrossing of tetraploid genotypes in subsequent generations restores heterozygosity. Tetraploid red clover generally attains up to 20% higher forage yields and is generally more tolerant to biotic and abiotic stresses compared to diploid red clover [6,9,10]. The downside of tetraploid cultivars is the often considerably lower seed yield [9,11,12].

The commercial success of any cultivar depends heavily on a reliable supply of competitively priced seed [11]. Yields of 500 kg seed/ha are considered satisfactory in large-scale productions, but such yields are often not attained [13]. In Western and Central Europe, diploid red clover cultivars typically produce around 400–500 kg seed/ha, depending on the harvest year, location, and cultivar [6]. In Oregon (USA), seed yields are usually higher: Between 600 to 1200 kg/ha [14].

Tetraploid cultivars produce in general 200–400 kg seed/ha [6,9], which is 20–50% less seed compared to diploids [2,9,15]. Numerous cultivars, especially tetraploids, but also diploids, are not commercialized in spite of their agronomic superiority [11]. In addition, various publications report that seed yield in red clover has decreased in the past decades [9,16–20]. In Sweden, seed yields of red clover have declined and became more variable since the 1980s [20]. In order to keep red clover on the market against competitive prices, seed yield needs to be improved, especially in tetraploid cultivars.

Given its relevance for breeders and farmers, in recent years, several studies have investigated diverse aspects of seed yield in red clover. It remains unclear, however, if unsatisfactory seed yields are due to poor cultural practices during seed production, inadequate pollination, fertility problems, or any other genetic problems. In this paper, we summarize the results of recent publications that shed new light on the complex process of seed yield in red clover and discuss possible directions of future research and venues to improve seed yield in red clover through breeding.

## 2. Reproductive Characteristics of Red Clover and Implications for Breeding

Red clover is an obligate out-crossing species with a strong gametophytic self-incompatibility controlled by a single S-locus and a severe inbreeding depression [2,6,21]. The diversity of S-alleles in red clover is huge: Populations are estimated to harbor between 143 and 193 unique S-alleles [22]. Due to the out-crossing character of the species, red clover populations display a high phenotypic and genotypic diversity, and intra-population variation is often larger than inter-population variation [6]. Breeding methods in red clover rely on open pollinated populations: Mass selection, family selection (or recurrent phenotypic selection), or polycross selection. These methods have helped to improve agronomic traits such as forage yield, winter hardiness, and resistance to certain diseases, but have been less successful for the improvement of seed yield [6,23].

The inflorescences of red clover are formed at the tips of the stems and consist of spherical heads composed of 50–200 individual flowers [24–26]. The length of the corolla tube is approximately 10 mm, with nectaries located at its base [25]. Flowers are generally pink, although pigmentation can range from white to purple [26]. The pistil and 10 stamina extend through the entire corolla tube, with the stigma and anthers enclosed by the keel petals [25]. Each ovary contains two ovules, but usually only one develops into a mature seed [25]. Red clover is insect-pollinated: A pollinator forcing its head into the corolla exerts pressure on the keel petals, which causes the stigma and anthers to "trip" and touch the insect [25]. Wild bumblebees (*Bombus* ssp.) and honeybees (*Apis mellifera* L.) are the most important pollinators of red clover, but different species and subspecies are important in different regions [25].

Bumblebees with long tongues such as *B. pascuorum* ssp., *B. ruderatus*, and *B. hortorum* ssp. are considered the most efficient pollinators of red clover, as their tongues can easily reach the nectar at the bottom of the corolla tube [19,24]. The effectiveness of long-tongued bumblebees became evident when red clover seed yields increased drastically after their introduction in New Zealand around 1885 [27]. Short-tongued bumblebees such as *B. terrestris* L. and *B. lucorum* L. are suboptimal, because they often fail to reach the nectar through the corolla tube [16,19,24,28,29]. Instead, short-tongued bumblebees can bite holes in the lower part of the corolla to "rob" the nectar without carrying out pollination, referred to as "nectar robbing" [24,25,28,29]. In this regard, Hawkins [30] found that seed yields of red clover were closely correlated to the abundance of long-tongued bumblebees, while the abundance of short-tongued bumblebees was not related to seed yield. Keeping in mind that bumblebee nests usually house fewer than 100 individuals, recommendations were made to grow red clover seed crops in small fields and in regions rich in bumblebee populations [31].

The ability of honeybees to pollinate red clover has long been questioned, as their short tongues render it difficult to reach the nectar [24]. Nonetheless, in hot and dry weather conditions, nectar rises sufficiently high in the corolla tube for honeybees to access it [32]. Also, when few alternative pollen and nectar sources are available, honeybees are observed as pollinators on red clover [12,24,33,34]. To conclude, honeybees are the main pollinators of red clover in warm climates such as Serbia [13], while bumblebees prevail in temperate climates such as Belgium [35] and in Nordic regions such as Denmark and Sweden [12,36]. Leafcutter bees (*Megachile rotundata* Fab.) are common pollinators of red clover in the western part of North America [37].

## 3. Steps Undertaken to Increase Seed Yield in Red Clover

### 3.1. Improvement of Agricultural Practices

In temperate and warm regions, red clover seed production fields are usually established in late summer or autumn, and harvested for seed in the two following years [11]. In most Nordic regions, red clover seed crops are seeded in spring, together with spring wheat, barley, or oat as a cover crop, and harvested for seed in the two following years [11].

Management practices that can impact seed yield are, for example, fertilization with phosphorus and potassium (depending on the soil nutrient level and exploitation of the crop) [2], fertilization with boron in B-deficient soils [38], irrigation in arid areas [39,40], and adequate weed and pest control [19]. Clover seed weevils (*Apion trifolii* and *Apion apricans*) and the lesser clover leaf weevil (*Hypera nigrirostris*) can cause heavy seed losses in some years and some locations [41]. Monitoring such major insect pests from the budding stage to the end of flowering allows timely insecticide application [42].

The most commonly applied practice in temperate and warm regions is pre-cutting in spring at the onset of flowering to synchronize flowering [2,43]. In addition, this lets flowering coincide with the peak in pollinator activity [44] and reduces vegetative biomass in the next cut, making seed harvest and threshing easier. In cold climates, such as most Nordic regions, pre-cutting in spring is not recommended due to the already short growing season, as it would excessively delay flowering and seed ripening [11].

A recently optimized practice is to apply plant growth regulators (PGRs) such as Trinexapac-ethyl—commercially registered as Palisade® EC or Moddus®—which inhibit gibberellic acid biosynthesis, resulting in shorter plants, firmer and more upraised stems, more intensive flowering (26–62% more flowers per square m) and easier harvesting [39]. When applied at stem elongation (BBCH 32), increases in seed yield of 9–21% have been reported [39,45,46]. Unfortunately, the effectiveness of Trinexapac-ethyl heavily depends on the accurate timing of the application, the weather conditions after application, and the crops' stress level [45].

Another common practice is to increase the number of pollinators by supplementing them [47]. The supplementation of pollinators is justified in warm climates, in large seed production fields, or in regions where bumblebee populations have declined [19,20,32,48]. Seed producers often prefer to supplement honeybees instead of bumblebees: Bumblebee nests are commercially available, but it is uneconomical to provide enough bumblebees to ensure adequate pollination. Moreover, in Europe, only *Bombus terrestris* L. is commercially available, yet this bumblebee is a poor pollinator of red clover as it often robs nectar [24,25]. Red clover seed producers typically supply two to three honeybee hives per hectare during flowering, sometimes more [49,50]. Seed yield increases up to 34 kg/ha are reported, although studies that statistically investigate the effect of supplementing honeybees on seed yield are not available [50]. The effect of supplementing honeybees is most likely region-dependent. Jevtic [51] reported that treating red clover with sugar syrup at full flowering attracted more honeybees, leading to 20% higher seed yields. Nonetheless, the effects of supplementing or attracting honeybees is often questioned, as honeybees can act as competitors for long-tongued bumblebees: Reduced numbers of long-tongued bumblebees have been reported on red clover when honeybee hives are supplemented [19,47].

### *3.2. Breeding for Higher Seed Yield*

Seed yield is, next to forage yield, disease resistance and persistence, a major breeding goal in red clover [6,23]. A typical breeding program of red clover relies on open pollinated populations, with selection of individual genotypes for vegetative traits in the first year(s) and selection for seed yield in the final year of the selection trial [6]. In spite of the breeders' efforts, seed yield in red clover has not improve substantially in the last decades [11,52]. Several factors can explain this. First, seed yield assessed on spaced plants is a poor predictor of seed production capacity in field conditions in forage crops [23]. The assessment of a genotype's breeding value for complex traits such as seed yield is difficult, as only one plant per genotype is evaluated. Instead, it would be more efficient to select for seed yield in plots of progenies from elite genotypes, or to assess the seed yield potential of breeding material early on by establishing seed production trials. Both options are rarely done because they would require substantially more time and effort. Second, strong selection for forage yield and persistence leaves perhaps little room for improvement of seed yield, since both traits are negatively correlated with seed yield [53,54].

Third, progress for seed yield (as for other traits) might be slowed down because in a typical breeding trial, the highest seed-yielding genotypes selected for the next breeding cycle have been pollinated to a large extent by non-selected genotypes with unsatisfactory seed yield. While in the first two cases, adaptation of the breeding scheme and selection criteria are the only option, it has been shown that marker-assisted parentage analysis enables to discriminate progeny plants that result from a cross between two high seed-yielding genotypes, offering opportunities to select the best progeny to advance to the next generation. For example, in diploid red clover, parentage analysis resulted in more efficient breeding for seed yield when applied in a family selection trial: Seed yields were 23% and 76% higher compared to traditional family selection in two successive harvest years [55]. Also, in tetraploid red clover, marker-assisted parentage analysis was able to identify progeny plants that were the result of pollination by a high seed-yielding genotype, but this did not help to improve seed yield further [56]. One of the main differences between both studies above was that in the former case, a first cycle selection trial was assayed, while in the latter case, a second cycle selection trial was tested, meaning that the tetraploid genotypes had undergone a first round of selection which might have reduced the variation available. The heritability of seed yield in this tetraploid population was very low [56], indicating that mainly non-additive genetic effects were responsible for the differences observed. Further exploration of this methodology in other tetraploid selection trials is required to formulate more definitive conclusions.

## 4. Factors Explaining Seed Yield in Red Clover

Figure 1 illustrates the process of seed formation in red clover and the various factors that affect it, including plant traits (under genetic control), management practices, and environmental factors.

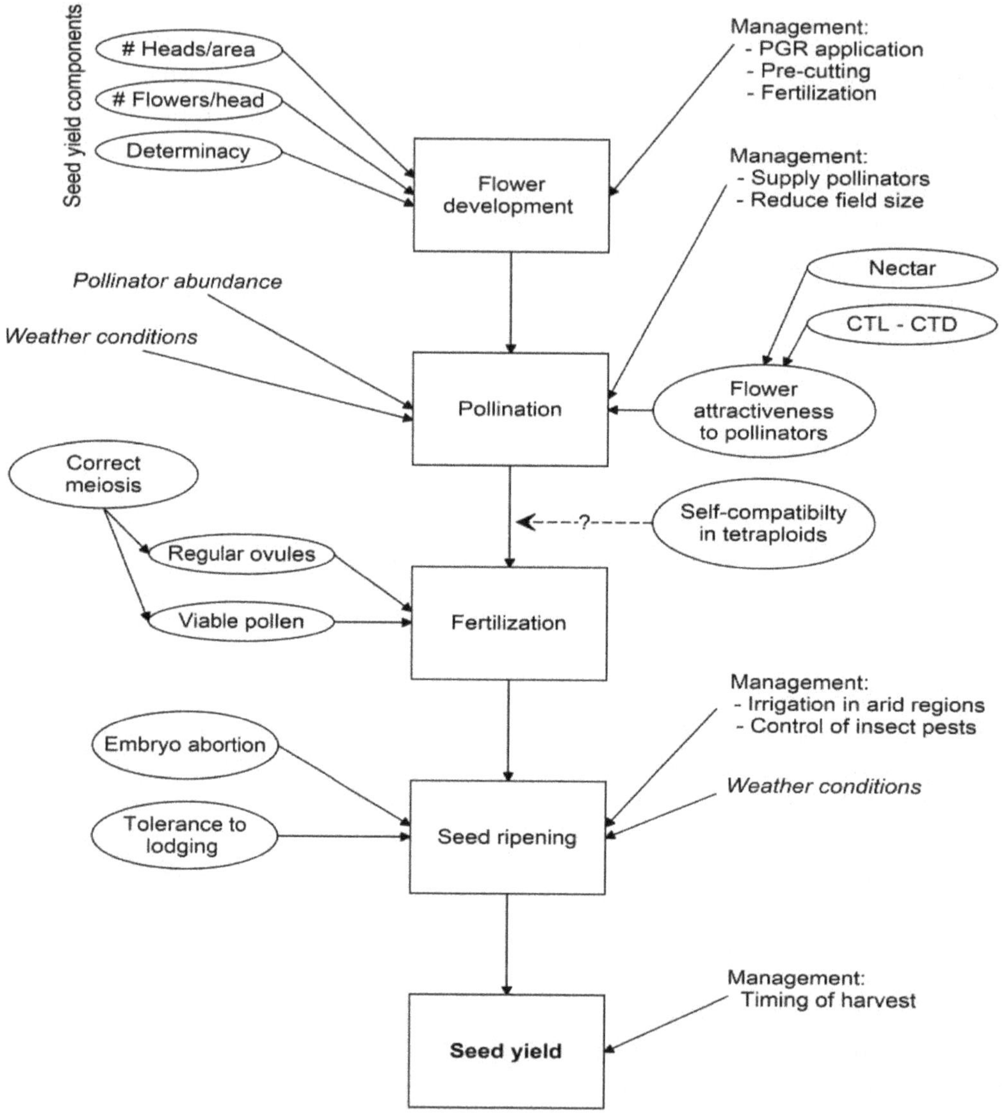


**Figure 1.** Process of seed formation in red clover with the underlying plant traits (circular boxes), management factors that can be applied to enhance seed yield, and environmental conditions (italic) that influence it. PGR = plant growth regulator; CTL = corolla tube length; CTD = corolla tube diameter.

### *4.1. Seed Yield Components*

Multiple studies have suggested that the number of flower heads—or head number (HN)—per plant or per unit area is the critical factor for seed yield in red clover [40,54,57]. Likewise, the QTL (quantitative trait loci) study of Herrmann [58] revealed that in diploid red clover, HN was the major seed yield component explaining seed yield per plant ($r = 0.89$), along with the seed number per head (SN/H) ($r = 0.47$). Dijkstra [59] showed that diploid plants had, on average, 7.4% two-seeded pods vs. 2.5% in tetraploids. There was a moderate correlation between seed yield per plant and the fraction of two-seeded pods in diploids ($r = 0.35$), but a poor correlation in tetraploids ($r = 0.08$).

A detailed analysis of seed yield components at the genotype level in diploid and tetraploid red clover was carried out recently [26,60,61]. The three studies pinpointed HN and SN/H as the most important components of seed yield in both diploids and tetraploids, with HN and SN/H explaining together up to 98% of the variation for seed yield per plant in tetraploids [60]. However, regression modeling revealed that seed yield per plant depends most on HN in diploids (62%), but on SN/H in tetraploids (62%) [26]. On average, tetraploid genotypes produced fewer seeds than diploids, which was largely explained by the lower SN/H (43.4 vs. 8.9 seeds/flower and 8930 vs. 1487 seeds/plant in diploids and tetraploids, respectively) [26,60]. In addition, tetraploid genotypes displayed fewer ripe flower heads at seed harvest than diploids (159.2 vs. 209.7 heads, or 73.1% vs. 78.3%, respectively), which indicates a slightly lower degree of branching and determinacy in tetraploids [26]. Amdahl [9] found similar results when studying seed yield components in dense stands of 12 tetraploid cultivars and populations from Scandinavia. A high correlation between seed yield $m^{-2}$ and the number of flower heads $m^{-2}$ was observed in two experiments ($r = 0.956$ and $r = 0.977$).

To conclude, recent studies have identified HN and SN/H as major determinants of seed yield in single plants and in dense stands of red clover [9,26,58,60,61]. Lower seed yield in tetraploids than in diploid genotypes is mostly explained by the lower seed number per head, and only partly by the lower number of ripe flower heads at seed harvest. The substantially lower seed number per head in tetraploids may indicate either inadequate pollination or problems with fertility. Both possibilities are discussed in the following sections.

### 4.2. Inadequate Pollination

Poor seed yields in red clover are often ascribed to inadequate pollination. It has been hypothesized that the declining abundance of bumblebees and wild pollinators in nature is partly responsible for the reduced seed yields observed recently [17]. In Denmark [48] and Sweden [20], for example, the abundance and species-richness of long-tongued bumblebees on red clover has severely declined since the 1940s, while short-tongued bumblebees have become more dominant. The change in pollinator communities in Sweden may explain the decline in seed yield and yield stability of red clover observed in recent years [20]. Pollination problems may be aggravated by a changed practice towards larger seed production fields, where pollination may become insufficient in the central part of the field due to reduced accessibility for the insects [31].

Other studies argue that the length of the corolla tubes can be the problem, especially in tetraploid red clover, given that tetraploid flowers are usually longer than diploid flowers. The nectar at the bottom of long corolla tubes would be less accessible for pollinators, resulting in inadequate pollination and low seed yield [25,62,63]. The "corolla hypothesis" likely originated in New Zealand where, before the introduction of European bumblebees around 1885, red clover often failed to set seed in the presence of the short-tongued native bees and honeybees [27,64,65]. The observation of nectar robbing spurred the corolla hypothesis [24,30,31]. Short-tongued bumblebees such as *Bombus terrestris* L. act as primary robbers and puncture the corolla tube to access the nectar [24]. Afterwards, secondary robbers such as honeybees, unable to bite through corolla tubes themselves, rob newly formed nectar through the bite holes as well [24]. Consequently, numerous flowers are not pollinated and deprived of nectar, thus rendered less attractive for long-tongued bumblebees [16,66].

There is some evidence for the corolla hypothesis. For example, Hawkins [67] observed that among eight red clover cultivars, cultivars with the shortest corolla tube length (CTL) were more frequently visited by bumblebees and produced more seeds than cultivars with the longest CTL. Bumblebee species with short tongues preferred to pollinate cultivars with shorter CTL, whereas corolla biting most frequently occurred in cultivars with longer CTL, presumably because less of their nectar had been removed [67]. However, most studies that support the corolla hypothesis only argue that the CTL in tetraploids is larger than in diploids [68,69]. For example, Bingefors [70] reported that the tetraploid cv. "Ulva" produced fewer seeds than its diploid ancestor cv. "Ultuna", which they attributed to the 25–30% lower flower head production per unit area and to the 0.74 mm longer CTL in "Ulva". Congruently, Bender [44] found that two tetraploid cultivars displayed 10% longer (and 30% wider) corolla tubes compared to their diploid ancestor cultivars (the ones that had been submitted to tetraploidization). In this case, it was demonstrated that higher proportions of long-tongued bumblebees (51.9% vs. 37.3%), lower proportions of short-tongued bumblebees (37.6% vs. 40.1%), and more frequent nectar robbing (11.1% vs. 4.6%) occurred on the tetraploid cultivars compared to their diploid ancestors [44].

There also exists a plethora of studies whose results do not support the corolla hypothesis. For example, Starling [71] observed no association between CTL and the seed number per head among 218 diploid genotypes derived from crosses between cultivars with diverging CTL: The cultivar with the shortest CTL actually displayed the lowest seed yield. Hawkins [72] stated that nectar robbing was not likely to be associated with the low seed setting of red clover, while Julén [73] stated that the tendency of bumblebees to rob nectar varies greatly between years and along the flowering period. Often, nectar robbing intensifies in late summer and autumn (personal observation), possibly because of few alternative nectar sources for pollinators in that period. Brodsgaard [16] observed that the CTL depended more on the location and management of the seed crop than on the ploidy level. The diploid cultivar "Rajah" had average CTLs of 8.71 mm in a conventional field vs. 8.82 mm in an organic field, while the CTL was 8.75 mm in the tetraploid cultivar "Kvarta" (conventional field). The degrees of corolla biting/nectar robbing were 3.6%, 13.4%, and 17.2% in the three fields, respectively [16].

In the above-mentioned studies, which suggest inadequate pollination as reason for low seed yield in tetraploid red clover, measurements were performed at the population level [44,68–70]. Yet, given the large genetic diversity within red clover cultivars, comparing average cultivar values often does not reveal hidden correlations between seed yield, flowering traits, and pollination preference, and may lead to faulty conclusions. Recent studies at the genotype level have provided different insights. Three recent studies found that tetraploid genotypes indeed have longer CTL than diploids [26,60,61], but a very large overlap in CTL between ploidy levels was also found. In addition, the corolla tube diameter (CTD) is larger in tetraploids than in diploids [26]. Importantly, these three studies found no association between CTL and seed yield. In fact, some genotypes with the largest CTL actually displayed the highest seed yields [60]. Hence, recent studies at the genotype level strongly contradict the corolla hypothesis [26,60,61]. Longer corolla tubes appear to be also wider, as observed by [26,74], which may allow pollinators to insert their heads deeper into the flower and reach the nectar in spite of the increased CTL. It is therefore not surprising that breeding for reduced CTL did not lead to great improvement in seed yield in the work of Hawkins [72]. Jablonski [75] reported increased seed yield in a population derived from a mutant genotype with short corolla tube. The population had a 24% shorter corolla than the cultivar "Nike", but also produced 29% more flower heads $m^{-2}$ [75]. It is therefore impossible to determine which factor (CTL or number of flower heads $m^{-2}$) contributed most to the increase in seed yield in this work.

In addition, no correlation has been found either between seed yield and other flowering traits such as the flower number per head (112 in diploids vs. 115 in tetraploids) or flower pigmentation [26]. Furthermore, Vanommeslaeghe [35] investigated whether pollinator preferences explained differences in seed yield between diploid and tetraploid red clover at the genotype level. Nectar production, nectar concentrations of glucose, fructose, sucrose, and total sugar in the nectar differed among genotypes with known differences in seed yield, but did not differ between ploidy levels, and were not associated with seed yield. In pollinator preference binary-choice experiments with *B. pascuorum* L., no bumblebee preference was recorded either for ploidy or seed yield, with the exception of a single tetraploid genotype that stood out for its high nectar production (10.3 μL per 10 flowers compared to 3.5 μL in the average tetraploid). The latter genotype was more attractive to *B. pascuorum* L., but actually displayed low seed yield, suggesting that factors other than pollination are responsible for its low seed yield [35].

Recent findings at the genotype level were confirmed by recent studies at field scale. A study in a seed production field of the tetraploid cv. "Avanti" confirmed the lack of correlation between CTL and seed yield when the analysis was done at the genotype level [74]. Jing [12] found similar visitation frequencies of honeybees on the diploid cv. "Rajah" and the tetraploid "Amos". However, the SN/H was 68.7% higher in "Rajah" than in "Amos", indicating a lower seed setting ability in the tetraploid cultivar under the same pollination conditions. Vanommeslaeghe [35] confirmed in a field experiment that *B. pascuorum* L. does not discriminate between diploid vs. tetraploid genotypes, nor between genotypes with high vs. low seed yield.

In conclusion, pollination in red clover under field conditions is complex, as it probably depends on the prevailing bumblebee/honeybee (sub)species, competing pollen and nectar sources, and local management strategies, which all differ from region-to-region and from year-to-year. For honeybees alone, there exist at least 29 subspecies [76]. Tongue length can vary substantially between subspecies, but even between colonies of the same subspecies, as observed by Brodsgaard [16] in "Buckfast" honeybees. Unfortunately, the prevailing pollinator subspecies and its tongue length is rarely reported in publications. While the corolla hypothesis may be valid in some regions or circumstances, it must not be generalized. Pollination can probably explain poor seed yield in some locations, e.g., where insufficient pollinators are present. However, neither pollinator preferences nor corolla tube dimensions can explain differences between genotypes, cultivars, or ploidy levels.

### *4.3. Fertility Problems*

Reduced fertility may lead to seed yield problems in red clover. Especially in tetraploid red clover, fertility problems often remain high in spite of ample breeding efforts, as is often observed in induced autotetraploids [77]. Reduced fertility may have various underlying reasons, including (1) deficient meiosis which may lead to (2) poor pollen quality, (3) embryo abortion during seed development, and possibly (4) altered self-incompatibility interactions in tetraploid red clover. In red clover, only a limited number of studies is available on fertility traits, and only few of them have investigated fertility traits in relation to seed yield at the genotype level.

Aberrations in meiotic division could lead to unviable pollen and ovules, with reduced fertility as a consequence [78]. Meiotic aberrations were observed in tetraploids by Büyükkartal in tetraploid red clover [79], but no link was made to seed yield. Vleugels [80] analyzed male meiosis among 19 diploid and 17 tetraploid genotypes with well-characterized and diverging seed yields: Tetraploid genotypes revealed more meiotic aberrations than diploids, and there was a trend of decreasing seed yield at increasing frequencies of meiotic aberrations. Analyzing a larger group of genotypes may allow to statistically prove an association between seed yield and meiotic aberrations. In certain low seed-yielding genotypes, high frequencies of meiotic aberrations (up to 56% aberrations) were observed, which suggests that in these genotypes, defective (male) meiosis may be a major cause of low seed yield [80].

Pollen quality was studied by Büyükkartal [81] in tetraploid red clover and compared between diploid and tetraploid cultivars [77] and genotypes [80] in later studies. Büyükkartal [81] found 57% pollen germination in a Turkish wild tetraploid red clover population. Grebrenisan [77] found that tetraploid cultivars produced pollen with lower viability and lower germination rates than diploid cultivars. Vleugels [80] observed a similar trend for pollen germination, although it is difficult to compare absolute values due to different culture media used in these studies. Vleugels [80] also noted that pollen viability was strongly influenced by the weather conditions during the three days before pollen harvest.

Embryo abortion is a commonly observed phenomenon in legumes, and has been the focus of various studies in forage legumes [80,82,83]. In diploid red clover, Povilaitis [84] found that 70% of ovules developed into mature embryos in high seed-yielding genotypes vs. 40% in low seed-yielding genotypes. In addition, flowers of high seed-yielding genotypes often raised two embryos, while flowers of low seed-yielding genotypes often contained no embryo at all. This indicates that low seed-yielding genotypes can simply support fewer developing seeds. In a more recent study in a tetraploid red clover population from Turkey, only 18% of the flowers raised one or two embryos, and merely 5.8% of the ovules developed into mature seeds [85]. Vleugels [80] studied embryo development in 11 genotypes (4 diploids, 7 tetraploids) with well-characterized, divergent seed yields. They noted high degrees of embryo abortion in most genotypes, but did not find a correlation between embryo abortion and seed yield. A breakdown of self-incompatibility is commonly observed in polyploid species [86], and has been described in tetraploid red clover as well [77,87–91]. Povilaitis [84] suggested that low seed-yielding red clover genotypes contain genetic factors that suppress pollen tube growth for both own and foreign pollen. Grebrenisan [77] studied pollen tube growth in detached red clover pistils. When the pistils were self-fertilized, a higher fraction of growing pollen tubes was observed in tetraploids than in diploids (12% vs. 3%, respectively). In both ploidy levels, however, self-fertilized pistils became more self-fertile when they were fertilized again with own pollen [77]. In other words, repeated self-pollination in the absence of other red clover pollen may inactivate self-incompatibility barriers. Vleugels [80] found that high seed-yielding genotypes self-fertilized to a large extent (83% on average), whereas no self-fertilization at all was detected in low seed-yielding tetraploids or in diploids. A subsequent study confirmed that self-fertilization is common in tetraploid red clover that has been pre-selected for seed yield and synchronous flowering [56]: 3 out of the 14 highest seed-yielding genotypes were self-fertile with individual selfing rates up to 20%. These observations suggest that self-fertility in tetraploid red clover acts similarly as in the autotetraploid alfalfa, where selfing rates average 11.8% in open-pollinated fields with individual selfing rates up to 52.2% [92].

As discussed by Vleugels [80], breeding for increased seed yield in tetraploid red clover may therefore lead to unintentional selection for increased self-fertility. Self-fertility can result in higher seed yield, but it is likely that population vigor will decrease in the long run due to inbreeding depression. All depends on the severity of the inbreeding depression in tetraploid red clover, which has not been thoroughly investigated yet. No information is available on the effect of inbreeding on forage yield in red clover, but some information exists on seed yield. Laczynska-Hulewicz [89] found reductions of 23–26% in seed yield of tetraploid red clover inbred lines. Vleugels [80] found similar results: Genotypes that were the result of self-fertilization yielded on average 20.5% less seed than genotypes that were the result of cross-pollination. A related, unresolved issue is how exactly self-incompatibility works in tetraploid red clover [22]. Self-incompatible genotypes cannot be fertilized by their own pollen (pollen carrying two overlapping S-alleles), but it is unclear if they can be fertilized by pollen with one overlapping S-allele (e.g., pollen from a related genotype). If not, numerous genotypes are excluded from fertilizing such a genotype, especially in tetraploid populations harboring few S-alleles [80].

To conclude, embryo abortion is common in tetraploid red clover, but additional research is needed to compare the degree of abortion between diploids and tetraploids. Aberrations in male meiosis tend to occur more frequently in tetraploid red clover, and tend to be negatively associated with seed yield, although more research is needed to prove such associations. Pollen germination is also generally lower in tetraploid red clover. Finally, self-fertility seems relatively common in tetraploid red clover and can contribute to increased seed yield. Selection for seed yield in tetraploid red clover may lead to co-selection for self-fertility, yet more research is needed to fully unravel the consequences of inbreeding for forage yield and seed yield.

### *4.4. Genomic Regions Associated with Seed Development and Seed Yield*

To date, only a few studies have investigated the genetics of seed yield in red clover. Herrmann [58] used a mapping population derived from a cross between two genotypes from the cultivars "Violetta" (high seed yield) and "Corvus" (poor seed yield) to identify genomic regions associated with seed yield. Thirty-eight QTLs related to eight seed yield components were identified and two genome regions containing four to five QTLs for different seed yield components were pinpointed as regions of interest for further QTL characterization. Recently, Kovi [52] used next-generation sequencing to analyze the transcriptome response during flowering and seed set in two tetraploid red clover genotypes with contrasting seed yield. They identified 34 differentially expressed genes related to pollination, flowering, and seed development, which can be considered as candidate genes for seed development in red clover [52]. The next step would be to perform loss-of-function analysis to understand the role of these candidate genes in seed development. Further efforts are therefore needed before genes affecting seed yield in red clover are identified and tools developed to exploit this information in practical breeding.

A second possible genetic factor is the effect of genetic diversity and heterozygosity on seed yield. The outcrossing nature of red clover suggests that increased heterozygosity leads to increased fitness, both at the genotype level and the population level. Hence, populations with a broad genetic basis and a high degree of heterozygosity are expected to attain higher seed yields than populations with less genetic diversity. This might be particularly relevant in tetraploid populations, because the tetraploid red clover genepool has probably undergone a bottleneck as tetraploidization efforts is usually successful for a limited number of plants [6]. To our knowledge, no studies have assessed the link between genetic diversity and fitness or seed yield at the population level.

In conclusion, several QTL regions and candidate genes for seed yield and seed development have been identified, which opens perspectives for the development of molecular markers for seed yield. Furthermore, a reference genome is now available for red clover, opening perspectives for genomic selection [93].

### 4.5. *Implications for Breeding*

Keeping the recent findings in mind, breeders aiming to improve seed yield in diploid or tetraploid red clover should select plants with numerous flower heads and high seed yield numbers per flower head. Flowering traits such as corolla tube dimensions and flower color can be omitted as breeding targets, but if a population could be created with substantially increased nectar production, it is likely to attract more pollinators [35]. Identifying genotypes with deficient meiosis in order to exclude them from crossings may positively influence seed yield as well. However, analyzing meiosis in a breeding program is presumably too labor- and cost-intensive. Marker-assisted parentage analysis can increase genetic gains for seed yield by excluding progeny from low seed-yielding genotypes.

Regarding tetraploid red clover, breeders have at least two options to improve seed yield. Selection for seed yield can take place at the diploid level, followed by chromosome doubling and at least three generations of intercrossing [94]. In this regard, Dijkstra [59] recommended to select for plants with two-seeded pods on the diploid level, followed by tetraploidization as a means to increase seed yield in tetraploids. Improving seed yield through additive genetic variance is relatively straightforward in diploids [94]. Heterozygosity will be restored when sufficient generations of intercrossing are performed [88]. Alternatively, breeders can perform multiple generations of selection and crossing among existing tetraploid populations. In this way, heterozygosity remains high at all times. However, more generations of crossing and selection may be required to achieve the same additive genetic gain than at the diploid level, since four (instead of two) beneficial alleles need to be accumulated on each locus related to seed yield [80]. Moreover, selecting for seed yield in tetraploid red clover may increase the risk to co-select for self-fertility [80], the effects of which are likely not beneficial for seed yield and general performance in the long term. Nonetheless, self-fertile genotypes could be identified and their seeds discarded, so that only seeds from self-incompatible genotypes are transferred to the next breeding cycle. Detection would be possible with marker-assisted parentage analysis [56], or by isolating a few flowering stems in a bag with pollinators while leaving the remaining flowers to cross-pollinate. Today, it is difficult to state which of these two breeding methods is most suitable to increase seed yield in tetraploid red clover. Amdahl [9] found that populations made by crossing existing tetraploids displayed higher seed yields (8.1% and 21.4% higher in two experiments) than tetraploid populations created by chromosome doubling. However, the fact that the diploid ancestor cultivars were not deliberately selected for seed yield, renders it difficult to compare both breeding methods.

## 5. Conclusions and Perspectives

Seed yield is a complex trait affected by various genetic, environmental, and agronomical aspects (Figure 1), some of which are not even discussed in this paper (e.g., disease resistance). However, all of the pieces of the puzzle together start to shed light on this complex process. The most important seed yield components are the number of flower heads per plant (in spaced plants) and the seed yield per flower head (in spaced plants and dense stands). Differences in seed yield between diploid and tetraploid red clover are mostly explained by the lower seed number per head and partly by the lower head number per plant at seed harvest.

The generally accepted statement that unsatisfactory seed yields in (tetraploid) red clover are mainly due to longer corolla tubes has been refuted by multiple studies. Corolla tube dimensions are not associated with seed yield, neither on a genotype basis nor in dense stands. Inadequate pollination can explain unsatisfactory seed yields in certain regions and years, e.g., where/when insufficient pollinators are present, but it cannot explain seed yield differences between genotypes, cultivars, or ploidy levels.

Recent studies seem to confirm that fertility problems may play an important role in seed yield, but more research is needed to confirm this. Meiotic aberrations tend to be more common in tetraploid red clover and in genotypes with low seed yield, with lower pollen germination as a consequence. A remarkable finding is the occurrence that high seed-yielding tetraploid genotypes are frequently self-compatible. Selection for seed yield in tetraploid red clover may perhaps lead to unintentional selection for self-fertility and a higher level of inbreeding. However, the consequences for inbreeding and the effects of inbreeding on seed yield and general performance in tetraploid red clover remain poorly understood, and should be the object of future research (currently underway at ILVO). Future research should also investigate how exactly self-incompatibility works in tetraploid red clover. An important question in this regard is whether self-incompatible genotypes can be fertilized by pollen with one overlapping S-allele (e.g., pollen from a related genotype).

So far, only few studies have investigated the genetics of seed yield in red clover. A transcriptome study revealed 34 candidate genes for seed development, which opens perspectives to develop markers for seed yield [52]. The recent construction of a reference genome for red clover is the first step towards genomic selection for seed yield [93]. Concerning breeding strategies, more research is needed to define which breeding method is the most suitable to increase seed yield in tetraploid red clover: Selection at the diploid level before tetraploidization, or selection at the tetraploid level. Elucidating the relation between genetic diversity and seed yield at the population level may help to uncover the most suitable breeding method.

**Author Contributions:** Conceptualization, T.V., I.R.-R., and G.C.; literature review, T.V. and H.A.; writing—original draft preparation, T.V. and H.A.; writing—review and editing, T.V., H.A., I.R.-R., and G.C.

**Acknowledgments:** The authors wish to thank multiple contact persons active in red clover breeding and seed production for providing valuable background information. Graminor would like to acknowledge the Norwegian Research Council for funding their research on seed yield of red clover.

## Abbreviations

| | |
|---|---|
| CTL | corolla tube length |
| CTD | corolla tube diameter |
| HN | head number per plant at seed harvest |
| SN/H | seed number per head at seed harvest |
| PGR | plant growth regulator |
| QTL | Quantitative trait locus |

## References

1. Annicchiarico, P.; Barrett, B.; Brummer, E.C.; Julier, B.; Marshall, A.H. Achievements and challenges in improving temperate perennial forage legumes. *Crit. Rev. Plant Sci.* **2015**, *34*, 327–380. [CrossRef]
2. Stockdale, E.A.; Lampkin, N.H.; Hovi, M.; Keatinge, R.; Lennartsson, E.K.M.; Macdonald, D.W.; Padel, S.; Tattersall, F.H.; Wolfe, M.S.; Watson, C.A. Agronomic and environmental implications of organic farming systems. *Adv. Agron.* **2003**, *70*, 261–327.

3. Taylor, N.L.; Quesenberry, K.H. *Red Clover Science*; Kluwer Academic Publishers: Dordrecht, The Netherlands, 1996.
4. DeWhurst, R.J.; Delaby, L.; Moloney, A.L.; Boland, T.; Lewis, E. Nutritive value of forage legumes used for grazing and silage. *Ir. J. Agric. Food Res.* **2009**, *48*, 167–187.
5. Huss-Danell, K.; Chaia, E.; Carlsson, G. $N_2$ fixation and nitrogen allocation to above and below ground plant parts in red clover-grasslands. *Plant Soil* **2007**, *299*, 215–226. [CrossRef]
6. Boller, B.; Schubiger, F.X.; Kölliker, R. Red clover. In *Handbook of Plant Breeding*; Boller, B., Ed.; Springer: Dordrecht, The Netherlands, 2010; pp. 439–455.
7. Sjödin, J.; Ellerström, S. Autopolyploid forage crops. In *Research and Results in Plant Breeding*; Olsson, G., Ed.; Svalöf: Stockholm, Sweden, 1986; pp. 102–113.
8. Meglic, V.; Smith, R.R. Self-incompatibility and seed set in colchicine-, nitrous oxide-, and sexually derived tetraploid red clover. *Crop Sci.* **1992**, *32*, 1133–1137. [CrossRef]
9. Amdahl, H.; Aamlid, T.S.; Ergon, A.; Kovi, M.R.; Marum, P.; Alsheikh, M.; Rognli, O.A. Seed yield of Norwegian and Swedish tetraploid red clover (*Trifolium pratense* L.) populations. *Crop Sci.* **2016**, *56*, 603–612. [CrossRef]
10. Hejduk, S.; Knot, P. Effect of provenance and ploidity of red clover varieties on productivity, persistence and growth pattern in mixture with grasses. *Plant Soil Environ.* **2010**, *56*, 111–119. [CrossRef]
11. Boelt, B.; Julier, B.; Karagic, D.; Hampton, J. Legume seed production meeting market requirements and economic impacts. *Crit. Rev. Plant Sci.* **2015**, *34*, 412–427. [CrossRef]
12. Jing, S. Pollination and Seed Setting of Diploid and Tetraploid Red Clover (*Trifolium pratense* L.). Master's Thesis, Aarhus University, Aarhus, Denmark, 2017.
13. Karagic, D.; Jevtic, G.; Terzic, D. Forage legumes seed production in Serbia. *Biotechnol. Anim. Husb.* **2010**, *26*, 133–148.
14. Anderson, N.P.; Garbaçik, C.J.; Chastain, T.G.; Elias, S. Boron effects on red clover seed production and quality. *Seed Prod. Res.* **2019**, *9*, 9–11.
15. Petrauskas, G.; Mikaliuniene, J.; Norkeviciene, E.; Statkeviciute, G.; Kemesyte, V. Breeding for improved seed yield of red clover. In *Breeding Grasses and Protein Crops in the Era of Genomics, Proceedings of the Joint Meeting of EUCARPIA Fodder Crops and Amenity Grasses Section and Protein Crops Working Group of Oil and Protein Crops Section, Vilnius, Lithuania, 11–14 September 2017*; Brazauskas, G., Statkeviciute, G., Jonaviciene, G., Eds.; Springer: Cham, Switzerland, 2018; pp. 96–100.
16. Brodsgaard, C.J.; Hansen, H. Pollination of red clover in Denmark. *DIAS Rep.* **2002**, *71*, 1–50.
17. Eckhart, T.; (Saatzucht Steinach, Steinach, Germany); Hartmann, S.; (Bavarian State Research Center for Agriculture, Freising, Germany). Personal communication, 2019.
18. Hejduk, S.; (Mendel University, Brno, Czech Republic). Personal communication, 2019.
19. Wermuth, K.H.; Dupont, Y.L. Effects of field characteristics on abundance of bumblebees (*Bombus* spp.) and seed yield in red clover fields. *Apidologie* **2010**, *41*, 657–666. [CrossRef]
20. Bommarco, R.; Lundin, O.; Smith, H.G.; Rundlöf, M. Drastic historic shifts in bumble-bee community composition in Sweden. *Proc. R. Soc. B* **2012**, *279*, 309–315. [CrossRef] [PubMed]
21. Julén, G. Rotklee, *Trifolium pratense* L. In *Handbuch der Pflanzenzüchtung*; Julén, G., Ed.; Paul Parey Publishing: Berlin, Germany, 1959; pp. 239–305.
22. Lawrence, M.J. Number of incompatibility alleles in clover and other species. *Heredity* **1996**, *76*, 610–615. [CrossRef]
23. Annicchiarico, P.; Boller, B.; Brummer, E.C.; Reheul, D. Improving the focus of forage breeding research. In *Breeding in a World of Scarcity*; Roldán-Ruiz, I., Baert, J., Reheul, D., Eds.; Springer: Cham, Switzerland, 2016; pp. 251–270.
24. Free, J.B. Leguminosae: Trifolium. In *Insect Pollination of Crops*, 2nd ed.; Academic Press: London, UK, 1993; pp. 271–297.
25. McGregor, S.E. Clover and some relatives. In *Insect Pollination of Cultivated Crop Plants*; McGregor, S.E., Ed.; USDA-ARS: Washington, DC, USA, 1976; pp. 162–235.
26. Vleugels, T.; Ceuppens, B.; Cnops, G.; Lootens, P.; van Parijs, F.R.D.; Smagghe, G.; Roldán-Ruiz, I. Models with only two predictor variables can accurately predict seed yield in diploid and tetraploid red clover. *Euphytica* **2016**, *209*, 507–523. [CrossRef]

27. Dunning, J.W. The importation of humble bees into New Zealand. *Trans. R. Entomol. Soc. Lond.* **1886**, *6*, 32–34.
28. Hawkins, R.P. Observations on the pollination of red clover by bees. *Ann. Appl. Biol.* **1961**, *49*, 55–65. [CrossRef]
29. Gurr, L. The role of bumblebees as pollinators of red clover and Lucerne in New Zealand: A review and prospect. *Proc. N. Z. Grassl. Assoc.* **1974**, *36*, 111–122.
30. Hawkins, R.P. A preliminary survey of red clover seed production. *Ann. Appl. Biol.* **1956**, *44*, 657–664. [CrossRef]
31. Hawkins, R.P. A survey of late-flowering and single cut red clover seed crops. *J. Natin. Inst. Agric. Bot.* **1958**, *8*, 450–461.
32. Goetze, G. Versuche zur Ausnutzung des Rotklees durch die Honigbienen. *Beitr. Agrarwiss.* **1948**, *2*, 35–47.
33. Rao, S.; Stephen, W.P. Bumble bee pollinators in red clover seed production. *Crop Sci.* **2009**, *49*, 2207–2214. [CrossRef]
34. Ruszkowski, A.; Sowa, S.; Biliński, M.; Kochańska, Z.; Galas, J.; Mostowska, I. The tongue length of the honeybee and the manner in which it forages on red clover. *Pszczel. Zesz. Nauk.* **1980**, *24*, 37–41.
35. Vanommeslaeghe, A.; Meeus, I.; Cnops, G.; Vleugels, T.; Merchiers, M.; Duquenne, B.; Roldán-Ruiz, I.; Smagghe, G. Influence of pollinator abundance and flower visitation on seed yield in red clover. *Arthropod Plant Interact.* **2018**, *12*, 339–349. [CrossRef]
36. Rundlöf, M.; Lundin, O.; Bommarco, R. Annual flower strips support pollinators and potentially enhance red clover seed yield. *Ecol. Evol.* **2018**, *8*, 7974–7985. [CrossRef]
37. Richards, K.W. Effectiveness of the alfalfa leafcutter bee *Megachile rotundata* Fab. to pollinate perennial clovers. *J. Apic. Res.* **2016**, *55*, 259–267. [CrossRef]
38. Stoltz, E.; Wallenhammer, A.C. Influence of boron in organic red clover (*Trifolium pratense* L.) seed production. *Grass Forage Sci.* **2014**, *69*, 285–293. [CrossRef]
39. Anderson, N.P.; Garbacik, C.J. Irrigation and trinexapac-ethyl effects on seed yield in first-and second-year red clover stands. *Agron. J.* **2016**, *108*, 1116–1123. [CrossRef]
40. Oliva, R.N.; Steiner, J.J.; Young, W.C. Red clover seed production: II. Plant water status on yield and yield components. *Crop Sci.* **1994**, *34*, 184–192. [CrossRef]
41. Kolařík, P.; Rotrekl, J. Regulation of the abundance of clover seed weevils, *Apion* spp. (Coleoptera: Curculionidae) in a seed stand of red clover (*Trifolium pratense* L.). *J. Entomol. Acarol. Res.* **2013**, *45*, e19. [CrossRef]
42. Lundin, O.; Rundlöf, M.; Smith, H.G.; Bommarco, R. Towards integrated pest management in red clover seed production. *J. Econ. Entomol.* **2012**, *105*, 1620–1628. [CrossRef] [PubMed]
43. Mihovsky, T.; Naydenova, G. Phenotypic analysis and heritability of seed production components in red clover (*Trifolium pratense* L.). *Bulg. J. Agric. Sci.* **2018**, *24*, 46–49.
44. Bender, A. An impact of polyploidization on the dimensions of red clover flowers, species' composition and number of insect pollinators. *Agraarteadus* **1999**, *4*, 9–23.
45. Chastain, T.G.; Anderson, N.P.; Garbacik, C.J.; Angsumalee, D.; Elias, S.G. Irrigation and trinexapac-ethyl effects on seed yield in a second-year red clover stand. In *2013 Seed Production Research Report*; Young, W.C., Ed.; Oregon State University: Salem, MA, USA, 2015; pp. 1–3.
46. Øverland, J.I.; Aamlid, T.S. Plant growth regulators and insect control in seed production of red clover (Trifolium pratense). In *Seed Production in the Northern Light, Proceedings of the 6th International Herbage Seeds Conference, Gjennestad, Norway, 18–20 June 2007*; Aamlid, T.S., Havstad, L.T., Boelt, B., Eds.; Bioforsk: Grimstad, Norway, 2007; pp. 226–230.
47. Rao, S.; Anderson, N.P. Comparison of bee pollinator abundance in red clover seed production fields with and without honey bee hives in the Willamette valley. In *2009 Seed Production Research Report*; Oregon State University: Corvallis, OR, USA, 2009; pp. 45–49.
48. Dupont, Y.L.; Damgaard, C.; Simonsen, V. Quantitative historical change in bumblebee (*Bombus* spp.) assemblages of red clover fields. *PLoS ONE* **2011**, *6*, e25172. [CrossRef]
49. Rincker, C.M.; Rampton, H.H. Seed production. In *Clover Science and Technology*; Taylor, N.L., Ed.; American Society of Agronomy: Madison, WI, USA, 1985; pp. 417–443.
50. Buchholz, L. Increasing red clover seed production by saturation of pollinators. In *Greenbook 2001*; Minnesota Department of Agriculture: Minnesota, WI, USA, 2001; pp. 13–14.

51. Jevtic, G.; Andelkovic, B.; Lugic, Z.; Mladenovic, M.; Nedic, N. The influence of the hive distance and the use of corn syrup on pollinator visits and red clover seed yield. *Biotechnol. Anim. Husb.* **2010**, *26*, 167–172.
52. Kovi, M.K.; Amdahl, H.; Alsheikh, M.; Rognli, O.A. De novo and reference transcriptome assembly of transcripts expressed during flowering provide insight into seed setting in tetraploid red clover. *Sci. Rep.* **2017**, *7*, 44383. [CrossRef]
53. Herrmann, D.; Boller, B.; Studer, B.; Widmer, F.; Kölliker, R. Improving persistence in red clover: Insights from QTL analysis and comparative phenotypic evaluation. *Crop Sci.* **2008**, *48*, 269–277. [CrossRef]
54. Steiner, J.J.; Smith, R.R.; Alderman, S.C. Red clover seed production: 4. Root rot resistance under forage and seed production systems. *Crop Sci.* **1997**, *37*, 1278–1282. [CrossRef]
55. Vleugels, T.; Cnops, G.; Roldán-Ruiz, I. Improving seed yield in red clover through marker assisted parentage analysis. *Euphytica* **2014**, *200*, 305–320. [CrossRef]
56. Vleugels, T.; Roldán-Ruiz, I.; Cnops, G. Marker-assisted parentage analysis reveals high individual selfing rates in tetraploid red clover genotypes selected for seed yield. *Plant Breed.* **2019**, *138*, 947–957. [CrossRef]
57. Montardo, D.P.; Dall'Agnol, M.; Crusius, A.F.; Paim, N.R. Path analysis for seed production in red clover (*Trifolium pratense* L.). *Rev. Bras. Zootec.* **2003**, *32*, 1076–1082. [CrossRef]
58. Herrmann, D.; Boller, B.; Studer, B.; Widmer, F.; Kölliker, R. QTL analysis of seed yield components in red clover (*Trifolium pratense* L.). *Theor. Appl. Genet.* **2006**, *112*, 536–545. [CrossRef] [PubMed]
59. Dijkstra, J. The importance of two-seeded pods in red clover (*Trifolium pratense* L.). *Euphytica* **1969**, *18*, 340–351. [CrossRef]
60. Amdahl, H.; Aamlid, T.S.; Marum, P.; Ergon, A.; Alsheikh, M.; Rognli, O.A. Seed yield components in single plants of diverse Scandinavian tetraploid red clover populations (*Trifolium pratense* L.). *Crop Sci.* **2017**, *57*, 108–117. [CrossRef]
61. Vleugels, T.; Roldán-Ruiz, I.; Cnops, G. Influence of flower and flowering characteristics on seed yield in diploid and tetraploid red clover. *Plant Breed.* **2015**, *134*, 56–61. [CrossRef]
62. Clifford, P.T.P.; Scott, D. Inflorescence, bumble bee and climate interactions in seed crops of a tetraploid red clover (*Trifolium pratense* L.). *J. Appl. Seed Prod.* **1989**, *7*, 38–45.
63. Tasei, J.N. Légumineuses fourragères et protéagineuses. In *Pollinisation et Productions Végétales*; Pesson, P., Louveaux, J., Eds.; INRA: Paris, France, 1984; pp. 261–308.
64. Pammel, L.H.; King, C.M. Pollination of clover. *Proc. Iowa Acad. Sci.* **1911**, *18*, 35–45.
65. Plath, O.E. The role of bumblebees in the pollination of certain cultivated plants. *Am. Nat.* **1925**, *59*, 441–451. [CrossRef]
66. Palmer-Jones, T.; Forster, I.W.; Clinch, P.G. Observations on the pollination of Montgomery red clover (*Trifolium pratense* L.). *N. Z. J. Agric. Res.* **1966**, *9*, 738–747. [CrossRef]
67. Hawkins, R.P. Factors affecting the yield of seed produced by different varieties of red clover. *J. Agric. Sci.* **1965**, *65*, 245–253. [CrossRef]
68. Bond, D.A. Variation between tetraploid red clover plants in corolla tube length and height of nectar. *J. Agric. Sci.* **1968**, *71*, 113–116. [CrossRef]
69. Skirde, W. Studies of morphology and secretion in diploid and tetraploid clovers. *Ann. Agric. Fenn.* **1963**, *2*, 90.
70. Bingefors, S.; Ellerstrom, S. Polyploidy breeding in red clover: The tetraploid variety Svalof's Ulva compared with some diploid and tetraploid varieties. *Z. Pflanz.* **1964**, *51*, 315–334.
71. Starling, T.M.; Wilsie, C.P.; Gilbert, N.W. Corolla tube length studies in red clover. *Agron. J.* **1950**, *42*, 25–32. [CrossRef]
72. Hawkins, R.P. Selection for height of nectar in the corolla tube of English singlecut Red Clover. *J. Argic. Sci. Camb.* **1971**, *77*, 347–350. [CrossRef]
73. Julén, G. Aspects on the breeding of tetraploid red clover with special reference to the seed setting problem. In Proceedings of the European Grassland Conference, Paris, France, 21–24 June 1954; European Productivity Agency; Organisation for European Economic Co-operation: Paris, France, 1954; pp. 69–72.
74. Vleugels, T.; Roldán-Ruiz, I.; Ceuppens, B.; Smagghe, G.; Cnops, G. Are corolla tube dimensions the reason for low seed yield in tetraploid red clover? In *Breeding in a World of Scarcity, Proceedings of the 2015 Meeting of the Section "Forage Crops and Turfgrasses" of Eucarpia, Ghent, Belgium, 13–17 September 2015*; Roldán-Ruiz, I., Baert, J., Reheul, D., Eds.; Springer: Dordrecht, The Netherlands, 2016; pp. 293–297.
75. Jablonski, B. A preliminary appreciation of agricultural and beekeeping value of a red clover population with a shortened flower tube. *Acta Hortic.* **2001**, *561*, 215–217. [CrossRef]

76. Han, F.; Wallberg, A.; Webster, M. From where did the Western honeybee (*Apis mellifera*) originate? *Ecol. Evol.* **2012**, *2*, 1949–1957. [CrossRef]
77. Grebrenisan, M.; Savatti, M. The comparative effects of the micro- and macrosporogenesis of the red clover with different levels of ploidy, in relation with its fertility. *Bull. Univ. Agric. Sci. Vet. Med. Agric.* **2011**, *68*, 138–143.
78. Mackiewicz, T. Low seed setting in tetraploid red clover (*Trifolium pratense* L.) in the light of cytoembryological analyses. *Genet. Pol.* **1965**, *6*, 5–39.
79. Büyükkartal, H.N. The reasons of sterility during pollen grain formation in the natural tetraploid *Trifolium pratense* L. *Int. J. Bot.* **2007**, *3*, 188–195.
80. Vleugels, T.; van Laere, K.; Roldán-Ruiz, I.; Cnops, G. Seed yield in red clover is associated with meiotic abnormalities and in tetraploid genotypes also with self-compatibility. *Euphytica* **2019**, *215*, 79. [CrossRef]
81. Büyükkartal, H.N. In vitro pollen germination and pollen tube characteristics in tetraploid red clover (*Trifolium pratense* L.). *Turk. J. Bot.* **2003**, *27*, 57–61.
82. Akhalkatsi, M.; Pfauth, M.; Calvin, C.L. Structural aspects of ovule and seed development and nonrandom abortion in *Melilotus officinalis* (Fabaceae). *Protoplasma* **1999**, *208*, 211–223. [CrossRef]
83. Wang, X.; Li, X.; Zhang, J.; Feng, G.; Zhang, S.; Huang, L.; Zhuo, R.; Jin, L. Characterization of nine alfalfa cultivars for differences in ovule numbers and ovule sterility. *Aust. J. Crop Sci.* **2011**, *5*, 447–452.
84. Povilaitis, B.; Boyes, J.W. Ovule development in diploid red clover. *Can. J. Bot.* **1960**, *38*, 507–523. [CrossRef]
85. Büyükkartal, H.N. Causes of low seed set in the natural tetraploid *Trifolium pratense* L. (Fabaceae). *Afr. J. Biotechnol.* **2008**, *7*, 1240–1249.
86. Marble, B.K. Polyploidy and self-compatibility: Is there an association? *New Phytol.* **2004**, *162*, 803–811. [CrossRef]
87. Drach, N.P.; Rogovchenko, V.M.; Skorobagat'ko, T.N.; Efimenko, G.M. Significance of self pollination in the reproduction of early red clover. *Sb. Nauchnykh Tr. Prikl. Bot. Genet. Sel.* **1986**, *103*, 31–34.
88. Elgersma, A. Effects of aneuploidy on seed production in autopolyploid forage crops. In *Ploidy and Chromosome Manipulation in Forage Breeding, Proceedings of the 17th Meeting of the Fodder Crops Section of EUCARPIA, Alghero, Italy, 14–18 October 1991*; Caredda, S., Veronesi, F., Bullitta, S., Eds.; Consiglio Nazionale delle Ricerche: Rome, Italy, 1991; pp. 124–131.
89. Laczynska-Hulewicz, T. Self-fertility and inbreeding in tetraploid red clover. *Genet. Pol.* **1963**, *4*, 97–119.
90. Vleugels, T.; Roldán-Ruiz, I.; Cnops, G. Comparing mitotic and meiotic tetraploid red clover plants for seed yield. In *Breeding Grasses and Protein Crops in the Era of Genomics, Joint Meeting of EUCARPIA Fodder Crops and Amenity Grasses Section and Protein Crops Working Group of Oil and Protein Crops Section, Vilnius, Lithuania, 11–14 September 2017*; Brazauskas, G., Statkeviciute, G., Jonaviciene, G., Eds.; Springer: Cham, Switzerland, 2018; pp. 39–45.
91. Williams, R.D.; Williams, W. Genetics of red clover (*Trifolium pratense* L.) compatibility III. The frequency of incompatibility S alleles in two non-pedigree populations of red clover. *J. Genet.* **1947**, *48*, 69–79. [CrossRef]
92. Riday, H.; Reisen, P.; Raasch, J.A.; Santa-Martinez, E.; Brunet, J. Selfing rate in an alfalfa seed production field pollinated with leafcutter bees. *Crop Sci.* **2015**, *55*, 1087–1095. [CrossRef]
93. De Vega, J.J.; Ayling, S.A.; Hegarty, M.; Kudrna, D.; Goicoechea, J.L.; Ergon, A.; Rognli, O.A.; Jones, C.; Swain, M.; Geurts, R.; et al. Red clover (*Trifolium pratense* L.) draft genome provides a platform for trait improvement. *Sci. Rep.* **2015**, *5*, e17394. [CrossRef] [PubMed]
94. Picard, J.; Berthaut, J. Fertility problems in autotetraploid red clover, an insight into the efficiency of selection at the diploid level. In Proceedings of the 10th International Grassland Congress, Helsinki, Finland, 7–16 July 1966; pp. 664–666.

# 5

# Do Phytomer Turnover Models of Plant Morphology Describe Perennial Ryegrass Root Data from Field Swards?

**Cory Matthew [1,*], Alec D. Mackay [2] and Arif Hasan Khan Robin [3]**

[1] Institute of Agriculture and Environment PN 433, Massey University, Private Bag 11-222, Palmerston North 4442, New Zealand
[2] AgResearch Grasslands Private Bag 11-008, Palmerston North 4442, New Zealand; alec.mackay@agresearch.co.nz
[3] Department of Genetics and Plant Breeding, Bangladesh Agricultural University, Mymensingh 02202, Bangladesh; gpb21bau@bau.edu.bd
* Correspondence: c.matthew@massey.ac.nz

Academic Editor: Les Copeland

**Abstract:** This study aimed to elucidate seasonal dynamics of ryegrass root systems in field swards. Established field swards of perennial ryegrass with white clover removed by herbicide and fertilised with nitrogen (N) to replace clover N fixation were subjected to lax and hard grazing management and root biomass deposition monitored using a root ingrowth core technique over a 13 month period. A previously published phytomer-based model of plant morphology that assumes continuous turnover of the root system was used to estimate mean individual root weight (mg) not previously available for field swards. The predicted root weights compared credibly with root data from hydroponic culture and the model output explained much of the seasonal variation in the field data. In particular, root deposition showed a seasonality consistent with influence of an architectural signal (*AS*) determined by plant morphology. This *AS* arises because it is theoretically expected that with rising temperatures and decreasing phyllochron in early summer, more than one leaf on average would feed each root bearing node. Conversely, in autumn the reverse would apply and root deposition is expected to be suppressed. The phytomer-based model was also able to explain deeper root penetration in summer dry conditions, as seen in the field data. A prediction of the model is that even though total root deposition is reduced by less than 10% under hard grazing, individual root weight is reduced proportionately more because the available substrate is being shared between a higher population of tillers. Two features of the field data not explained by the phytomer based model, and therefore suggestive of hormonal signaling, were peaks of root production after summer drought and in late winter that preceded associated herbage mass rises by about one month. In summary, this research supports a view that the root system of ryegrass is turning over on a continuous basis, like the leaves above ground. The phytomer based model was able to explain much of the seasonal variation in root deposition in field swards, and also predicts a shift of root deposition activity, deeper in summer and shallower in winter.

**Keywords:** perennial ryegrass; root dynamics; field swards; phytomer; phyllochron

---

## 1. Introduction

Historic studies of grass sward dynamics have tended to focus separately on behavior of leaves [1,2], tiller populations [3], or roots [4]. Compartmentalisation of sward dynamics studies in this way probably occurred, in part, as a pragmatic reaction to the fact that such studies are extremely time consuming, and focusing on one aspect of sward behavior helps ensure data quality. However, the disadvantage of single sward attribute studies is they can give the impression plant organs respond independently to external stimuli. Some attempts have been made to examine interrelationships between shoot traits, both conceptually [5,6] and from experimental data [7,8]. One study that attempted to synthesise a picture of root–shoot interactions of field swards of perennial ryegrass [9] achieved this by coordination of above- [10] and below-ground [11] measurements in separate studies of the same lax- or hard-grazed perennial ryegrass swards.

Root systems were described by Davidson [12] as 'the forgotten component of pastures', but that comment may represent a perception rather than fact, considering that a monograph by Troughton [13] two decades earlier had cited over 800 references. Moreover, there has been a steady stream of ongoing studies. A literature search in Web of Science by the authors identified just over 100 papers from 1995 to 2015 with both 'root' and 'ryegrass' in the title. Approximately 20 of these explore some aspect of the physical morphology and turnover of the root system. For example, one series of reports of New Zealand research compared root development in wild and bred ryegrass populations [14], in 200 genotypes of a ryegrass QTL mapping population [15] and in cloned ryegrass plants with or without fungal endophyte [16]. Apart from those which examined shoot:root ratio, all of these studies have looked at roots in terms of their rate of arrival or population density in specific soil layers without consideration of the underlying plant dynamics regulating root production. By contrast, research elsewhere has tended to focus on physiological function of roots, such as factors affecting abiotic stress resistance of roots or the various roles of root exudates [17,18].

Pursuit of these two emphases of researching root mass or density in defined soil layers or the physiological function of roots, while clearly valuable, has left a large gap in the knowledge about perennial ryegrass root systems. Specifically, we now understand from organising data on the growth of single leaves [19,20] how the phytomer-based plant morphology contributes to leaf tissue generation and loss in ryegrass swards over a defoliation interval, and there has been recent extension of this understanding to hydroponically grown ryegrass and wheat plants [21–24]. However, there is no study known to the authors that has explored quantitatively, the relationship between root presence in soil and phytomer-level events of the single tillers that comprise the sward.

The present paper arises from the first author deciding to submit for publication the remainder of a data set from a large field experiment studying root-shoot relations in perennial ryegrass swards, conducted almost 30 years ago, and only partly published at the time [9]. The experiment in question used an 'ingrowth core' technique adapted from European research [25] to measure seasonality of new root production (kg DM/ha/day) in lax- or hard-grazed ryegrass plots rotationally grazed by sheep. Here we add to the previously published data for 0–250 mm soil depth, data for the 250–600 mm soil layer and seasonal root and shoot production data. We draw on emerging awareness of phytomer-level organisation of the grass root system [21–24,26–29] to develop a phytomer-oriented mechanistic model; and attempt to reconcile the annual cycle of field data collected from December 1986 to January 1988 with the output from the phytomer based model. The aim is to show the quantitative link between root appearance rates measured in units of mass flow typically used for herbage accumulation (kg DM/ha/day) in the field swards of perennial ryegrass described above on the one hand, and root formation events at each phytomer on individual tillers on the other hand. It should be noted that a parallel concept for leaf turnover above ground is now well established in research targeted at informing extension advice to farmers [20,30], and so extending this approach to include understanding of the root system is overdue. The development of a theoretical framework for phytomer-based turnover of the root system represents a step change from previous thinking which has generally assumed that roots are replaced annually [31,32]. One feature that needs inclusion in the calculations is a representation of the proposed 'architectural signal' (*AS*) [26,33], whereby root:shoot ratio potentially changes during the year, reflecting variation in the number of leaves available to 'feed' developing roots at different times of the year. The model should also be capable of providing insight into the effects of lax and hard grazing on root development.

## 2. Materials and Methods

### 2.1. Field Plots

The experiment was conducted from 1986 to 1988 [9–11]. Briefly, white clover was removed using herbicide from established sheep-grazed perennial ryegrass swards at Palmerston North, New Zealand. The soil at the site is a fine silt loam formed from a blanket of wind-blown loess at least 2 m deep over marine sediments of an uplifted former shore platform. The experiment comprised 4 grazing managements including "lax" (L) (approx. 2000 and 3000 kg DM/ha, post- and pre-grazing herbage mass, respectively) and hard (H) (approx. 800 and 1500 kg DM/ha, post- and pre-grazing) which were maintained in 4 replicates in a Latin square design on 100 $m^2$ plots from December 1986 to February 1988, so as to collect root data over an annual cycle. Four additional H and four L plots were used for L-to-H and H-to-L crossover grazing management treatments from November 1987 and their data are not reported here. Within this period new root growth was allowed to accumulate for approximately 40 days into core-holes of 75 mm diameter and approx. 700 mm depth, filled with builders' sand [11] (online version of [11] contains photographs at pp. 33–34). Cores were then destructively harvested, root ingrowth determined as kg DM/ha/day ash free DM by weight loss on combustion of samples at 650 °C [34], and root length as km/ $m^2$ using a grid intersect method [35], and a new set of ingrowth cores established. Mean root diameter was estimated from the ratio of root length:root mass recovered from ingrowth cores at each harvest [11]. For simplicity of presentation, in this paper and focus on the morphology of root mass deposition in the field, we present the root mass data and a relative root length, obtained by adjusting the root mass deposition for diameter differences between harvest dates.

The cycle of ingrowth core placement and harvest was repeated 12 times from October 1986 to May 1988 and data for 9 of those harvests are presented here to achieve one year of continuous observation of root formation rates for the 1988 calendar year. Grazing occurred 'as needed' to keep herbage mass of plots within predetermined target values above, as far as possible. L plots were grazed 16 times H plots 19 times during the experiment, with grazing interval 40 to 60 days in autumn-winter and as short as 15 days in spring-summer. Ryegrass tiller population density was determined at each root harvest using a coring method [36]. Herbage mass was determined 11 times during the experiment on dates approximately coordinated with root harvesting, using four 0.1 $m^2$ quadrats (20 cm × 50 cm) cut to ground level, but these data were not coordinated with grazing and so herbage accumulation could not be determined from them. For December to August, herbage accumulation was estimated from weather and soil fertility data for the site using the model "Grow" [37] and for the remainder of the experiment herbage accumulation data was obtained from combining leaf elongation and tiller density data for 2 or 3 representative time periods within each interval between grazing events [9]. Root ingrowth (kg DM/ha/day) and herbage accumulation were plotted together to elucidate the relationship between them. Nitrogen fertilizer was applied at 9 kg N/ha, approximately twice monthly during the experiment, a rate judged from the literature [38] to replace N fixation by clover rhizobia.

### 2.2. Mechanistic Modeling of Phytomer Root Production

We selected Equation (5) of Matthew et al. [26] for use in this study, which, when reorganized to suit the data collected in this experiment can be expressed (with reconciliation of units below):

$$F_r = T \times P \times AS \times R_n \times W_r \times UC/(b \cdot x \cdot f) \tag{1}$$

$$(\text{kg/ha}\cdot\text{day}) = (\text{no./m}^2) \times (P/\text{day}) \times (\text{day/day}) \times (\text{roots}/P) \times (\text{mg/root}) \times (\text{kg/mg}) \times (\text{m}^2/\text{ha})$$

where $Fr$ denotes mass flux of root formation (kg DM/ha/day); $T$ denotes tiller population density per $m^2$; $P$ is the number of phyllochrons per day (units: $\text{days}^{-1}$); $AS$ (architectural signal) is the ratio between the phyllochron at the time of leaf formation and the phyllochron at the time of root formation, denoted by [26] as $I_{t-d}/I_t$; $R_n$ denotes the average number of roots produced by each phytomer on the tiller axis (assuming steady state turnover); $W_r$ denotes the average weight of a single root (mg $\text{root}^{-1}$);

*UC* is a unit correction (0.01) to convert units of mg/m$^2$ on the right hand side of Equation (1) to kg DM/ha/day on the left hand side; *b* is a calibration coefficient indicating any difference from unity in the comparative rates of root colonization of refilled cores and undisturbed soils; and *f* is an adjustment to represent the proportion of tillers which are large and/or dominant within the population and responsible for a majority of sward root formation activity while another sub-population of subservient daughter tillers produce little new root biomass. The coefficient *b* was determined by using a fibre optic viewing device (Ultrafine Technology, London, UK) to count and compare root arrival events in 3 pairs of 25 mm diameter plastic observation tubes with one core of each pair installed in the centre of a root ingrowth core and the other 50 cm away in undisturbed soil. This calibration was performed for 6 successive cycles of ingrowth cores over a period of approximately 8 months. Equation (1) was solved for $W_r$ to estimate the mean dry weight per root (mg) required to generate the observed root ingrowth in refilled cores. Of the required entities for this calculation, *Fr*, *T* and *b* were measured in the field experiment as described above, *P* and *DF* were calculated, as described below, and $R_n$ had to be assumed based on values reported in the literature [21,26]. The coefficient *f* is an estimate at this stage, taken from visual observation of tiller status in ryegrass swards similar to those used in the experiment.

To calculate *P*, a thermal time methodology was used with the base temperature and leaf appearance interval for ryegrass taken from subsequent New Zealand studies [39,40] as being 2.0 °C and 101 °C· d, respectively. Daily maximum and minimum temperature data for Palmerston North weather station NAN3238, approximately 1 km distant from the experiment site were downloaded from the 'Cliflo' online database of the New Zealand National Institute of Water and Atmospheric Research Ltd. (Auckland, New Zealand). The thermal time (°C· d/day) was then calculated for each day from installation to harvest of each cycle of root ingrowth cores and the thermal time per day averaged for each root harvest cycle. The ryegrass thermal time phyllochron of 101 °C· d was then divided by these values to obtain an estimated value in days for the average ryegrass leaf appearance interval within each of the 9 root harvest periods.

*AS* is a theoretical morphogenetic influence on *Wr* [26,33] which arises because leaf formation at any particular phytomer on the tiller axis precedes root formation at the same phytomer by about 5 phyllochrons (505 °C· d thermal time) on average [21,41]. Within that delay between leaf and root formation at the same phytomer the phyllochron can change because of seasonal factors like day length. This means that in spring fractionally more than one leaf-bearing phytomer feeds each root bearing phytomer, while in autumn the reverse would occur [33]. Because 5 phyllochrons (30–65 days) was broadly similar to the harvesting interval of root ingrowth cores (approx. 40 days), in the interests of simplicity *AS* was taken as the ratio between the mean value of *P* for a given root harvest and for the following harvest.

Correlations of *P* and *AS* (and their reciprocals) with root deposition and root mean diameter data were explored. In order to explore whether *P* was acting simply as a surrogate variable for day length and/or insolation, a cosine curve was constructed based on a 360° cycle over 365 days from the 22 December 1986, and data correlation with this seasonal cosine curve was also examined.

The experiment site is prone to summer drought with a mean November to April plant water deficit of 258 mm (range 106–432 mm) for the years 2002–2011 [42]. For the summer of 1986/1987, 22.4 mm rainfall was recorded at the nearby weather station between installing and harvesting first cycle of ingrowth cores (indicative of water deficit stress); and 133.5 mm between installing and harvesting the second cycle (indicative of water deficit alleviation).

### 2.3. Statistical Analysis

Field data from the H and L plots of each harvest were analyzed using a repeated measures option in Proc GLM of SAS (SAS Institute Inc., Cary, NC, USA), in order to obtain relevant standard errors for comparing annual means of H and L grazing management treatments, or to compare the harvest mean of H and L treatments at different harvest dates. Calculations of mean root weight (mg) using

Equation (1), were performed in MicrosoftXL. Pearson correlation coefficients to assess similarities of seasonal trends for *P*, *AS*, and other entities, were calculated using Minitab version 10.51 (Minitab Inc., 2081 Enterprise Drive, State College, PA, USA).

## 3. Results

### *3.1. Root and Herbage Biomass Fluxes*

Root biomass deposition differed little between L and H grazing management regimes, but showed an approximately logarithmic decline with increasing soil depth. Marked seasonal variation, ranging from a low of less than 2 kg DM/ha/day in June to a high of 13.5 kg DM/ha/day in November was also recorded (Table 1). The 250–600 mm soil depth could not be harvested from June to October because of a high water table, physically preventing core extraction. This should not have compromised root deposition data as no evidence of root formation below the water table was observed in this period. When plotted with herbage accumulation, root biomass deposition was found to be about 15% of above ground herbage accumulation, with a similar seasonality, but with root deposition bursts in late winter and following autumn rain, preceding the corresponding above ground events by approximately one month. However, the highest recorded values of $F_r$ occurred during an early summer burst of root growth, in November (Figure 1). The early summer root growth would coincide with flowering, and this is explored further below.

Tiller population density was always significantly lower in L swards (mean 5270 tillers/m$^2$) than in H swards (mean 7820 tillers/m$^2$) and showed summer peaks and a winter low (Table 1). Mean herbage mass averaged over the whole experiment was 2700 and 1095 kg DM/ha for L and H grazing treatments, respectively. This gives an indicative mean tiller weight of 51 mg DM/tiller for L swards, and 14 mg/tiller for H swards.

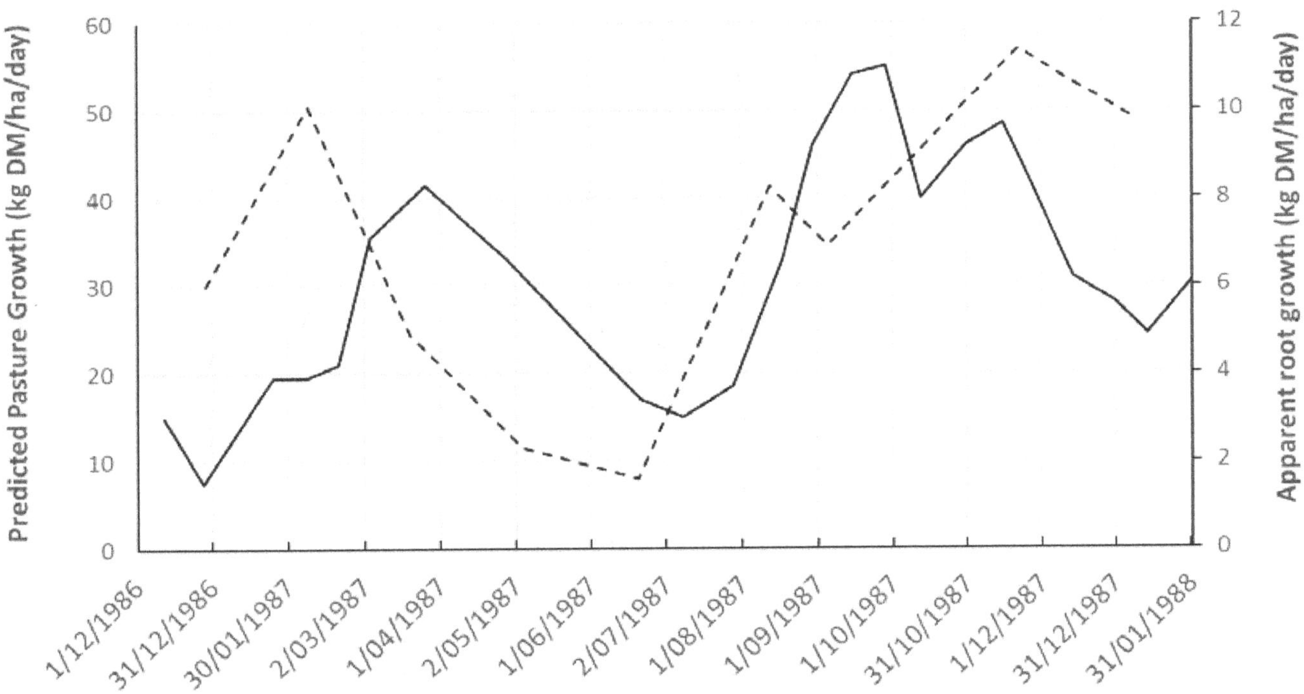


**Figure 1.** Seasonality of new root deposition (– – –) ($F_r$) in perennial ryegrass swards at Palmerston North, plotted with herbage accumulation rate (————) estimated by computer modeling from weather data, as described by Butler et al. [37]. Peaks of root growth occurred after summer rain (January–February 1987), in late winter (August 1987), and approximately coincident with flowering in early summer (November 1987).

**Table 1.** Apparent root deposition as measured by ingrowth cores (kg DM/ha/day ash-free DM), mean root diameter (mm), relative root length (RRL), and tiller population density for three soil depths and two grazing managements determined from ingrowth core samples for the period December 1986 to January 1988. Dates shown are those for the median of each cycle of ingrowth core placement and harvest.

| | | | Harvest Number and Median Date of Root Ingrowth Core Capture | | | | | | | | | | | |
|---|---|---|---|---|---|---|---|---|---|---|---|---|---|---|
| Trait | Grazing | Soil Depth | 2 | 3 | 4 | 5 | 6 | 7 | 8 | 9 | 10 | MEAN | S.E. [1] | S.E. [2] |
| | | (mm) | 25 December 1986 | 3 February 1987 | 13 March 1987 | 1 May 1987 | 10 June 1987 | 2 August 1987 | 26 September 1987 | 14 November 1987 | 29 December 1987 | | | |
| Root deposition (kg DM/ha/day) | L | 0–70 | 4.3 | 7.6 | 3.1 | 2.3 | 1.2 | 6.1 | 4.2 | 6.8 | 5.9 | 4.6 | 0.4 | 0.9 |
| | | 70–250 | 1.8 | 3.3 | 1.4 | 0.9 | - | 2 | 2 | 4.6 | 2.8 | 2.2 | 0.3 | 0.6 |
| | | 250–600 | 1.1 | 0.6 | 0.4 | 0.1 | -[3] | - | - | 2.4 | 1.4 | 1.1 | 0.3 | 0.4 |
| | | 0–600 | 7.2 | 11.5 | 4.9 | 3.3 | 1.8 | 8.1 | 6.2 | 13.8 | 10.1 | | | |
| | H | 0–70 | 3.8 | 6 | 3.6 | 0.9 | 0.9 | 5.8 | 4.8 | 7.8 | 7.7 | 4.6 | 0.4 | 0.9 |
| | | 70–250 | 2.1 | 3.4 | 1.8 | 0.4 | 0.4 | 2.6 | 2.8 | 3.5 | 3.1 | 2.2 | 0.3 | 0.6 |
| | | 250–600 | 1.9 | 0.8 | 0.6 | 0.1 | -[3] | - | - | 1.8 | 3 | 1.4 | 0.3 | 0.4 |
| | | 0–600 | 7.8 | 10.2 | 6 | 1.4 | 1.3 | 8.4 | 7.6 | 13.1 | 13.8 | | | |
| | Mean | 0–70 | 4 | 6.8 | 3.3 | 1.6 | 1.1 | 6 | 4.5 | 7.3 | 6.8 | 4.6 | 0.3 | 0.6 |
| | | 70–250 | 1.9 | 3.4 | 1.6 | 0.6 | 0.5 | 2.3 | 2.4 | 4 | 2.9 | 2.2 | 0.2 | 0.4 |
| | | 250–600 | 1.3 | 0.6 | 0.5 | 0.1 | -[3] | - | - | 2.2 | 2.2 | 1.2 | 0.2 | 0.3 |
| Mean diameter (mm) | Mean | | 0.24 | 0.25 | 0.27 | 0.22 | 0.3 | 0.31 | 0.21 | 0.18 | 0.22 | 0.24 | 0.011 | 0.011 |
| Relative root length [4] | Mean | | 0.81 | 1.15 | 0.49 | 0.32 | 0.12 | 0.59 | 1.09 | 2.7 | 1.74 | 1 | n.d. [5] | n.d. |
| Tiller population density (Tillers·m$^{-2}$) | L | | 4183 | 8257 | 5615 | 4624 | 3853 | 5284 | 4624 | 4404 | 6606 | 5270 | 622 | 289 |
| | H | | 5725 | 11,009 | 7927 | 7266 | 6606 | 6606 | 6606 | 7486 | 11,119 | 7820 | | |

Notes: [1] Standard error for testing differences between H and L grazing; [2] Standard error for testing seasonal differences; [3] No root deposition as water table rose to approx. 300 mm soil depth at this time; [4] Denotes root length recovered from ingrowth cores on a scale where the annual mean = 1.0; [5] Not determined.

**Table 2.** Mean values of phyllochron (P), architectural signal (*AS*) and modelled weight per root ($W_r$, mg) using Equation (1) for the 9 ingrowth core cycles from December 1986 to January 1988. Dates shown are those for the mid-point of each ingrowth core cycle. *AS* is a numerical index quantifying a theoretical tendency, predicted by plant architecture considerations [26], for one root to be fed by fewer leaves in autumn and more leaves in early summer.

| | Grazing | Harvest Number and Median Date of Root Ingrowth Core Capture | | | | | | | | | |
|---|---|---|---|---|---|---|---|---|---|---|---|
| | | 2 | 3 | 4 | 5 | 6 | 7 | 8 | 9 | 10 | MEAN |
| | | 25 December 1986 | 3 February 1987 | 13 March 1987 | 1 May 1987 | 10 June 1987 | 2 August 1987 | 26 September 1987 | 14 November 1987 | 29 December 1987 | |
| P [1] (days) | | 6.4 | 6.6 | 7.4 | 9.4 | 13.1 | 13.2 | 10.5 | 7.6 | 6.6 | 8.8a |
| *AS* | | 1.03 | 0.97 | 0.89 | 0.79 | 0.72 | 0.99 | 1.26 | 1.38 | 1.15 | |
| $W_r$ | L | 1.1 | 0.9 | 0.6 | 0.6 | 0.6 | 1.9 | 1.4 | 2.3 | 1.0 | 1.2 |
| | H | 0.8 | 0.6 | 0.5 | 0.2 | 0.2 | 1.6 | 1.2 | 1.3 | 0.8 | 0.8 |

Note: [1] Total thermal time for the 1987 calendar year was 4174 °C·d; at 101 °C·d/leaf this is 43.1 leaves/year.

### 3.2. Determination of Phyllochron (P) and Rainfall Pattern

The accumulated thermal time from the National Institute of Water and Atmospheric Research (NIWA) weather data for the 1987 calendar year was 4174 °C· d, which converts to 41.3 phyllochrons per year. Mean phyllochron for root ingrowth periods ranged from 6.4 days in December/January to a little over 13 days in June/July (Table 2). Notable features in the seasonal pattern of thermal time units per day are that the summer maximum and winter minimum follow the longest and shortest days, respectively, by about one month, and that the period of most rapid temperature rise following winter in this particular year, which could potentially create *AS* responses, occurred in late September, October and November (Figure 2).

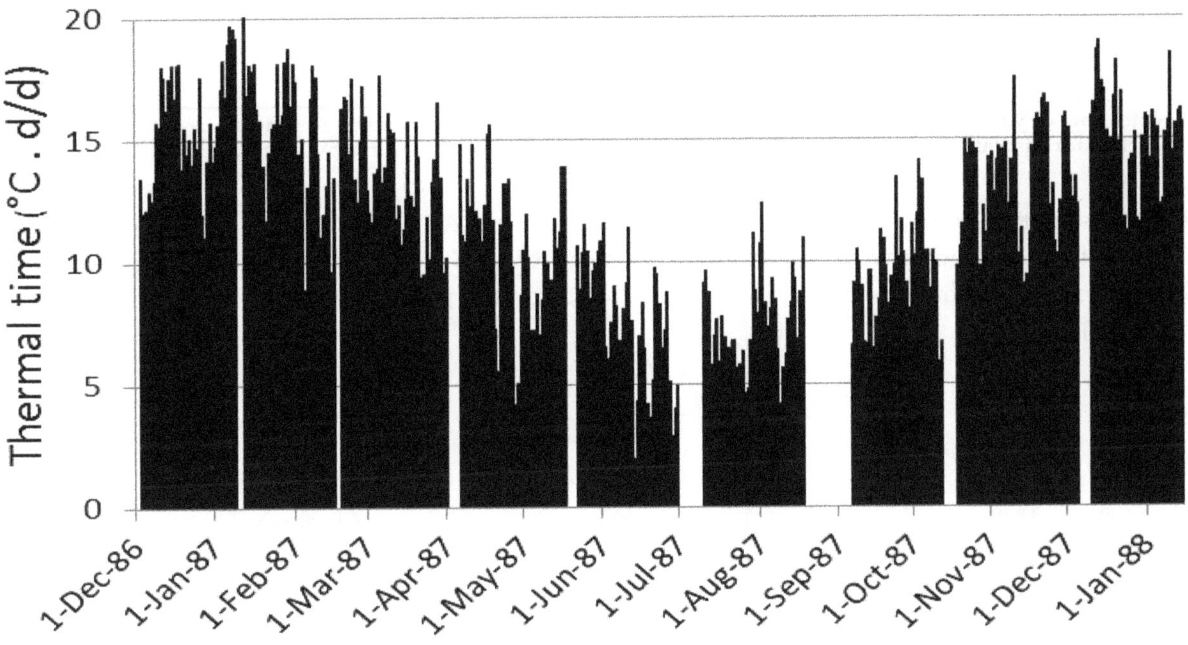


**Figure 2.** Seasonal variation in daily thermal time (°C· d/day) for each of 9 ingrowth core placement and harvest cycles completed between December 1986 and January 1988.

### 3.3. Determination of Architectural Signal (AS), and Root Dry Weight Per Phytomer ($W_r$)

Based on values reported in the literature [21,26] $R_n$ was taken as 1.3; based on comparing counts of root arrivals at the centre of ingrowth cores and in nearby undisturbed soil as described above, $b$ was taken as 0.5 [11]; and based on dissection of ryegrass plants from a field pasture, $f$ was taken as 0.8. Inclusion of these values in Equation (1), together with ingrowth core root deposition data from Equation 1 yielded $W_r$ values varying seasonally from 0.6 mg in winter to 2.3 mg in early summer in L-grazed swards with corresponding values of 0.2 mg and 1.6 mg for H-grazed swards (Table 2).

### 3.4. Relationship between Root Traits and Seasonal Change in Phyllochron (P) and Architectural Signal (AS)

An unexpected finding in the root data was a tendency for negative correlation between $F_r$ and root diameter. While the correlation between $F_{r\ (0\text{–}600)}$ and mean root diameter was non-significant ($p = 0.203$), the correlation between $F_{r\ (70\text{–}600)}$ and mean root diameter neared significance ($r = -0.591$, $p = 0.094$) and the correlation between *AS* and mean root diameter was statistically significant ($r = -0.700$, $p = 0.036$).

Correlation between thermal time elapsed during each root ingrowth core sampling cycle, as defined by *P* (or its reciprocal) and the root deposition and diameter traits, was typically not significant, while for *AS* (or its reciprocal) there was normally a strong correlation. The cosine function was significantly correlated with the root traits, but less so than *AS* (Table 3).

**Table 3.** Correlations (and statistical probabilities) between root deposition in the 3 soil depths of the ingrowth cores, total root deposition, and root diameter, phyllochron (*P*) and architectural signal (*AS*) or their reciprocals, and a cosine function representing annual seasonal change in insolation.

| | $F_{r(0–70)}$ | $F_{r(70–250)}$ | $F_{r(250–600)}$ | $F_{r(0–600)}$ | **Diameter** |
|---|---|---|---|---|---|
| *P* | −0.409 | −0.475 | −0.690 | −0.521 | 0.587 |
| | (0.275) | (0.196) | (0.040) | (0.150) | (0.097) |
| 1/*P* | 0.443 | 0.485 | 0.708 | 0.547 | −0.476 |
| | (0.232) | (0.186) | (0.033) | (0.127) | (0.195) |
| *AS* | 0.758 | 0.830 | 0.620 | 0.804 | −0.700 |
| | (0.018) | (0.006) | (0.075) | (0.009) | (0.036) |
| 1/*AS* | −0.816 | −0.856 | −0.609 | −0.842 | 0.642 |
| | (0.007) | (0.003) | (0.002) | (0.004) | (0.062) |
| Cos day of year [1] | 0.660 | 0.720 | 0.807 | 0.760 | −0.614 |
| | (0.053) | (0.029) | (0.009) | (0.018) | (0.078) |

[1] Note: Day of year was multiplied by 360/365 to create a 360° annual cycle beginning 22 December to form a covariate for insolation, to test environmental factors associated with root ingrowth into refilled cores.

Visual inspection of seasonal alignment of $F_r$, *RRL* and *AS*, indicates that the November $F_r$ peak in root deposition coincides with the seasonal peak in the *AS* signal, and that increased $F_r$ associated with decreased root diameter at this time combines to create greater seasonal response magnitude in *RRL* (Figure 3).

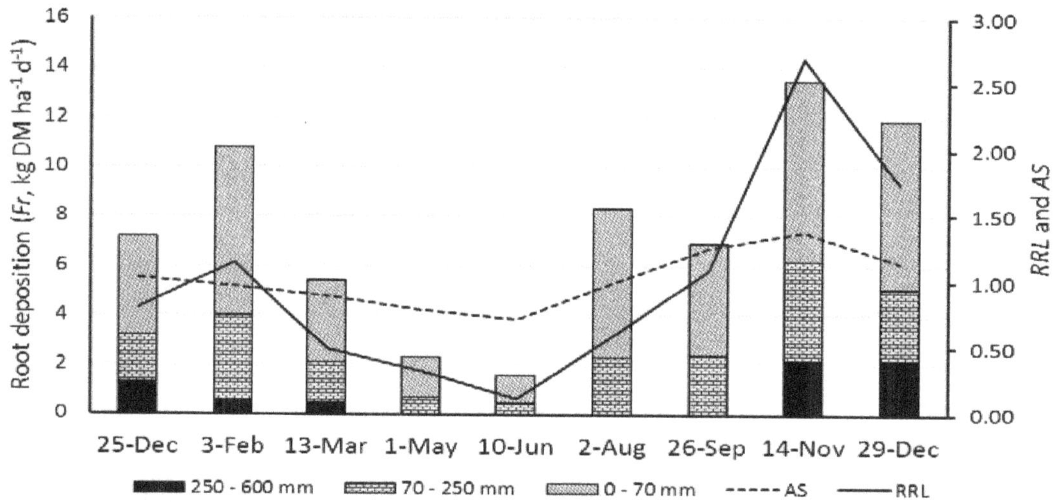


**Figure 3.** Seasonal pattern of root deposition in 3 soil depths and relative root length (*RRL*) for mean of Lax- and Hard-grazed swards at Palmerston North, New Zealand. The hypothetical 'architectural signal' (*AS*) arising from seasonal increase or decrease in the number of leaves expected to feed each root is also shown, and correlates with root deposition significantly better than thermal time or a cosine curve representing insolation (Table 3).

## 4. Discussion

### *4.1. Ingrowth Core Data Provides Insight Complementary to that from Other Techniques*

Understanding of the field behavior of the root system of grasses remains partial. The difficulty in developing suitable measurement techniques is one factor in this situation. Window methods [43,44] can give information about rates of root tip arrival and some information about proliferation at the two-dimensional surface of the window. This can be regarded as a random sample of undisturbed soil and can provide data on traits like root tip elongation rate or interactions with soil fauna, but it is difficult to build up a complete three-dimensional picture of the whole root or quantify mass flow using this technique alone. Core break methods can yield information on root length density within the soil profile but have been found to be of limited accuracy [45]. Root length density data does allow inference about nutrient and water capture, not possible by other methods [46]. Destructive sampling in field swards provides limited insight into root dynamics because of the slow decomposition rate of dead roots, resulting in a high proportion of dead roots in samples that are difficult to distinguish from live roots [9,11]. Destructive sampling of plants grown in deep pots or pipes for a limited period of time provides valuable information on inherent genetic differences in root biomass distribution with soil depth and root:shoot biomass allocation ratios, making this technique potentially valuable for plant breeding purposes [14–16]. Studying plants grown in hydroponic culture allows building of understanding on whole root characteristics, or root:shoot relations [21,47], but leaves uncertainties about extrapolation of results to the field, since the hydroponic environment tends to be unnatural in many respects, including usually the light and temperature regime, and lack of physical resistance to root elongation.

The particular advantage of ingrowth cores is that they allow comparison between experiment factors (in this case L and H grazing management and season) of mass flow of root DM. While installation of ingrowth cores creates a disturbance to root growth, this approach does allow inference and insight complementary to that obtained from techniques described above, and fills a gap in present knowledge by providing information about the seasonal cycle of root development and deposition in grazed field swards.

### *4.2. Signals Determining Seasonality of Root Deposition ($F_r$) and Root Diameter*

The finding that field swards will generate a flush of root growth both at the end of winter and after a summer drought event (Figure 1) would appear to indicate a temporary increase in substrate allocation to roots. The seasonal timing of these two root growth events does not match the postulated *AS* and would appear to arise from root:shoot ratio control by the plant, several models for which were described by Wilson [48]. In short, there is evidence from the field data of root-shoot 'cross-talk' that ensures seasonal shoot growth flushes are preceded by root deposition activity that will presumably deliver N and other nutrients required by the leaves. Such responses, especially post-drought root responses would be variable from year to year and this may partly explain variation in reported seasonal patterns of root deposition. This point is also relevant to both extension advice to farmers on pasture husbandry and to modeling of pasture herbage accumulation. Farmers can be advised to avoid high grazing pressure when root system recovery is occurring, and when modeling pasture herbage accumulation based on temperature or radiation a correction term can be included for the temporary period of preferential root allocation.

With respect to the proposed 'architectural signal' termed here *AS* and first identified theoretically almost two decades ago [26], this is the first evaluation, to the authors' knowledge, of field data for indication of the expression of such a signal. Despite theoretical calculations showing an *AS could* exist, the *actual* existence of the signal, can not necessarily be assumed, because various plant internal compensations such as seasonal differences in area of individual leaves or their photosynthesis rates could neutralize it. Since the identification of the theoretical *AS* signal, one question has been: 'How could it be measured?' In this study, the finding that root biomass deposition rates of field swards correlate more strongly with the estimated value of the *AS*, than thermal time as represented by *P*, or a cosine curve representing insolation, provides circumstantial evidence for, but not proof of the operation of the *AS* as a factor in determining seasonality of root growth in these swards. The fact that total root deposition correlates with the *AS* indicates that root:shoot partitioning is altered seasonally by the *AS*, not just the weight of individual roots. If *AS* reflects the relative mass of tiller source and sink tissues, this is logical, but confirmation from further research would be desirable. By way of comparison with *P*-values assumed here based on thermal time studies for ryegrass in New Zealand, longer P-values of 9 and 35 days for summer and winter, respectively, have been reported from a colder UK climate [49]

Prior to making the thermal time calculation from the temperature data (Figure 2), we intuitively expected the *AS* signal earlier in the spring and autumn than indicated Figure 3. It can be seen that there was indeed a peak in $F_r$ coinciding with the early summer peak in *AS*, and there was also low root growth coincident with an *AS* less than 1.0 in June (Figure 3). Logically, if root allocation is driven by a regular pattern of root initiation and shoot:root allocation of DM on senescence of successive leaves (Figure 4), except perhaps for some variation during the flowering period [50], $F_r$ should correlate best with seasonal changes in thermal time. To the contrary, correlations between $F_r$ in the various soil depth horizons and *P* or its reciprocal were generally non-significant, while correlations between $F_r$ and *AS* or its reciprocal were mostly strongly significant, even though the *AS* clearly does not account for the 'late winter' and 'drought recovery' root flushes.

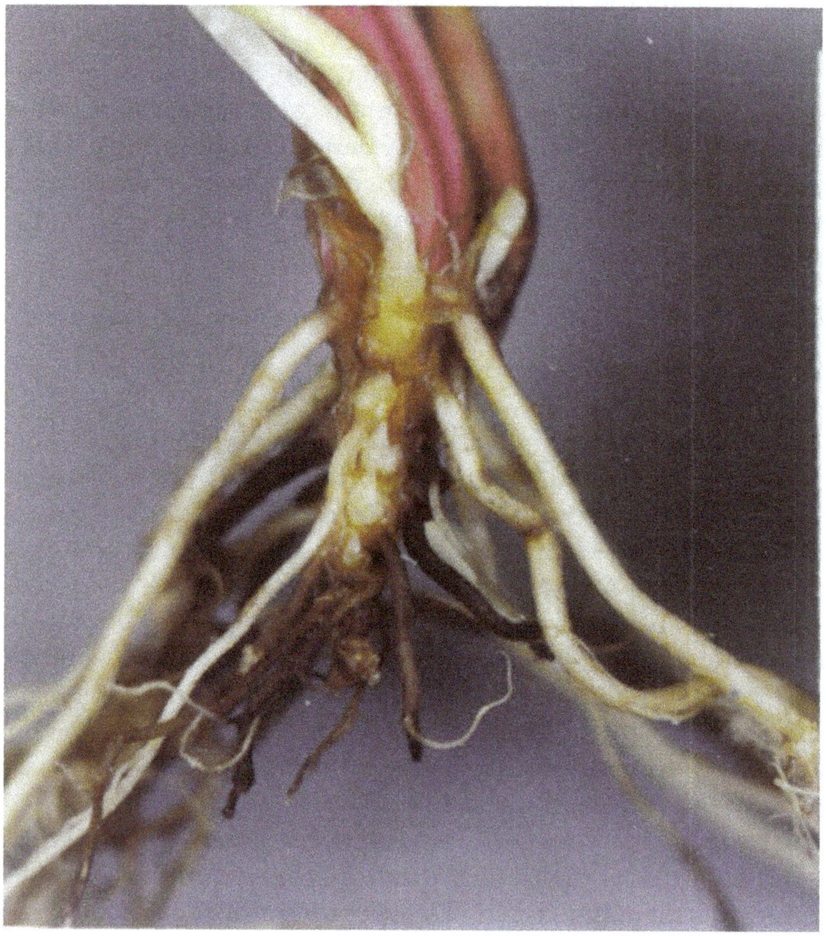

**Figure 4.** A ryegrass tiller base with 6 visible tillers or buds, representing approximately 12 phytomers on the tiller axis, since alternate tiller buds appear alternately on opposite sides. The young white roots on the uppermost root bearing phytomer are typically short and unbranched, having only recently been initiated, and branching begins one or two phyllochrons later [24]. At around 12 phytomers distance from the uppermost root, roots are typically highly branched and dead or dying [21]. An animation showing roots and leaves in 'steady state' turnover on the ryegrass tiller axis, as in hydropnic culture can be viewed at Supplementary file.

Previously, root diameter reduction in spring was assumed to relate to the appearance of large numbers of new tillers, many of these formed at the base of flowering tillers [9,11]. It is possible, however, that more rapid elongation results in narrower diameter roots and vice versa. This point is deserving of future study.

### 4.3. Reconciliation of $F_r$ Data from Field Swards with Phytomer Turnover Models of Root Formation

In hydroponic culture where almost every phytomer on the tiller axis produces one or more roots in an approximately steady state turnover, roots of ryegrass typically cease dry weight increase and main axis elongation about five phyllochrons after initiation [21]. This cessation presumably arises from the accumulation of younger roots higher on the tiller axis capturing the basipetal flow of photosynthate from leaves before it reaches roots more distant from the youngest root-bearing phytomer [21,24]. In this context, Equation (1) is not intended to imply that all the substrate from one leaf is allocated to root formation at a particular phytomer, but rather that over the period of root elongation the sum of partial contributions from the contributing leaves is equivalent to the total contribution of one leaf.

It is useful at this point to review the morphology of a ryegrass tiller base (Figure 4), and to note that live roots are present over a span of approximately 10–12 phytomers at any one time with young roots forming above, and older roots dying below [21,41]. A similar behavior is observed in wheat and rice [22–24,51].

An inference from Figure 4 is that ryegrass roots will be 'architecturally' limited in depth penetration by the root tip elongation attainable within the 'feeding period' after initiation (as mentioned above, typically five phyllochrons in hydroponic culture). It is a hypothesis for further study that the root elongation period may be extended either if a daughter tiller has initiated at a particular phytomer and is feeding substrate to the region of the tiller axis where that root is attached, or if new root formation is temporarily suppressed. This latter point is a very interesting insight from our data. Both an earlier field study in New Zealand [31], and a UK glasshouse study [52] found that that new root initiation does not occur in ryegrass in dry summer conditions. It seems logical then, that suspension of new root initiation would provide for supply of substrate moving basipetally to continue to reach existing roots for an extended period allowing those roots to penetrate below 250 mm soil depth in summer (Figure 3).

The use of the field root deposition rate data ($F_r$) to estimate individual root dry weight, $W_r$ is somewhat hypothetical, but instructive even so, as it gives a preliminary evaluation, subject to refinement, of the morphological details of root system formation. Of the parameters assumed for this calculation in this paper, a mean value for $R_n$ can be determined by low power microscopy of tiller bases collected from a sward [26], and the same would be true for $f$. The value for $b$ used here was obtained by calibration during a subsequent experiment [11], though the calibration work was time consuming. It is intuitively likely that low nutrient levels in the builders sand used to fill the ingrowth cores might have reduced root colonization and resulted in a lower value for $b$. This might be overcome in future experiments either by adding a small amount of slow release fertilizer to the material used to refill root ingrowth cores or by refilling cores with the same soil material bored out and using some kind of sleeve to delineate the ingrowth core for future harvest, as is common practice in other studies.

Early discussion of ryegrass root systems assumed an annual root replacement event [31,53] and this was at least partially supported by the pattern of new root emergence through an annual cycle [32]. This does not reconcile well with the picture of phytomer-based steady state turnover implied by hydroponic studies [21], nor by the morphology seen in Figure 4. However, the data presented here go a considerable way to showing how a phytomer based plant morphology can deliver an annual cycle of root development under field conditions. Several signals operate to modify the underlying steady state turnover generated by new phytomer addition on the tiller axis. There is evidence for seasonal variation in root:shoot allocation (Figure 1; [50]), $P$ varies seasonally (Figure 2) and root deposition rates would be expected to be influenced. There is also evidence that an $AS$ further modifies seasonality of root deposition with suppression of root deposition in mid-winter and enhancement in early summer, possibly with associated effects on root diameter. Lastly, where summer moisture deficit suppresses new root initiation, existing roots should receive additional substrate and penetrate deeper.

Considering that ryegrass tillers in hydroponic culture [21] had a mean dry weight of 425 mg with 191 mg root dry weight, comprising on average 15.2 live roots per tiller weighing 12.6 mg, the $W_r$ values of 0.2 to 2.3 mg DW root$^{-1}$ (Table 2) seem low but credible and indicate that these calculations are worth developing in future studies for the insight they can offer on root dynamics. A perspective that emerges from this calculation is that although $F_r$ is little changed under hard grazing, dividing the smaller substrate pool among a larger tiller population density results in a proportionately larger drop in allocation to individual roots, in H-grazed swards, especially in winter, unless there is some compensation within the tiller. For example, a decrease in $R_n$, in H-grazed swards could neutralize the $W_r$ difference between L and H swards obtained from Equation (1) here. Further investigation would be desirable, since this point is important when providing husbandry recommendations to farmers, as an extension opinion sometimes promoted is that hard grazing to enhance tiller density will be beneficial [54]. While hard grazing does indeed increase tiller density in most situations (Table 1), the implications of defoliation severity for root weight reduction (Table 2), and sward leaf area index reduction [55] also need to be considered.

## 5. Conclusions

Root deposition rate as determined by an ingrowth core technique was about 15% of above ground herbage accumulation and showed a very similar seasonal pattern.

Peaks of root deposition after summer drought and in late winter preceded above ground herbage accumulation by about one month and likely relate to conventional root:shoot signaling.

Data support the existence of a plant morphology-driven 'architectural signal' partly determining root deposition rate through seasonal changes in the number of leaves feeding sites where roots are being formed.

The architectural signal theoretically results in root formation being boosted in rising temperatures and suppressed in falling temperatures and field data correlate significantly with the theoretical signal.

A model based on the phytomer structure of the tiller axis produced credible values for the weight of individual roots in field swards as affected by season and grazing intensity. A tentative insight from this model for confirmation by further research is that hard grazing will limit individual root size, especially in winter.

The root system of perennial ryegrass is dynamic, with an inbuilt tendency to shift deposition activity deeper in summer and shallower in winter.

**Acknowledgments:** The first author warmly thanks Alex C.P. Chu and John Hodgson for their guidance as Supervisors. The field study of ryegrass root deposition over a 12 month cycle was labor intensive and was assisted by more than a dozen people at different times [11]. We thank the C. Alma Baker Trust, The Massey University Research Fund, The Massey University Agricultural Research Foundation, and the T.R. Ellett Agricultural Research Trust for funding support. We thank Cécile Duranton for preparation of Figure 1.

**Author Contributions:** This work was a part of the PhD study of Cory Matthew with Alec Mackay as supervisor and active participant. Arif Hasan Khan Robin contributed expert knowledge on root turnover in the compilation of the manuscript.

## References

1. Parsons, A.J.; Leafe, E.L.; Collett, B.; Stiles, W. The physiology of grass production under grazing. I. Characteristics of leaf and canopy photosynthesis of continuously-grazed swards. *J. Appl. Ecol.* **1983**, *20*, 117–126. [CrossRef]
2. Parsons, A.J.; Leafe, E.L.; Collett, B.; Penning, P.D.; Lewis, J. The physiology of grass production under grazing. II. Photosynthesis, crop growth and animal intake of continuously-grazed swards. *J. Appl. Ecol.* **1983**, *20*, 127–139. [CrossRef]
3. Korte, C.J. Tillering in 'Grasslands Nui'perennial ryegrass swards 2. Seasonal pattern of tillering and age of flowering tillers with two mowing frequencies. *N. Z. J. Agric. Res.* **1986**, *29*, 629–638. [CrossRef]

4. Garwood, E.A. Seasonal variation in appearance and growth of grass roots. *Grass Forage Sci.* **1967**, *22*, 121–129. [CrossRef]
5. Chapman, D.F.; Lemaire, G. Morphogenetic and structural determinants of plant regrowth after defoliation. In Proceedings of the XVII International Grassland Congress, Palmerston North, New Zealand, 8–21 February 1993; pp. 95–104.
6. Bahmani, I.; Hazard, L.; Varlet-Grancher, C.; Betin, M.; Lemaire, G.; Matthew, C.; Thom, E.R. Differences in tillering of long-and short-leaved perennial ryegrass genetic lines under full light and shade treatments. *Crop Sci.* **2000**, *40*, 1095–1102. [CrossRef]
7. Van Loo, E.N. On the Relation between Tillering, Leaf Area Dynamics and Growth of Perennial Ryegrass (*Lolium perenne* L.). Ph.D. Thesis, Wageningen University, Wageningen, The Netherlands, May 1993.
8. Sartie, A.M.; Matthew, C.; Easton, H.S.; Faville, M.J. Phenotypic and QTL analysis of herbage production-related traits in perennial ryegrass (*Lolium perenne* L.). *Euphytica* **2011**, *182*, 295–315. [CrossRef]
9. Matthew, C.; Xia, J.; Chu, A.; Mackay, A.; Hodgson, J. Relationship between root production and tiller appearance rates in perennial ryegrass (*Lolium perenne* L.). In *Plant Root Growth—An Ecological Perspective*; Atkinson, D., Ed.; British Ecological Society Special Publication, Blackwell Scientific Publications: London, UK, 1991; pp. 281–290.
10. Xia, J.X. The Effects of Defoliation on Tissue Turnover and Pasture Production in Perennial Ryegrass, Prairie Grass and Smooth Bromegrass Pasture. Ph.D. Thesis, Massey University, Palmerston North, New Zealand, 1991.
11. Matthew, C. A Study of Seasonal Root and Tiller Dynamics in Swards of Perennial Ryegrass (*Lolium perenne* L.). Ph.D. Thesis, Massey University, Palmerston North, New Zealand, 1993. Available online: http://mro.massey.ac.nz/xmlui/bitstream/handle/10179/3289/02_whole.pdf (accessed on 11 June 2016).
12. Davidson, R.L. Root systems—The forgotten component of pastures. In *Plant Relations in Pastures*; Wilson, J.R., Ed.; CSIRO: Melbourne, Australia, 1978; pp. 86–94.
13. Troughton, A. *The Underground Organs of Herbage Grasses*; Bulletin No. 44; Commonwealth Bureau of Pastures and Field Crops: Hurley, UK, 1957.
14. Crush, J.R.; Nichols, S.N.; Easton, H.S.; Ouyang, L.; Hume, D.E. Comparisons between wild populations and bred perennial ryegrasses for root growth and root/shoot partitioning. *N. Z. J. Agric. Res.* **2009**, *52*, 161–169. [CrossRef]
15. Crush, J.R.; Easton, H.S.; Waller, J.E.; Hume, D.E.; Faville, M.J. Genotypic variation in patterns of root distribution, nitrate interception and response to moisture stress of a perennial ryegrass (*Lolium perenne* L.) mapping population. *Grass For. Sci.* **2007**, *62*, 265–273. [CrossRef]
16. Crush, J.R.; Popay, A.J.; Waller, J. Effect of different *Neotyphodium* endophytes on root distribution of a perennial ryegrass (*Lolium perenne* L.) cultivar. *N. Z. J. Agric. Res.* **2004**, *47*, 345–349. [CrossRef]
17. Cartes, P.; Jara, A.A.; Pinilla, L.; Rosas, A.; Mora, M.L. Selenium improves the antioxidant ability against aluminium-induced oxidative stress in ryegrass roots. *Ann. Appl. Biol.* **2010**, *156*, 297–307. [CrossRef]
18. Xu, W.H.; Liu, H.; Ma, Q.F.; Xiong, Z.T. Root exudates, rhizosphere Zn fractions, and Zn accumulation of ryegrass at different soil Zn levels. *Pedosphere* **2007**, *17*, 389–396. [CrossRef]
19. Bircham, J.S.; Hodgson, J. The influence of sward condition on rates of herbage growth and senescence in mixed swards under continuous stocking management. *Grass For. Sci.* **1983**, *38*, 323–331. [CrossRef]
20. Donaghy, D.J.; Fulkerson, W.J. Plant-soluble carbohydrate reserves and senescence—Key criteria for developing an effective grazing management system for ryegrass-based pastures: A review. *Aust. J. Exp. Agric.* **2001**, *41*, 261–275.
21. Robin, A.H.K.; Matthew, C.; Crush, J. Time course of root initiation and development in perennial ryegrass—A new perspective. *Proc. N. Z. Grassl. Assoc.* **2010**, *72*, 233–239.
22. Robin, A.H.K.; Uddin, M.J.; Bayazid, K.N. Polyethylene glycol (PEG)-treated hydroponic culture reduces length and diameter of root hairs of wheat varieties. *Agronomy* **2015**, *5*, 506–518. [CrossRef]
23. Robin, A.H.K.; Uddin, M.J.; Afrin, S.; Paul, P.R. Genotypic variations in root traits of wheat varieties at phytomer level. *J. Bangladesh Agric. Univ.* **2014**, *12*, 45–54. [CrossRef]
24. Robin, A.H.K.; Matthew, C.; Uddin, M.J.; Bayazid, K.N. Salinity-induced reduction in root surface area and changes in major root and shoot traits at the phytomer level in wheat. *J. Exp. Bot.* **2016**. [CrossRef] [PubMed]
25. Steen, E. Variation in root growth in a grass ley studied with a mesh bag technique. *Swed. J. Agric. Res.* **1984**, *14*, 93–97.

26. Matthew, C.; Yang, J.Z.; Potter, J.F. Determination of tiller and root appearance in perennial ryegrass (*Lolium perenne*) swards by observation of the tiller axis, and potential application in mechanistic modelling. *N. Z. J. Agric. Res.* **1998**, *41*, 1–10. [CrossRef]
27. Bos, H.J.; Neuteboom, J.H. Morphological analysis of leaf and tiller number dynamics of wheat (*Triticum aestivum* L.): Responses to temperature and light Intensity. *Ann. Bot.* **1998**, *81*, 131–139. [CrossRef]
28. Hitch, P.A.; Sharman, B.C. Vascular pattern of festucoid grass axes, with particular reference to nodal plexi. *Bot. Gaz.* **1971**, *132*, 38–56. [CrossRef]
29. Etter, A.G. How Kentucky bluegrass grows. *Ann. MO. Bot. Gard.* **1951**, *38*, 293–375. [CrossRef]
30. Henessy, D.; O'Donovan, M.; French, P.; Laidlaw, A.S. Factors influencing tissue turnover during winter in perennial ryegrass-dominated swards. *Grass For. Sci.* **2008**, *63*, 202–211. [CrossRef]
31. Caradus, J.R.; Evans, P.S. Seasonal root formation of white clover, ryegrass and cocksfoot in New Zealand. *N. Z. J. Agric. Res.* **1977**, *20*, 337–342. [CrossRef]
32. Jacques, W.A. Root development in some common New Zealand pasture plants. IX. The root replacement pattern in perennial ryegrass (*Lolium perenne*). *N. Z. J. Sci. Technol. A* **1956**, *38*, 160–165.
33. Matthew, C.; Van Loo, E.N.; Thom, E.R.; Dawson, L.A.; Care, D.A. Understanding shoot and root development. In Proceedings of the XIX International Grassland Congress, Sao Pedro, Brazil, 11–21 February 2001; pp. 19–27.
34. Bohm, W. Methods of studying root systems. In *Ecological Studies—Analysis and Synthesis*; Billings, W.D., Golley, F., Lange, O.L., Olson, J.S., Eds.; Springer-Verlag: Berlin, Germany, 1979; Volume 33.
35. Tennant, D. A test of a modified line intersect method of estimating root length. *J. Ecol.* **1975**, *63*, 995–1001. [CrossRef]
36. Mitchell, K.J.; Glenday, A.C. The tiller population of pasture. *N. Z. J. Agric. Res.* **1958**, *1*, 305–318.
37. Butler, B.M.; Matthew, C.; Heerdegen, R. The greenhouse effect—What consequences for seasonality of pasture production. *Weather Clim.* **1990**, *10*, 55–60.
38. Hoglund, J.H.; Crush, J.R.; Brock, J.L.; Carran, R.A. Nitrogen fixation in pasture. 12. General discussion. *N. Z. J. Exp. Agric.* **1979**, *1*, 45–51.
39. Moot, D.J.; Scott, W.R.; Roy, A.M.; Nicholls, A.C. Base temperature and thermal time requirements for germination and emergence of temperate pasture species. *N. Z. J. Agric. Res.* **2000**, *43*, 15–25. [CrossRef]
40. Black, A.D.; Moot, D.J.; Lucas, R.J. Seedling development and growth of white clover, Caucasian clover, and perennial ryegrass sown in field and controlled environments. *Proc. N. Z. Grassl. Assoc.* **2002**, *64*, 197–204.
41. Yang, J.Z.; Matthew, C.; Rowland, R.E. Tiller axis observations for perennial ryegrass (*Lolium perenne*) and tall rescue (*Festuca arundinacea*): Number of active phytomers, probability of tiller appearance, and frequency of root appearance per phytomer for three cutting heights. *N. Z. J. Agric. Res.* **1998**, *41*, 11–18. [CrossRef]
42. Matthew, C.; van der Linden, A.; Hussain, S.; Easton, H.S.; Hatier, J.-H.B.; Horne, D.J. Which way forward in the quest for drought tolerance in perennial ryegrass. *Proc. N. Z. Grassl. Assoc.* **2012**, *74*, 195–200.
43. Sackville Hamilton, C.A.G.; Cherrtt, J.M. The development of clover and ryegrass root systems in a pasture and their interactions with the soil fauna. In *Plant Root Growth—An Ecological Perspective*; Atkinson, D., Ed.; British Ecological Society Special Publication, Blackwell Scientific Publications: London, UK, 1991; pp. 291–301.
44. Evans, P.S. The effect of repeated defoliation to three different levels on root growth of five pasture species. *N. Z. J. Agric. Res.* **1973**, *16*, 31–34. [CrossRef]
45. Bland, W.L. Estimating root length density by the core break method. *Soil Sci. Soc. Am. J.* **1989**, *53*, 1595–1597. [CrossRef]
46. De Willigen, P.; van Noordwijk, M. Roots, Plant Production and Nutrient Use Efficiency. Ph.D. Thesis, Wageningen University, Wageningen, The Netherlands, 1987.
47. Gastal, F.; Saugier, B. Relationships between nitrogen uptake and carbon assimilation in whole plants of tall fescue. *Plant Cell Environ.* **1989**, *12*, 407–418. [CrossRef]
48. Wilson, J.B. A review of evidence on the control of root:shoot ratio in relation to models. *Ann. Bot.* **1988**, *61*, 433–449.
49. Davies, A. Structure of the grass sward. In *Animal Production from Temperate Grassland*; Gilsenan, B., Ed.; Irish Grassland and Animal Production Association: Dublin, Ireland, 1977; pp. 36–44.
50. Troughton, A. Further studies on the relationship between root and shoot systems of grasses. *J. Br. Grassl. Soc.* **1960**, *15*, 4–47. [CrossRef]

51. Robin, A.H.K.; Saha, P.S. Morphology of lateral roots of twelve rice cultivars of Bangladesh: Dimension increase and diameter reduction in progressive root branching at the vegetative stage. *Plant Root* **2015**, *9*, 34–42. [CrossRef]
52. Troughton, A. Production of root axes and leaf elongation in perennial ryegrass in relation to dryness of the upper soil layer. *J. Agric. Sci.* **1980**, *95*, 533–538. [CrossRef]
53. Stuckey, I.H. Seasonal growth of grass roots. *Am. J. Bot.* **1941**, *28*, 486–491. [CrossRef]
54. Edwards, G.R.; Chapman, D.F. Plant responses to defoliation and relationships with pasture persistence. In *Pasture Persistence Symposium*; Mercer, C.F., Ed.; New Zealand Grassland Association: Dunedin, New Zeaalnd, 2011; pp. 99–108.
55. Matthew, C.; Lemaire, G.; Sackville Hamilton, N.R.; Hernandez-Garay, A. A modified self-thinning equation to describe size/density relationships for defoliated swards. *Ann. Bot.* **1995**, *76*, 579–587. [CrossRef]

# 6

# Using Ecophysiology to Improve Farm Efficiency: Application in Temperate Dairy Grazing Systems

**David F. Chapman**

DairyNZ, c/o P.O. Box 85066, Lincoln University, Lincoln 7647, New Zealand; david.chapman@dairynz.co.nz

Academic Editor: Cory Matthew

**Abstract:** Information on the physiological ecology of grass-dominant pastures has made a substantial contribution to the development of practices that optimise the amount of feed harvested by grazing animals in temperate livestock systems. However, the contribution of ecophysiology is often under-stated, and the need for further research in this field is sometimes questioned. The challenge for ecophysiolgists, therefore, is to demonstrate how ecophysiological knowledge can help solve significant problems looming for grassland farming in temperate regions while also removing constraints to improved productivity from grazed pastures. To do this, ecophysiological research needs to align more closely with related disciplines, particularly genetics/genomics, agronomy, and farming systems, including systems modelling. This review considers how ecophysiological information has contributed to the development of grazing management practices in the New Zealand dairy industry, an industry that is generally regarded as a world leader in the efficiency with which pasture is grown and utilised for animal production. Even so, there are clear opportunities for further gains in pasture utilisation through the refinement of grazing management practices and the harnessing of those practices to improved pasture plant cultivars with phenotypes that facilitate greater grazing efficiency. Meanwhile, sub-optimal persistence of new pastures continues to constrain productivity in some environments. The underlying plant and population processes associated with this have not been clearly defined. Ecophysiological information, placed in the context of trait identification, grounded in well-designed agronomic studies and linked to plant improvements programmes, is required to address this.

**Keywords:** temperate pastures; dairy grazing management; pasture regrowth; herbage utilisation; farm systems

---

## 1. Introduction

The overview of the contribution of research into the physiological ecology of forage plants presented by Da Silva *et al.* [1] illustrates the critical role played by this scientific discipline in quantifying the dynamics of leaf area expansion, light capture, and sward structural and compositional changes (including fluxes of plant organic reserves) as they affect herbage accumulation in grass-dominant pastures. This information has been instrumental in developing sward management practices that maximise the intake potential of grazing animals while also sustaining high rates of net herbage accumulation. Hodgson [2] weaved the underlying science of plant and pasture growth processes into a clear account of the grazing management principles and practices that optimise the sustainable yield of herbage under different grazing methods operating in a range of environments. The science, principles and practices were further developed in Lemaire *et al.* [3].

Despite the impressive body of knowledge summarised in these two volumes and this special issue, the contribution of ecophysiology to grassland development around the world is often under-stated and the value of further ecophysiological research has sometimes been questioned. For example, Bryant [4], in summarising optimum feed management practices for New Zealand dairy farms, concluded that over and above the implementation of current grazing recommendations, the effect of grazing management on production was small, and proposed this " ... must surely signal the end of a long era in grazing management research ... ". This is a bold statement, considering that maintenance of the global competitive advantage of the New Zealand dairy industry rests heavily on achieving continued improvements in the amount of feed harvested from grazed, perennial ryegrass-dominant pasture. Recent analyses indicate that New Zealand dairy farmers have lost their previous position as the lowest-cost milk producers in the world [5] and, seemingly, also lost their focus on achieving good control of pasture growth and utilisation [6]. These trends suggest that a resurgence in "grazing management research" is required: or, at least, that questions are asked about how well the ecophysiogical knowledge already available is being used to reinforce management advice to dairy farmers (e.g., [7]).

The New Zealand dairy industry is built upon the combined natural elements of fertile soils, high rainfall (usually evenly distributed throughout the year), and a temperate maritime climate which allows large amounts of pasture to be grown to feed high-producing animals. It is internationally recognised as a model for how a globally competitive industry can be developed from a sustainable natural resource base without the use of price subsidies or other protectionist market policies. At the farm level, there is a close positive relationship between the amount of pasture consumed and operating profit per hectare [8], as there is in other countries with largely pasture-based dairy industries [9]. Continued gains in the amount of pasture utilised are essential for future growth of the industry at both the global and the farm levels.

The aim of this review is to use the New Zealand dairy industry as a case study to: illustrate the important contribution of ecophysiological knowledge to the development of efficient grazing management practices; show how those practices are implemented (or, as seen in some surveys, not implemented) on dairy farms; identify where existing and new ecophysiological information can be used to help improve pasture performance further; and consider how the discipline of ecophysiology needs to link to other disciplines to increase the rate of gain in pasture productivity being achieved on-farm.

## 2. Critical Ecophysiological Principles and Their Translation to Grazing Management Practice

The starting point for designing efficient grazing management systems is the pasture regrowth curve. Regrowth is a result of each grass tiller producing new leaves to replace leaves that were removed by grazing, thus steadily increasing the pasture mass produced over time. The typical regrowth interval in dairy grazing systems is between 20 and 60 days, varying depending on temperature, radiation intensity, and water and nutrient availability. The critical ecophysiological processes supporting the accumulation of mass have been well documented in the research reviewed in this special issue [1,10,11]. They include: the emergence and expansion of new leaves from the tiller apical meristem to replace lamina area and increase light interception and carbon (C) assimilation; the mobilisation of some stored C and nitrogen (N) to support new leaf production and Rubisco synthesis, with subsequent replenishment of those stores once the energy and N status of the whole plant recovers; changes in the relative allocation of available C between the shoot and root systems, initially to promote new leaf growth, and subsequently to resume root growth so that the capacity to assimilate nutrients and water remains in balance with the increase in C assimilation capacity of the canopy; tiller initiation and emergence and consequent effects on sward tiller density; changes in leaf morphology including the specific and absolute area of the new leaves produced as the plant responds to its changing energy status; and the turnover of older leaf material entering senescence including re-translocation of some constituents of old leaves, especially N, to support new growth.

The underlying drivers are light interception, whole-plant energy status, and the maintenance of homeostatic growth such that the balance between root and shoot (including reproductive development) maximises the long-term ability of grass plants to compete for light and other resources and so sustain their presence within the grassland community [12]. The high degree of spatial and temporal variation in weather conditions (influencing temperature, rainfall, and total radiation receipt), nutrient distribution (including the effects of excreta return) and defoliation (which is spatially variable irrespective of whether swards are continuously or intermittently grazed, just at different scales [13]), combined with the diversity of species-specific traits governing resource capture and use found in mixed swards, makes for a highly complex system which farmers must manage. Furthermore, efficient livestock farming requires that as much herbage as possible is harvested to feed animals. But this very process removes the leaf area required for plants to sustain growth and sets up a fundamental conflict between the needs of the pasture and the needs of the animal [14]. It also sets in train repeated cycles of leaf removal and replacement, the outcome of which can be measured in terms of, for example, the total amount of herbage harvested per unit area per year, or (of more direct importance to farm profitability) the amount of animal product (milk, or meat) derived from grazed pasture per year. The degree to which these repeated cycles are managed to achieve a sustainable balance between herbage harvest rates and herbage growth rates becomes the pivot point upon which farm business success depends.

The complexity of the processes described above is daunting for scientists. As the reviews published in this special edition attest, despite decades of research into the critical ecophysiological processes, our knowledge of the interactions is far from complete. The challenge is much more daunting for farmers. A key goal for scientists working in the field of grassland ecophysiology must, therefore, be to make the "complex seem simple". To achieve this, ecophysiological principles must be translated into easily implemented practices and decision rules (with accompanying management information resources) for farmers and their advisers. Grassland farming efficiency in the UK, New Zealand, Australia and other countries has benefitted enormously from work of this nature [15]. Examples include: development of sward height targets in the UK [2,16]; grazing rotation guidelines in New Zealand and Australia [15,17]; use of the leaf stage indicator for perennial ryegrass [18], and the compilation of comprehensive grazing decision rules [19]. Most of these apply to temperate zones, but the adaptation of the same ecophysiological principles to $C_4$ species is making great strides towards improving the efficiency of tropical grassland farming [1].

Even in the temperate grassland world, further efficiency gains are well within reach for farmers if new ecophysiological information can be gathered, and existing ecophysiological knowledge is re-visited, to identify how improvements in pasture harvest rates can be achieved. The clearest opportunity is to capture more of the potential for herbage accumulation available from the pasture regrowth curve in intermittent (rotational) defoliation systems. The following sections relate the ecophysiological processes of light capture, herbage accumulation, and tissue turnover in grass-based pastures during regrowth to current grazing management recommendations in the New Zealand dairy industry. They draw on evidence from on-farm studies to illustrate how opportunities for improved biological efficiency, and greater profitability, are being foregone in this industry which is often considered the benchmark against which other pasture-based livestock industries should measure their potential for improvement.

## 3. Managing the Pasture Regrowth Curve

### *3.1. The Ecophysiological Basis of Current Grazing Management Recommendations for New Zealand Dairy Farms*

Pasture yield was first related to light interception in the 1950s. Brougham [20,21] showed that, when the pasture canopy intercepts about 95% of the available light, the rate of accumulation of pasture mass declines to quickly reach a ceiling yield that varies according to seasonal differences in light intensity and the leaf area index (LAI) that can be sustained. In the 1970s and 1980s, the flow of mass through plants and the canopy as pastures regrow up to ceiling yield was measured and modelled [22,23]. The principles uncovered in these studies are fundamental to the way pastures are now managed in intensive pasture-based livestock systems.

Mass flux processes during regrowth were comprehensively described by Parsons *et al.* [23]. This analysis reconciled the apparent discrepancies in pasture herbage accumulation rates observed between continuous and intermittent defoliation by relating regrowth dynamics to LAI. Using the common sigmoidal pasture regrowth curve, Parsons *et al.* [23] showed how net instantaneous and average growth rates can be derived from this curve and used to identify the optimum time to harvest during the regrowth cycle to maximise herbage yield over repeated cycles of defoliation. The optimum is the point at which average growth (Equation (1)) reaches its maximum value during regrowth [14].

$$\text{Average growth rate} = (W - W_0)/t \tag{1}$$

where $W$ = pasture mass at any time $t$ during regrowth, and $W_0$ = pasture mass at the start of the regrowth period, also known as "residual" pasture mass.

Farmers cannot measure average growth rate. However, ecophysiological information provides an indicator that can easily be adopted by farmers, termed the "leaf stage" rule which is a measure of the readiness for grazing during regrowth from the plant perspective. Mass flux dynamics reveals that maximum average growth rate occurs at the point when the rate of senescence of old leaf material causes the instantaneous growth rate (Equation (2)) to fall below the average growth rate for the first time since the start of the regrowth period (see for example Figure 3.9 in [14]).

$$\text{Instantaneous growth rate (kg}\cdot\text{DM/ha/day)} = dW/dt \tag{2}$$

where $W$ = pasture mass and $t$ = time, in units of one day

Tissue turnover studies (e.g., [22,24,25]) further reveal that the relative rates of new leaf growth and senescence of old leaf material are fixed by the pattern of leaf appearance and longevity on the ryegrass tiller. Perennial ryegrass is often termed a "three leaf" plant because it generally sustains a maximum of three live leaves on a tiller [26]. Hence, after grazing, once the third new leaf has been produced, the first leaf which was produced after grazing will start to die. The sequence of new leaf production and the death of older leaves is fundamental to the overall regrowth process. When the mass of old leaves that is dying and disappearing from the bottom of the pasture is equal to the mass of the new leaf being produced at the top of the pasture, the pasture reaches ceiling yield. New leaves are still being produced at this point, but the mass of new leaf added to the canopy is cancelled out by the mass of leaf that is dying and decaying: hence the net rate of pasture growth is zero. This will generally occur at, or after, three new leaves have been produced, unless the post-grazing residual is very low or very high, in which case it will occur later or earlier, respectively, as discussed below.

From this, Fulkerson and Donaghy [18] proposed that a simple assessment of the number of leaves produced since the previous grazing event carried out on a random sample of 10 to 20 tillers per paddock allows farmers to track plant physiological state during regrowth. They recommended that grazing be implemented when tillers have regrown between two and three new leaves since the previous defoliation, reasoning that this will provide the balance between rates of new leaf production, senescence of old leaves, maintenance of non-structural carbohydrate reserves to assist regrowth and plant survival during periods of stress required to maximise long-term pasture harvest rates. Subsequently, tools for farmers have been developed from this concept (e.g., [27,28]) and the leaf stage method has been adopted by dairy farmers in New Zealand, Australia, and Ireland.

In addition to assisting decision-making regarding when to graze during regrowth, the type of analysis performed by Parsons *et al.* [23] also helps define the optimum residual state to which pasture should be defoliated to maximise pasture harvest rates. They showed that when pasture is defoliated to a very low residual LAI, the initial rate of regrowth is retarded since the first leaf produced post-defoliation is restricted in size due to the limited amount of energy available to invest in the lamina of that leaf. Since leaf emergence rate and leaf elongation duration are relatively insensitive to defoliation severity within the range normally applied in farming systems [10], the plant is unable to compensate for the loss of C assimilation capacity (despite the mobilisation of carbohydrate reserves plus changes in specific leaf area [11]) and the first leaf is necessarily restricted in size. Subsequent leaves are larger as C assimilation capacity increases and the full extent of compensatory effects (such as cessation of C export to roots to support new leaf growth [29]) is realised. In this situation, the point of maximum average growth will be delayed compared with a pasture where a higher LAI (and, therefore, a larger source of current assimilate for investment in new leaf growth) is left post-defoliation. Practically, the length of the regrowth cycle should be extended in response to severe defoliation to maximise pasture harvest, but the loss of yield incurred in the early stage of regrowth cannot be recovered and total pasture harvest over time will be less than for a system where a higher residual LAI is left.

Conversely, when a high LAI is left after defoliation, the rate of senescence will increase earlier in the regrowth cycle compared with a pasture where a lower LAI is left post-defoliation, since the residual LAI inevitably includes older leaf tissue nearing the end of its natural life. In this case, since the timing of maximum average growth rate is dependent on the relative rates of new leaf growth and senescence, it will be reached earlier, and the appropriate response in practice is to reduce the length of the regrowth interval to ensure yield is not foregone later in the regrowth cycle.

When translated into guidelines for farmers, these principles lead to the recommendation that pastures be grazed when between two and three new leaves have been produced since the previous grazing, or at approximately 2600–3200 kg· DM/ha pre-grazing pasture mass, leaving a spatially uniform residual pasture mass of between 1500 and 1700 kg· DM/ha [19,30]. Ideally, the rotation length (grazing interval) applied will be an emergent property of these rules as they are adapted to changing leaf appearance rates (e.g., [27]). In practice, it is often necessary to over-ride the theoretical optimum grazing interval to balance whole farm feed supply and demand over an annual cycle, for example ensuring that targets for the average amount of herbage mass across the farm are met at the beginning and end of each lactation [31]. This is an acceptable trade-off, provided farmers are aware they are "breaking the rules", know why they are doing it, and know how to restore optimum grazing practices once the other needs have been met [32].

### 3.2. *Opportunities for Improving Pasture Harvest Rates*

The analysis above shows that, as well as tracking the accumulation of herbage mass during regrowth, the sigmoidal growth curve holds additional information that is valuable for improving the efficiency of grazing management. However, this information has seldom been used explicitly to test and further develop grazing recommendations. In part, this is due to the difficulty of accurately measuring the accumulation of mass during regrowth, given the temporal variation in growth conditions that typically occurs over periods of 3–6 weeks, the spatial heterogeneity in herbage mass that is ubiquitous in pastures (especially when grazed), and the relatively low precision of techniques for measuring pasture mass. Nonetheless, examples of sigmoidal regrowth curves can be found in the literature (e.g., [33–35]).

Figure 1 shows three regrowth curves for a perennial ryegrass pasture at Lincoln, Canterbury, New Zealand, areas of which were mown to residual pasture masses of 1150, 1500 or 1850 kg· DM/ha in early autumn (28 March). Before mowing, the pasture had been grazed as per "normal" farm practice: about 10–12 grazings per year, to a post-grazing residual between 1480 and 1750 kg· DM/ha. Relative to the pasture state maintained before the comparison was initiated, the three starting residuals represent over-grazing (1150 kg· DM/ha—hereafter called the "over" treatment), optimal grazing (1500 kg· DM/ha—"target", as per the recommendations listed above), or under-grazing (1850 kg· DM/ha—"under"). The pastures were allowed to regrow for 45 days. At least once per week during regrowth, pasture mass was measured using a calibrated rising plate meter, light interception was assessed with a hand-held ceptometer, and leaf stage was monitored by counting the number of new leaves that had emerged since defoliation on a sample of 20 tillers per treatment. Pasture botanical composition and dead matter content were assessed 28 days after the start of the regrowth period, by hand dissection of four sub-samples per treatment cut to ground level.

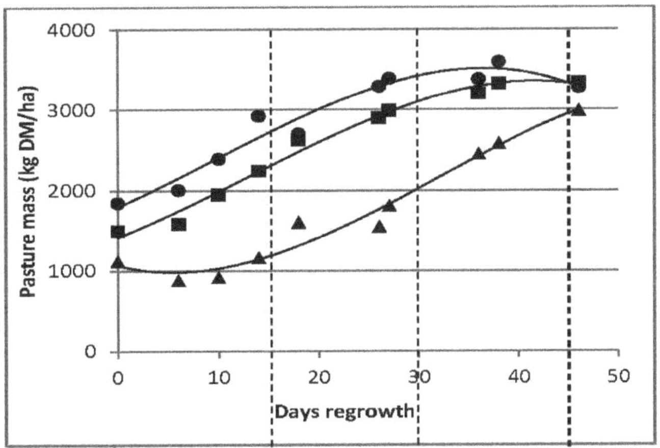


| Leaf stage: | 0 | 1.0 | 2.0 | 3.0 |
|---|---|---|---|---|

| Light interception (%): | | | | | |
|---|---|---|---|---|---|
| 'Over' | 18 | 48 | 77 | 90 | 97 |
| 'Target' | 57 | 82 | 95 | 99 | 100 |
| 'Under' | 77 | 89 | 97 | 99 | 100 |

**Figure 1.** A single example of the regrowth of perennial ryegrass pasture following defoliation to three levels: 1150 kg· DM/ha (triangles, representing "over-grazing"), 1500 kg· DM/ha (squares, representing the "target" for residual state) or 1850 kg· DM/ha (circles, representing "under-grazing") kg· DM/ha. Regrowth commenced on 28 March. Timing of the full emergence of the first, second and third new leaves after defoliation, and light interception, are shown below the x-axis.

During the regrowth period in April and May, rainfall was well above the long-term monthly averages, therefore plant available water was never limiting. The average daily air temperature was relatively consistent at around 11.5 °C, although day length was declining. The relatively favourable and consistent growth conditions meant that the data fitted the expected sigmoidal growth pattern. This is not always the case, and attempts to confirm the underlying ecophysiology with field measurements are often confounded by short-term variability in growth conditions. Also, the data presented in Figure 1 pertain to only one regrowth cycle in one environment. There is no implication in what follows that regrowth in all circumstances follows the classical pattern; rather, the intention is to show how the ecophysiological information presents opportunities to re-appraise grazing practices in the pursuit of improved pasture harvest efficiency, and thereby also increase farm profitability.

In this example, the interval between appearance of successive new leaves on tillers was consistent throughout the regrowth period at about 15 days. There was no difference between the treatments in the rate of leaf emergence, which is consistent with other, more-comprehensive studies [36,37]: unless grazing/cutting is extremely severe, the rate at which new leaves is produced is not affected, but the size (length, and weight) of new leaves can be reduced by severe defoliation [29,36]. Because leaf emergence rate was consistent across time and treatments in this example, the data illustrate how light interception, herbage accumulation and senescence change during regrowth, in accordance with changes in plant energy status. From this, it is possible to explore how management affects these interactions and estimate the costs in biological efficiency, and in profit per hectare, of failure to achieve the targets embedded in the recommendations listed above.

### 3.2.1. Targets for Residual Pasture State

In the example shown in Figure 1, ceiling yield of ~3500 kg· DM/ha was reached in the "under" and "target" treatments after about 38 and 45 days regrowth respectively. At this point, virtually all of the light available was intercepted by the pasture canopy (as shown by the numbers for light interception along the bottom of Figure 1). In the "over" treatment, net herbage accumulation in the first 15 days of regrowth when the first new leaf was produced was minimal, indicating that this first leaf was much smaller than the first new leaf produced in the other treatments. After seven days of regrowth, only 18% of total light was intercepted by the pasture canopy in the "over" treatment, whereas in the "target" and "under" treatments, 57%–77% of light was being intercepted. Therefore, in the "over" treatment, plants had relatively little energy to invest in growing the first new leaf, and its size was restricted. As leaf size and light interception increased, the rate of herbage accumulation accelerated in this treatment such that the pasture approached ceiling yield after 45 days regrowth.

Parsons *et al.* [23] demonstrated that, to maximise the amount of pasture harvested per year in intermittent defoliation systems, the optimum time to graze during regrowth is the point when the maximum average growth rate of the pasture is reached. At this point, the optimum balance between the amount of new leaf being produced, and the amount of old leaf dying, has been reached. Going beyond this point means that the rate of leaf death increases and the efficiency with which further mass is added to the pasture declines. It is this declining efficiency which defines the maximum average growth rate as the optimal point to graze the pasture again.

Figure 2 plots the average growth rate for the "over", "target" and "under" treatments shown in Figure 1. The arrows in Figure 2 indicate when maximum average growth rate is reached for the "under" and "target" treatments. The maximum average growth rate was reached after about 18 days in the "under" treatment indicating that to optimise pasture harvest when a high residual is left after grazing requires a relatively short grazing interval. The maximum average growth rate for the "target" treatment was reached about 30 days after regrowth commenced, whereas the average growth rate curve for the "over" treatment had still not reached a maximum after 45 days regrowth.

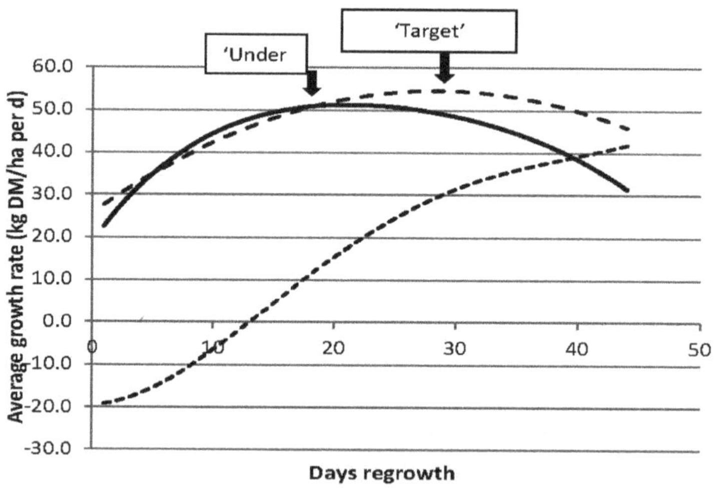


**Figure 2.** Average growth rate during regrowth for pastures mown to about 1150 ("Over", dotted line), 1500 ("Target", dashed line) or 1850 ("Under", solid line) kg· DM/ha. Arrows indicate the point when the maximum average growth rate was reached during regrowth in the "under" and "target" treatments. The curves, and the point of maximum average growth rate, are derived from the fitted curves shown in Figure 1.

Importantly, the growth rate achieved at the point of maximum average growth rate is higher for the "target" treatment (about 55 kg· DM/ha per day on the vertical axis in Figure 2) than for the "under" treatment (about 51 kg· DM/ha/day; see also Figure 3.10 in [14]) although the difference is small. The total net amount of pasture grown at the point when maximum average growth rate was reached in the "target" and "under" treatments was 2110 and 1260 kg· DM/ha respectively, or an average of around 70 kg· DM/ha per day in both treatments. As noted above, maximum average growth rate was not attained in the "over" treatment: the highest growth rate reached was just over

40 kg· DM/ha per day, and after 45 days regrowth (the three leaf stage) pastures in this treatment had accumulated 2370 kg· DM/ha at a mean growth rate of 52 g· DM/ha per day. The "over" treatment can never catch up with the "target" treatment because of the lag in regrowth that occurred after defoliation (Figure 1). This is fundamentally a problem caused by over-grazing (residual pasture mass/leaf area is too low). There is an important interaction here between the residual pasture state post-grazing and the length of the regrowth period; both of which can be controlled by grazing management decisions.

Continuing with the example shown in Figures 1 and 2 net herbage accumulation in the "under" and "target" treatments would be similar if the pasture were grazed again around the 2-leaf stage, but grazing after this point would result in relatively more herbage accumulating in the "target" treatment ( Table 1). Conversely, net herbage accumulation in the "over" treatment would be much lower than in the "target" treatment if grazing occurred after two new leaves had emerged since the last grazing, but the gap between these treatments would narrow as the interval to the next grazing stretched out ( Table 1). When averaged across the three scenarios shown in Table 1, the "under" and "over" treatments resulted in 240–465 kg· DM/ha lower herbage accumulation compared with the "target" treatment. Using the Dairy NZ Forage Value Index figure for the economic value (EV) of additional pasture grown in autumn in Canterbury ($0.30 per kg· DM, [38]), this equates to between $70 and $140/ha potential profit foregone. This is the potential economic cost of missing the target range for the optimum residual state of the pasture. This loss will obviously compound if the target is missed frequently.

**Table 1.** Estimated effect of different regrowth intervals on the net amount of herbage accumulated (kg· DM/ha) during regrowth of pastures mown to about 1150 ("Over"), 1500 ("Target") or 1850 ("Under") kg· DM/ha residual pasture mass. Data apply to just a single regrowth event, in autumn, in Canterbury New Zealand, using the results shown in Figure 1.

| Residual Pasture State | Timing of Next Grazing | | |
|---|---|---|---|
| | 2 Leaves (30 Days) | 2.5 Leaves (37.5 Days) | 3 Leaves (45 Days) |
| "Under" grazing | 2095 | 2260 | 2065 |
| "Target" grazing | 2100 | 2440 | 2595 |
| "Over" grazing | 1415 | 1945 | 2375 |

How realistic are these estimates? Using cutting management applied to perennial ryegrass pastures from September to April, Lee *et al.* [39] measured about 1.2 t· DM/ha less pasture yield from pastures managed to a consistent residual of 1100 kg· DM/ha compared with a consistent residual of 1500 kg· DM/ha. This equates to around $360/ha lost operating profit, based on an EV of $0.30 per kg· DM. When cut to a consistent residual of 2300 kg· DM/ha, there was minimal effect on pasture yield. In a grazing experiment, Garcia and Holmes [40] observed a statistically significant 20% reduction in pasture growth from residuals of less than 1300 kg· DM/ha compared to residuals of greater than 1500 kg· DM/ha. Mean total annual pasture yields in their study were around 12 t· DM/ha: hence a 20% reduction in yield equates to around 2.4 t· DM/ha in lost feed production, or potentially $600/ha lost operating profit. Similar to Lee *et al.* [39], Garcia and Holmes [40] did not observe any negative effect on pasture yield from higher pasture residuals of up to 2300 kg· DM/ha. The difference in herbage accumulation that emerges from the analysis of the regrowth curves in Figure 1 is not supported by these more-comprehensive studies but, as noted above: (1) the difference herbage accumulation over one regrowth cycle is relatively small; and (2) much depends on the interval between grazing that was applied because of the interaction that exists between post-grazing pasture state and regrowth interval. Nonetheless, negative effects of high residuals on pasture quality and sward structure would be expected, which could seriously restrict the efficiency of pasture utilisation and animal production. In the example shown in Figure 1, dead matter comprised 30% of the total pasture mass in the "under" treatment after 28 days regrowth, whereas it comprised only 12% of total mass in the "target" treatment and 2% in the "over" treatment.

### 3.2.2. Targets for Timing of Defoliation

When pasture regrows from a consistent post-grazing residual mass of between 1500 and 1750 kg· DM/ha, there is usually some curvature in the plot of herbage mass *versus* time (e.g., Figure 3), reflecting an increase in mass of the successive leaves. Using data from two years of a dairy grazing study where good control of pasture residual state was achieved, Chapman *et al.* [35] calculated that the first, second and third leaves produced after defoliation contribute 25%, 35% and 40% respectively of the total available pasture mass at the next grazing. This pattern is consistent with the theoretical analysis of Parsons *et al.* [23] which was further developed into the leaf stage grazing rule by Fulkerson and Donaghy [18], as described above. It provides a proxy from which the impacts of missing the optimum regrowth interval on the efficiency of pasture harvest, and farm profit, can be calculated without the need for empirical regrowth curve data. The only additional information needed to do this is average seasonal net herbage accumulation rates and leaf emergence intervals for a given environment.

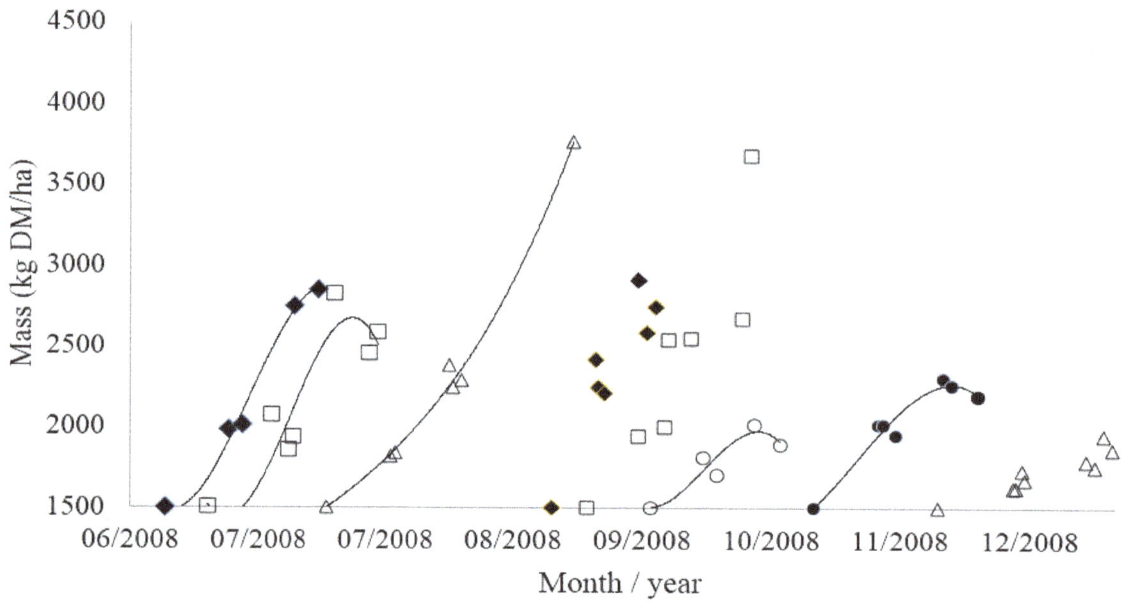


**Figure 3.** Regrowth curves for perennial ryegrass lamina tissue, derived from data for percentage of total tiller weight comprised by lamina, total herbage mass at the time of sampling, and observed leaf appearance rate data. Fitted lines are third-order polynomials, and are presented only where the polynomial described more of the variation in the data than a linear regression. From [35].

Table 2 shows the possible cost of erring towards a shorter regrowth interval (grazing at 1½ or 2 leaves) compared with a longer interval (3 leaves). It applies the 25:35:40 ratio seasonally (except for October–November, when reproductive growth is present and the difference in mass between successive leaves largely disappears [35]), along with typical pasture growth rates and leaf appearance intervals for irrigated dairy pastures in Canterbury. It also uses the DairyNZ Forage Value Index seasonal economic values previously described. The analysis shows that, for example in February-March, if growth rates are 55 kg· DM/ha per day, and the leaf appearance interval is 10 days, an additional 330 kg· DM/ha could be harvested if pastures were grazed at the three-leaf stage compared with the two-leaf stage. With an EV of $0.25 per kg· DM, the total value for the season of the additional DM grown on the longer rotation is estimated at $83. Across the full year, the analysis suggests around $260/ha additional profit could be achieved using a longer defoliation interval, aligned with the three-leaf stage of regrowth.

**Table 2.** Estimated difference in dry matter grown and economic value (EV) of pasture for a dairy farm in Canterbury, New Zealand, where grazing occurs consistently at the 1.5 or 2-leaf stage of regrowth compared with the 3-leaf stage of regrowth.

| Period | Growth Rate (kg· DM/ha/d) | Leaf Appearance Interval (d) | Contribution of 1st:2nd:3rd Leaves to Total DM (%) [1] | DM grown (kg/ha) When Grazed at | | | Difference in DM Grown (kg/ha) | | EV for Season ($/kg· DM) [2] | EV of the Difference ($/ha) | |
|---|---|---|---|---|---|---|---|---|---|---|---|
| | | | | 1.5-Leaf | 2-Leaf | 3-Leaf | 1.5- v. 3-Leaf | 2- v 3-Leaf | | 1.5- v. 3-Leaf | 2- v 3-Leaf |
| February–March | 55 | 10 | 25:35:40 | 2810 | 2970 | 3300 | −500 | −330 | 0.25 | −125 | −85 |
| April–May | 26 | 15 | 25:35:40 | 860 | 940 | 980 | −120 | −40 | 0.31 | −35 | −10 |
| June–July | 0 | | | 0 | 0 | 0 | 0 | 0 | | | |
| August–September | 34 | 15 | 25:35:40 | 1120 | 1220 | 1280 | −150 | −50 | 0.42 | −65 | −20 |
| October–November | 95 | 7.5 | 30:35:35 | 7200 | 7410 | 7550 | −360 | −140 | 0.29 | −105 | −40 |
| December–January | 95 | 10 | 25:35:40 | 4850 | 5130 | 5700 | −860 | −570 | 0.18 | −155 | −105 |
| *Totals* | | | | *16,840* | *17,670* | *18,810* | *−1990* | *−1130* | | *−485* | *−260* |

[1] From [35]; [2] From [38].

Again, it is necessary to ask if these estimates are realistic. In Tasmania, Turner *et al.* [41] observed 20% greater herbage accumulation (18.0 *versus* 15.0 t· DM//ha) over 12 months when irrigated pastures were defoliated (using cutting) at the 3-leaf stage compared with the 1.5-leaf stage. This margin was consistent for the three different cultivars used in the study. The proxy approach shown in Table 2 predicts a 12% difference (18.8 *versus* 16.8 t· DM/ha) for pastures defoliated (by grazing) at the 3-leaf stage *versus* the 1.5-leaf stage respectively in Canterbury. These observations align well, and indicate that an additional $260–$480/ha in operating profit may be available for dairy farmers in Canterbury by consistently re-grazing at the point of maximum average growth rate (3-leaf stage) compared with earlier stages (2- and 1.5-leaf, respectively) of regrowth ( Table 2).

3.2.3. Pasture Quality

One possible draw-back of this strategy is that the nutritive value of the feed eaten could be lower when pasture is grazed at the later stages of regrowth. It is well known that the composition of herbage of grass-dominant pastures changes during regrowth. As the leaves of both temperate and tropical grasses age, their digestibility declines due to increasing indigestible cell wall content and decreasing cell wall digestibility [41–44]. Thus, overall fibre content of herbage increases, while fibre digestibility declines, as regrowth proceeds. Countering these trends, grazing later in the regrowth cycle offers at least two advantages. Firstly, the crude protein (CP) concentration of old leaves is lower than young leaves meaning that the total CP content of grass herbage can be diluted by delaying grazing to later in the regrowth cycle [45–47]. The CP concentration of well-fertilised, ryegrass-dominant pastures frequently exceeds the concentration required to meet animal protein requirements [48]. Seventy to eighty percent of the excess N consumed is excreted in urine [49], resulting in high urinary N loadings in urine patches which provide a ready source of nitrate that can be leached below the pasture root zone eventually accumulating in receiving waters such as streams and lakes. In NZ and other temperate countries with intensive agricultural industries, this nutrient contamination of freshwater bodies is a significant environmental threat which is subject to increasingly stringent regulation at the source, the farm.

Secondly, the content of water soluble carbohydrates (WSC) increases during regrowth [46,50–52] with positive implications for the WSC:CP ratio of herbage and, hence, the efficiency of utilisation of protein in the rumen [53]. Furthermore, the expression of higher WSC content in cultivars bred for this trait is aided by a longer regrowth interval [54,55]. Therefore, changes in herbage composition associated with grazing later in the regrowth cycle can have positive benefits for managing some of the externalities of grassland farming. In this example, the effects of greater expression of the WSC trait and dilution of the total leaf CP concentration are additive and have been demonstrated to offer important potential gains in control of nitrate leaching [56].

*3.3. Implementing Grazing Management on Farm*

Considering the combined effects of over-grazing pasture and grazing too early in the regrowth cycle, as analysed in Section 3.2., it can be postulated that farmers could readily increase pasture harvest rates by 1–2 t· DM/ha per year and profit by $300–600/ha per year through accurate implementation of well-documented grazing decision rules. Of course, such benefits will only be realised if optimal grazing targets are currently not being achieved on-farm. Despite a long history of focussing on grazing management and a reputation for achieving high levels of pasture utilisation on New Zealand dairy farms, there is evidence of wide variation in grazing management practices with potentially large negative consequences for pasture productivity. In a study where pre-grazing pasture mass, leaf stage and post-grazing pasture mass were measured every two weeks on seven dairy farms in the lower North Island, New Zealand for 10 months, McCarthy *et al.* [6] observed that 49% of grazing events occurred at or before the 2-leaf stage, and only 27% occurred between the 2- and 3-leaf stage. The recommended post-grazing herbage mass target was achieved in only 48% of grazing events, with individual farms showing a tendency toward consistently either under-grazing or over-grazing relative to the target.

It must be noted that, on seasonal calving dairy farms that rely mostly on pasture for feed, there are times of the year when the theoretical best practice for management of pasture regrowth does not result in the optimal overall outcome for the farm system. For example, the application of long regrowth intervals in late autumn and winter when pasture growth rates are relatively low helps to increase average herbage mass across the farm and ensure sufficient pasture is available to bridge the feed gap between calving and the acceleration of spring pasture growth [31]. Here, extended regrowth intervals are implemented via strip grazing of paddocks at a time when cows are not lactating and can be fed a restricted diet sufficient to maintain, or achieve modest increases in, body condition score. The result is that some of the feed grown in late autumn or winter (a period of relatively low feed demand) is "transferred" to late winter-early spring (a period of high feed demand, but variable and often low pasture growth). In many situations, this will result in low post-grazing herbage mass on grazed strips, and push regrowth beyond the 3-leaf stage on strips that are withheld from grazing for long periods (up to 100 days): however, the "cost" of doing this is generally less than the cost of purchasing supplements to fill the early lactation feed gap, therefore overall farm profit is improved. This is an example of a situation where the "rules" of best management for regrowth can be bent for a good reason [32].

The effects of these management practices will be reflected to some extent in the survey findings referred to above [6]. Nonetheless, the results indicated that there is substantial room for improvement in grazing management on New Zealand dairy farms, and prompted (among others) two responses. Firstly, a renewed focus on the fundamental principles of grazing management emerged in extension material distributed to all New Zealand dairy farmers (e.g., [7]). Secondly, recognising that part of the problem is related to relatively low adoption of pasture monitoring practices such as weekly assessment of pasture mass in each paddock and the generation of a whole-farm feed profile for assisting short-term grazing management decisions (e.g., [57]), farm system simulation was used to estimate the profit foregone as a result of having incomplete information on which to make pasture allocation decisions. This analysis [58] showed that, compared with a situation where no information was available for the herbage mass of individual paddocks, farm operating profit could be increased by about $385/ha per year if pasture mass was measured for each paddock each week, even if the error associated with the measurement was ±15%. That is, having some information is substantially better than having no information. The profit gains estimated by Beukes *et al.* [58] are similar to those arrived at in the analysis in Sections 3.2.1 and 3.2.2 above. Both approaches point toward large gains in profit being available in many New Zealand dairy farms by application of well-proven management practices that have strong ecophysiological foundations.

## 4. Future Opportunities for Application of Ecophysiological Information

### *4.1. The Extended Ryegrass Phenome and Its Implications for Grazing Management Practices*

One objection to the idea of implementing longer regrowth intervals (as discussed in Section 3.2, above) is that the pasture mass at the time of grazing will be greater, making it more difficult for animals to graze down to the target residual pasture state. Grazing management recommendations developed in the NZ dairy industry in the 1970s and 1980s emphasised the need for high stocking rates to control pasture residual state and maintain high feed quality (e.g., [4]). At this time, many dairy pastures were likely based on old ecotypes of perennial ryegrass. Pasture renewal rates were low, and cultivar choices were restricted to a handful of mid-season flowering types with strong winter and spring growth, and relatively poor summer growth. They were also characterised by relatively high sheath/pseudostem to leaf ratios which would have posed a sward structural barrier to grazing into the lower sward layers [59]. Irrespective of whether pastures were old or recently renewed, the ryegrass populations in them would have been highly infected with the "standard" or "wild type" ryegrass endophyte *Epichloë festucae* var. *lolii* (formerly *Neotyphodium lolii*, [60]). This endophyte produces high concentrations of the alkaloids lolitrem and ergovaline, especially in the summer/autumn dry conditions of northern New Zealand [61,62] where the majority of New Zealand's dairy cows were farmed. These alkaloids are concentrated in the base of the leaf sheath. They induce aversive responses in ruminants, which probably limit grazing intensity, *i.e.*, it was necessary to "force" animals to graze low into the canopy. Hence, high stocking rates were necessary to "graze out" pastures effectively.

The influence of the practical guidelines for grazing that were developed in the 1970s and 1980s can still be seen in current grazing management recommendations, for example in the guideline that pre-graze pasture mass does not exceed 3200 kg· DM/ha. Yet the types of pastures used for milk production today are markedly different. Firstly, very few "old" pastures now remain in the industry, after farming has intensified in the traditional areas and expanded into non-traditional areas mainly through the conversion of sheep and beef farms to dairying. The latter change in land use has always been accompanied by pasture renewal. Secondly, the role of endophytes in the ecology of grazed pastures is now well-documented [63–65] and endophyte strains that do not produce the alkaloids that are toxic to animals have been commercialised [64]. Thirdly, the genetic diversity of cultivars available in the NZ market is now much greater [66]. Breeding objectives during the 1990s and 2000s emphasised later flowering [67], using Mediterranean germplasm and polyploidy to delay maturity, reduce the magnitude of the traditional mid-spring peak in growth, and increase growth rates in summer and autumn. Some of these traits, especially tetraploidy, are well-documented as facilitating greater dry matter intake [68] allowing better control of post-grazing sward state. Milk production benefits are likely from the use of such material [69], but this has not yet been confirmed in rigorous grazing systems experiments [70].

More targeted research is needed to quantify exactly how different modern cultivars are compared with older cultivars in terms of grassland productivity [70], and therefore some caution is required in when questioning whether current guidelines still result in best possible pasture harvest rates. Nonetheless, the emergence of greater genetic diversity through the development of an "extended phenome" of perennial ryegrass (combining genetic variation in the ryegrass plant and the endophytic fungus) may open opportunities for farmers to grow and utilise more pasture, thereby controlling the costs of milk production by reducing/avoiding the need to purchase expensive supplements. This opportunity has not yet been investigated in targeted research, even though the theoretical and practical bases by which it can be realised are well-documented (as described in the previous section). One of the barriers is the absence of quantitative information on the amount of phenotypic variation in specific shoot traits that exists in current perennial ryegrass cultivars, and their impact on grazing efficiency, animal intake and milk production. Ecophysiological explanations for the effects of management and environment on the traits and processes of interest are essential. This information must be allied with field-scale grazing studies to provide the information that farmers require to correctly match genotype, management and environment for better biological efficiency and profit. Early inroads are being made into this field. For example, Wims *et al.* [71] identified an important interaction between cultivar and growth stage (vegetative *versus* reproductive) which influenced relative milk production of dairy cows and concluded that sward structural differences between cultivars are more important during reproductive development, while differences in digestibility are more important during vegetative growth. There is more to be learnt in this field of plant phenotypic variation and pasture-animal interactions, which offers gains for pasture-based livestock producers in all temperate regions.

### *4.2. Pasture Persistence*

As the intensification and expansion trends in the New Zealand dairy industry described above have proceeded, a growing concern has emerged regarding the persistence of newly sown pastures. This issue was the subject of a special symposium held in 2011 [72] to review the state of knowledge regarding the extent, seriousness, and causes of the problem. The symposium clearly demonstrated that it is a multi-faceted problem, involving many interacting causal factors which can be broadly grouped under the headings of environment, management and genetics (perennial ryegrass plus fungal endophyte). The symposium also highlighted how little is known about the underlying plant growth and population processes that are likely to be involved. This is surely a prime opportunity for ecophysiological research: for, as with the example discussed in the previous section, ecophysiological information is critical to the solution.

Unravelling the problem first requires a clear definition of what is meant by "persistence". Parsons *et al.* [73] argued that it is best defined as the persistence of the expected yield advantage of a new pasture compared with the pasture that was in place before. In this context, "poor persistence" is not just a function of physical survival of plants of the sown species/cultivar but can also be caused by changes in the expression of yield traits over time which erode early advantages. From the perspective of ecophysiological investigations, this broader view of the problem invokes the need for: agronomic information on the relative yields of "old" and "new" pastures to define when there is a persistence "failure" and the magnitude of the failure; well-constructed control treatments (both positive and negative) so that the incidence of persistence failure can be confirmed; a stronger focus on the fundamental plant traits, rather than on the cultivar (given that there is strong overlap between cultivars in traits which frequently confounds attempts to elucidate key factors); and more emphasis on community and ecosystem processes to complement the traditional whole-plant focus of ecophysiology.

In the case of pasture persistence, previous research into the problem has addressed only some of these requirements, creating difficulties in pin-pointing critical processes and reconciling apparently conflicting results. Often the focus has been on ryegrass tiller density in pastures as an indicator of persistence. However, the existence of a relationship between tiller density and the persistence of an expected yield advantage has not been proven, apart from the obvious situation where perennial ryegrass populations are eliminated by, for example, major insect pest damage to pastures lacking the protection available from novel endophytes in regions vulnerable to pest outbreaks [74]. More progress has been made in relating changes in ground score [75] to trends in dry matter yield of ryegrass cultivars in Ireland [76]. Ground score change has subsequently been incorporated into the Irish Pasture Profit Index which ranks ryegrass cultivars according to their economic worth [77]. Similar information for New Zealand cultivars and their yield trends is not available, although initial analysis of one long-term data set comparing yields of 25 perennial ryegrass cultivars over 8 years in a summer-dry environment indicates no relationship between ground score and long-term yield differences [78].

At the population level, it is obvious that persistence is dependent on maintenance of a population of tillers. Individual tillers have a finite life span [79] so it is logical to look for tillering traits that might explain why some pastures "persist" better than others. However, for the density of a population of perennial ryegrass to remain stable, each tiller need leave only one replacement [80]. Perennial ryegrass is quite capable of doing this, since plants can rapidly initiate new tillers under favourable conditions [10,81]. There is no evidence that failure to replace tillers that die naturally or in response to short-term growth stresses, limits ryegrass "persistence" in temperate regions where soil moisture deficits are absent or relatively mild and/or short in duration [82].

Well-targeted ecophysiogical research remains critical to resolving the pasture persistence problem. However, the discussion above highlights that it must be closely-aligned with agronomic investigations (so that, for example, yield differences in response to treatments can be confirmed, and plant characteristics can be related to differences in pasture yield), and genetics/genomics (so that the focus can shift to traits, rather than cultivars, and opportunities for identifying markers for persistence can be exploited). There is also a distinct gap in knowledge of what changes, if any, occur in the genotypic and phenotypic structure of ryegrass populations over time, how fast those changes occur and what traits they favour (the direction of change) and, indeed, if these changes ultimately have any bearing on the persistence of yield. Hence, stronger connections between ecophysiology and population ecology are also needed.

## 5. Conclusions

The discipline of plant physiological ecology is fundamentally important for the advancement of the biological efficiency and sustainability of pasture-based livestock systems throughout the world. Ecophysiological information has already contributed to the development of important management guidelines and tools that are used by farmers in temperate regions. Several opportunities for further developing those guidelines and tools are available using the existing body of ecophysiological information: if taken, those opportunities will allow farmers to increase pasture harvest rates at little extra cost and therefore support improvements in farm profit.

By contrast, ecophysiological information has made less of a contribution to gains in forage plant breeding, even though several credible routes have been proposed (as reviewed by [74]). Closer links between ecophysiology and plant genetics/genomics (including population genetics) will help focus effort on the right traits and determine when/if meaningful progress is being made. These two discipline areas are naturally aligned, but seldom deployed together. Likewise, the horizons for ecophysiology also need to encompass well-constructed agronomic investigations, as well as farm systems experimentation and systems modelling. There are few examples where the links between these various disciplines have been forged. Until these types of links are developed, progress in improving the efficiency of pasture-based agriculture will be limited, and the discipline of ecophysiology may struggle to retain its relevance.

## References

1. Da Silva, S.C.; Sbrissia, A.F.; Pereira, L.E.T. Ecophysiology of $C_4$ forage grasses—Understanding plant growth for optimising their use and management. *Agriculture* **2015**, *5*, 598–625. [CrossRef]
2. Hodgson, J. *Grazing Management: Science into Practice*; Longman Scientific and Technical: Harlow, Essex, UK, 1990.
3. Lemaire, G., Hodgson, J., Moraes, A., Nabinger, C., Carvalho, P.C.F., Eds.; *Grassland Ecophysiology and Grazing Ecology*; CAB International: Wallingford, UK, 2000.
4. Bryant, A.M. Optimum stocking and feed management practices. In Proceedings of the 42nd Ruakura Farmers Conference, Hamilton, New Zealand, June 1990; MAF: Hamilton New Zealand, 1990; pp. 55–59.
5. IFCN. *Dairy Report 2015*; International Farm Comparison Network: Kiel, Germany, 2015.
6. McCarthy, S.; Hirst, C.; Donaghy, D.; Gray, D.; Wood, B. Opportunities to improve grazing management. *Proc. N. Z. Grassl. Assoc.* **2014**, *76*, 75–80.
7. DairyNZ. *Grazing Management: The Root to Success*; Technical Series, Issue 23; DairyNZ: Hamilton, New Zealand, 2014.
8. Savage, J.; Lewis, C. Applying science as a tool for dairy farmers. *Proc. N. Z. Grassl. Assoc.* **2005**, *67*, 61–66.
9. Dillon, P.; Roche, J.R.; Shalloo, L.; Horan, B. Optimising financial returns from grazing in temperate pastures. In *Utilisation of Grazed Grass in Temperate Animal Systems*; Wageningen Academic Publishers: Wageningen, The Netherlands, 2005; pp. 131–147.
10. Gastal, F.; Lemaire, G. Defoliation, shoot plasticity, sward structure and herbage utilization in pasture: Review of the underlying ecophysiological processes. *Agriculture* **2015**, *5*, 1146–1171. [CrossRef]
11. Irving, J. Carbon assimilation, biomass partitioning and productivity in grasses. *Agriculture* **2015**, *5*, 1116–1134. [CrossRef]
12. Lambers, H.; Chapin, F.S.; Pons, T.L. *Plant Physiological Ecology*; Springer-Verlag: New York, NY, USA, 1998.
13. Chapman, D.F.; Parsons, A.J.; Cosgrove, G.P.; Barker, D.J.; Marotti, D.M.; Venning, K.J.; Rutter, S.M.; Hill, J.; Thompson, A.N. Impacts of spatial patterns in pasture on animal grazing behaviour, intake, and performance. *Crop Sci.* **2007**, *47*, 399–415. [CrossRef]
14. Parsons, A.J.; Chapman, D.F. The principles of pasture growth and utilisation. In *Grass: Its Production and Utilisation*; Hopkins, A., Ed.; Blackwell Science: Oxford, UK, 2000; pp. 31–89.
15. Clark, D.A.; Kanneganti, V.R. Grazing management systems for dairy cattle. In *Grass for Dairy Cattle*; Cherney, J.H., Cherney, D.J.R., Eds.; CABI: Wallingford, UK, 1998; pp. 311–334.

16. Agricultural Development and Advisory Service. *Grassland Management Calendar. Dairy Cows. Continuous Grazing*; HMSO: London, UK, 1987.
17. Milligan, K. Comments on the grazing management session. *Proc. N. Z. Grassl. Assoc.* **1984**, *45*, 207–208.
18. Fulkerson, W.J.; Donaghy, D.J. Plant soluble carbohydrate reserves and senescence—Key criteria for developing an effective grazing management system for ryegrass-based pastures: A review. *Aust. J. Exp. Agric.* **2001**, *41*, 261–275. [CrossRef]
19. MacDonald, K.A.; Glassey, C.B.; Rawnsley, R.P. The emergence, development and effectiveness of decision rules for pasture-based dairy systems. In Proceedings of the 4th Australasian Dairy Science Symposium, Lincoln, New Zealand, 31 August–2 September 2010; pp. 233–238.
20. Brougham, R.W. Effect of intensity of defoliation on regrowth of pasture. *Aust. J. Agric. Res.* **1956**, *7*, 377–387. [CrossRef]
21. Brougham, R.W. Pasture growth rate studies in relation to defoliation. *Proc. N. Z. Soc. Ann. Prod.* **1957**, *17*, 46–55.
22. Bircham, J.S.; Hodgson, J. The influence of sward conditions on rate of herbage growth and senescence in mixed swards under continuous stocking management. *Grass Forage Sci.* **1983**, *38*, 323–331. [CrossRef]
23. Parsons, A.J.; Johnson, I.R.; Harvey, A. Use of a model to optimize the interaction between frequency and severity of intermittent defoliation and to provide a fundamental comparison of the continuous and intermittent defoliation of grass. *Grass Forage Sci.* **1988**, *43*, 49–59. [CrossRef]
24. Grant, S.A.; Barthram, G.T.; Torvell, L.; King, J.; Smith, H.K. Sward management, lamina turnover and tiller population density in continuously stocked *Lolium perenne*-dominated swards. *Grass Forage Sci.* **1983**, *38*, 333–344. [CrossRef]
25. Parsons, A.J.; Leafe, E.L.; Collett, B.; Stiles, W. The physiology of grass production under grazing. 1. Characteristics of leaf and canopy photosynthesis of continuously grazed swards. *J. Appl. Ecol.* **1983**, *20*, 117–126. [CrossRef]
26. Davies, A. Tissue turnover in the sward. In *Sward Measurement Handbook*, 2nd ed.; Hodgson, J., Baker, R.D., Davies, A., Laidlaw, A.S., Leaver, J.D., Eds.; The British Grassland Society: Reading, UK, 1981; pp. 183–215.
27. Rawnsley, R.P.; Snare, T.A.; Lee, G.; Lane, P.A.; Turner, L.R. Effects of ambient temperature and osmotic stress on leaf appearance rate. In Proceedings of the 4th Australasian Dairy Science Symposium, Lincoln, New Zealand, 31 August–2 September 2010; pp. 345–350.
28. Shannon, P. Feeding pastures for profit—An innovative and practical approach to understanding and managing grazing based feeding systems. In Proceedings of the 4th Australasian Dairy Science Symposium, Lincoln, New Zealand, 31 August–2 September 2010; pp. 185–189.
29. Richards, J.H. Physiology of plants recovering from defoliation. In Proceedings of the XVII International Grassland Congress, Palmerston North, New Zealand, 8–21 February 1993; New Zealand Grassland Association: Palmerston North, New Zealand, 1993; pp. 85–94.
30. Lee, J.; Hedley, P.; Roche, J. *Grazing Management Guidelines for Optimum Pasture Growth and Quality*; Technical Series, Issue 5; DairyNZ: Hamilton, New Zealand, 2011.
31. Holmes, C.W.; Brookes, I.M.; Garrick, D.J.; MacKenzie, D.D.S.; Parkinson, T.J.; Wilson, G.F. *Milk Production from Pasture. Principles and Practices*; Massey University: Palmerston North, New Zealand, 2002.
32. Chapman, D.; McCarthy, S.; Kay, J. Hidden dollars in grazing management: Getting the most profit from your pastures. In Proceedings of the South Island Dairy Event, Invercargill, New Zealand, 23–25 June 2014; pp. 21–36.
33. Piggot, G.J.; Baars, J.A.; Waller, J.E.; Farrell, C.A. Initial development of the "pasture growth rate" concept for estimating pasture growth on farms across Northland. In Proceedings of the 16th Agronomy Society of New Zealand Conference, Auckland, New Zealand, 1–5 December 1986; Agronomy Society of New Zealand: Palmerston North, New Zealand, 1986; pp. 65–69.
34. Piggott, G.J. Herbage accumulation patterns of pastures at various sites in Northland. In Proceedings of the 18th Agronomy Society of New Zealand Conference, Havelock North, New Zealand, 1988; Agronomy Society of New Zealand: Palmerston North, New Zealand, 1988; pp. 81–84.
35. Chapman, D.F.; Tharmaraj, J.; Agnusdei, M.; Hill, J. Regrowth dynamics and grazing decision rules: Further analysis for dairy production systems based on perennial ryegrass (*Lolium perenne* L.) pastures. *Grass Forage Sci.* **2012**, *67*, 77–95. [CrossRef]

36. Grant, S.A.; Barthram, G.T.; Torvell, L. Components of regrowth in grazed and cut *Lolium perenne* swards. *Grass Forage Sci.* **1981**, *36*, 155–168. [CrossRef]
37. Gautier, H.; Varlet-Grancher, C.; Hazard, L. Tillering responses to the light environment and to defoliation in populations of perennial ryegrass (*Lolium perenne* L.) selected for contrasting leaf length. *Ann. Bot.* **1999**, *83*, 423–429. [CrossRef]
38. Chapman, D.F.; Bryant, J.R.; Olayemi, M.E.; Edwards, G.R.; Thorrold, B.S.; McMillan, W.H.; Kerr, G.A.; Judson, G.; Cookson, T.; Moorhead, A.; *et al.* An economically based evaluation index for perennial and short-term ryegrasses in New Zealand dairy farm systems. *Grass Forage Sci.* **2016**. in press. [CrossRef]
39. Lee, J.M.; Donaghy, D.J.; Roche, J.R. Effect of defoliation severity on regrowth and nutritive value of perennial ryegrass (*Lolium perenne* L.) dominant swards. *Agron. J.* **2008**, *100*, 308–314. [CrossRef]
40. Garcia, S.C.; Holmes, C.W. Seasonality of calving in pasture-based dairy systems: Its effects on herbage production, utilisation and dry matter intake. *Aust. J. Exp. Agric.* **2005**, *45*, 1–9. [CrossRef]
41. Agnusdei, M.G.; Di Marco, O.N.; Nenning, F.R.; Aello, M.S. Leaf blade nutritional quality of rhodes grass (*Chloris gayana*) as affected by leaf age and length. *Crop Pasture Sci.* **2011**, *62*, 1098–1105. [CrossRef]
42. Groot, J.C.J.; Neuteboom, J.H. Composition and digestibility during ageing of Italian ryegrass leaves of consecutive insertion levels. *J. Sci. Food Agric.* **1997**, *75*, 227–236. [CrossRef]
43. Groot, J.C.J.; Neuteboom, J.H.; Deinum, B. Composition and digestibility during ageing of consecutive leaves on the main stem of Italian ryegrass plants, growing undisturbed or regrowing after cutting. *J. Sci. Food Agric.* **1999**, *79*, 1691–1697. [CrossRef]
44. Wilson, J.R. Variation of leaf characteristics with level of insertion on a grass tiller. I. Development rate, chemical composition and dry matter digestibility. *Aust. J. Agric. Res.* **1976**, *27*, 343–354. [CrossRef]
45. Minson, D.J. *Forage in Ruminant Nutrition*; California Academic Press: Millbrae, CA, USA, 1990.
46. Sinclair, K.; Fulkerson, W.J.; Morris, S.G. Influence of regrowth time on the fresh quality of prairie grass, perennial ryegrass and tall fescue under non-limiting soil nutrient and moisture conditions. *Aust. J. Exp. Agric.* **2006**, *46*, 45–51. [CrossRef]
47. Turner, L.R.; Donaghy, D.J.; Lane, P.A.; Rawnsley, R.P. Effect of defoliation management, based on leaf stage, on perennial ryegrass (*Lolium perenne* L.), prairie grass (*Bromus willdenowii* Kunth.) and cocksfoot (*Dactylis glomerata* L.) under dryland conditions. 2. Nutritive value. *Grass Forage Sci.* **2006**, *61*, 175–181. [CrossRef]
48. Pacheco, D.; Waghorn, G.C. Dietary nitrogen—Definitions, digestion, excretion and consequences of excess for grazing ruminants. *Proc. N. Z. Grassl. Assoc.* **2008**, *70*, 107–116.
49. Kebreab, E.; France, J.; Beever, D.E.; Castillo, A.R. Nitrogen pollution by dairy cows and its mitigation by dietary manipulation. *Nut. Cycl. Agroecosys* **2001**, *60*, 275–285. [CrossRef]
50. Donaghy, D.J.; Fulkerson, W.J. Priority for allocation of water-soluble carbohydrate reserves during regrowth of *Lolium perenne*. *Grass Forage Sci.* **1998**, *53*, 211–218. [CrossRef]
51. Donaghy, D.J.; Turner, L.R.; Adamczewski, K.A. Effect of defoliation management on water-soluble carbohydrate energy reserves, dry matter yields, and herbage quality of tall fescue. *Agron. J.* **2008**, *100*, 122–127. [CrossRef]
52. Rawnsley, R.P.; Donaghy, D.J.; Fulkerson, W.J.; Lane, P.A. Changes in the physiology and feed quality of cocksfoot (*Dactylis glomerata* L.) during regrowth. *Grass Forage Sci.* **2002**, *57*, 203–211. [CrossRef]
53. Miller, L.A.; Moorby, J.M.; Davies, D.R.; Humphreys, M.O.; Scollan, N.D.; MacRae, J.C.; Theodorou, M.K. Increased concentration of water-soluble carbohydrate in perennial ryegrass (*Lolium perenne* L.): Milk production from late-lactation dairy cows. *Grass Forage Sci.* **2001**, *56*, 383–394. [CrossRef]
54. Turner, L.R.; Donaghy, D.J.; Pembleton, K.; Rawnsley, R.P. Longer defoliation interval ensures expression of the "high sugar" trait in perennial ryegrass cultivars in cool temperate Tasmania, Australia. *J. Agric. Sci.* **2015**, *153*, 995–1005. [CrossRef]
55. Rasmussen, S.; Parsons, A.J.; Xue, H.; Newman, J.A. High sugar grasses—Harnessing the benefits of new cultivars through grazing management. *Proc. N. Z. Grassl. Assoc.* **2009**, *71*, 167–175.
56. Gregorini, P.; Beukes, P.C.; Bryant, R.H.; Romera, A.J. A brief overview and simulation of the effects of some feeding strategies on nitrogen excretion and enteric methane emission from grazing dairy cows. In Proceedings of the 4th Australasian Dairy Science Symposium, Lincoln, New Zealand, 31 August–2 September 2010; pp. 29–43.

57. Van Bysterveldt, A.; Christie, R. Dairy farmer adoption of science demonstrated by a commercially focussed demonstration farm. In Proceedings of the 3rd Australasian Dairy Science Symposium, Melbourne, Australia, November 2007; pp. 535–540.
58. Beukes, P.C.; McCarthy, S.; Wims, C.M.; Romera, A.J. Regular estimates of paddock pasture mass can improve profitability on New Zealand dairy farms. *J. N. Z. Grassl.* **2015**, *77*, 29–34.
59. Hodgson, J. Influence of sward characteristics on diet selection and herbage intake by the grazing animal. In *Nutritional Limits to Animal Production from Pasture*; Hacker, J.B., Ed.; CAB International: Wallingford, UK, 1982; pp. 153–166.
60. Leuchtmann, A.; Bacon, C.W.; Schardl, C.L.; White, J.F., Jr.; Tadych, M. Nomenclatural realignment of *Neotyphodium* species with genus Epichloë. *Mycologia* **2014**, *106*, 202–215. [CrossRef] [PubMed]
61. Hume, D.E.; Cooper, B.M.; Pankhurst, K.A. The role of endophyte in determining the persistence and productivity of perennial ryegrass, tall fescue and meadow fescue in Northland. *Proc. N. Z. Grassl. Assoc.* **2009**, *71*, 145–150.
62. Thom, E.R.; Clark, D.A.; Waugh, C.D. Growth, persistence and alkaloid levels of endophyte-infected and endophyte-free ryegrass pastures grazed by dairy cows in northern New Zealand. *N. Z. J. Agric. Res.* **2009**, *42*, 241–253. [CrossRef]
63. Popay, A.J.; Hume, D.E. Endophytes improve persistence by controlling insects. In *Pasture Persistence Symposium*; Mercer, C.F., Ed.; New Zealand Grassland Association: Dunedin, New Zealand, 2011; pp. 149–156.
64. Thom, E.R.; Popay, A.J.; Hume, D.E.; Fletcher, L.R. Evaluating the performance of endophytes in farm systems to improve farmer outcomes—A review. *Crop Pasture Sci.* **2012**, *63*, 927–943. [CrossRef]
65. Hume, D.E.; Sewell, J.C. Agronomic advantages conferred by endophyte infection of perennial ryegrass (*Lolium perenne* L.) and tall fescue (*Festuca arundinaceae* Schreb.) in Australia. *Crop Pasture Sci.* **2014**, *65*, 747–757. [CrossRef]
66. Stewart, A.V. Genetic origins of perennial ryegrass (*Lolium perenne*) for New Zealand pastures. In Breeding for Success: Diversity in Action, Proceedings of the 13th Australasian Plant Breeding Conference, Christchurch, New Zealand, 18–21 April 2006; Mercer, C.F., Ed.; pp. 11–20.
67. Lee, J.M.; Matthew, C.; Thom, E.R.; Chapman, D.F. Perennial ryegrass breeding in New Zealand: A dairy industry perspective. *Crop Pasture Sci.* **2012**, *63*, 107–127. [CrossRef]
68. Gowen, N.; O'Donovan, M.; Casey, I.; Rath, M.; Delaby, L.; Stakelum, G. The effect of grass cultivars differing in heading date and ploidy on the performance and dry matter intake of spring calving dairy cows at pasture. *Anim. Res.* **2003**, *52*, 321–336. [CrossRef]
69. O'Donovan, M.; Delaby, L. A comparison of perennial ryegrass cultivars differing in heading date and grass ploidy with spring calving dairy cows at two stocking rates. *Anim. Res.* **2005**, *54*, 337–350. [CrossRef]
70. Chapman, D.F.; Edwards, G.R.; Stewart, A.V.; McEvoy, M.; O'Donovan, M.; Waghorn, G.C. Valuing forages for genetic selection: Which traits should we focus on? *Anim. Prod. Sci.* **2015**, *55*, 869–882. [CrossRef]
71. Wims, C.M.; McEvoy, M.; Delaby, L.; Boland, T.M.; O'Donovan, M. Effect of perennial ryegrass (*Lolium perenne* L.) cultivars on the milk yield of grazing dairy cows. *Animal* **2013**, *7*, 410–421. [CrossRef] [PubMed]
72. Mercer, C.F., Ed.; *Pasture Persistence Symposium. Grassland Research and Practice Series No. 15*; New Zealand Grassland Association: Dunedin, New Zealand, 2011.
73. Parsons, A.J.; Edwards, G.R.; Newton, P.C.D.; Chapman, D.F.; Caradus, J.R.; Rasmussen, S.; Rowarth, J.S. Past lessons and future prospects: Plant breeding for yield and persistence in cool-temperate pastures. *Grass Forage Sci.* **2011**, *66*, 153–172. [CrossRef]
74. Popay, A.J.; Thom, E.R. Endophyte effects on major insect pests in Waikato dairy pasture. *Proc. N. Z. Grassl. Assoc.* **2009**, *71*, 121–126.
75. Camlin, M.S.; Stewart, R.H. The assessment of persistence and its application to the evaluation of mid season and late perennial ryegrass cultivars. *J. Br. Grassl. Soc.* **1978**, *33*, 275–282. [CrossRef]
76. Cashman, P.; O'Donovan, M.; McEvoy, M.; Shalloo, L. Quantifying ground score change on perennial ryegrass swards exposed to different grazing regimes. In Proceedings of the Agricultural Research Forum, Tullamore, Ireland, 10–11 March 2014; p. 89.

77. McEvoy, M.; McHugh, N.; O'Donovan, M.; Grogan, D.; Shalloo, L. Pasture profit index: Updated economic values and inclusion of persistency. In *EGF at 50: The Future of European Grasslands*; Hopkins, A., Collins, R.P., Fraser, M.D., King, V.R., Lloyd, D.C., Moorby, J.M., Robson, P.R.H., Eds.; European Grassland Federation: Aberystwyth, UK, 2014; pp. 843–845.
78. Chapman, D.F.; Muir, P.D.; Faville, M.J. Persistence of dry matter yield among New Zealand perennial ryegrass (*Lolium perenne* L.) cultivars: Insights from a long-term data set. *J. N. Z. Grassl.* **2015**, *77*, 177–184.
79. Matthew, C.; Agnusdei, M.G.; Assuero, S.G.; Sbrissia, A.F.; Scheneiter, O.; da Silva, S.C. State of knowledge in tiller dynamics. In Proceedings of the XXII International Grassland Congress, Sydney, Australia, 15–19 September 2013; pp. 1041–1044.
80. Edwards, G.R.; Chapman, D.F. Plant responses to defoliation and relationships with pasture persistence. In *Pasture persistence symposium*; Mercer, C.F., Ed.; New Zealand Grassland Association: Dunedin, New Zealand, 2011; pp. 39–46.
81. Matthew, C.; Assuero, S.G.; Black, C.K.; Sackville-Hamilton, N.R. Tiller dynamics in grazed swards. In *Grassland Ecophysiology and Grazing Ecology*; Lemaire, G., Hodgson, J., Moraes, A., de Carvalho, P.C.F., Nabinger, C., Eds.; CAB International: Wallingford, UK, 2000; pp. 127–150.
82. Chapman, D.F.; Edwards, G.R.; Nie, Z.N. Plant responses to climate and relationships with pasture persistence. In *Pasture Persistence Symposium*; Mercer, C.F., Ed.; New Zealand Grassland Association: Dunedin, New Zealand, 2011; pp. 99–107.

# 7

# Defoliation, Shoot Plasticity, Sward Structure and Herbage Utilization in Pasture: Review of the Underlying Ecophysiological Processes

**François Gastal [1,2,*] and Gilles Lemaire [2]**

[1] INRA, Unité Expérimentale Fourrages Environnement Ruminants, INRA-FERLUS, CS 80006, Lusignan 86600, France

[2] INRA, Unité de Recherche Prairies et Plantes Fourragères, INRA-URP3F, CS 80006, Lusignan 86600, France; gilles.lemaire.inra@gmail.com

* Author to whom correspondence should be addressed; Francois.Gastal@lusignan.inra.fr

Academic Editor: Cory Matthew

**Abstract:** Sward structure affects herbage growth, pasture species dynamics, and herbage utilization. Defoliation management has a major impact on sward structure. In particular, tiller size-tiller density compensations allow for the maintenance of herbage growth. Tiller size and tiller density are determined by several major morphogenetical components. Defoliation affects these morphogenetical components, depending on its frequency and its intensity, through several direct and indirect physiological and environmental processes. Due to the implications of leaf area removal, defoliation has a direct effect on the mobilization of C and N reserves and their supply to growing leaves. In addition, defoliation has an indirect effect on leaf and tiller morphogenesis, due to its impact on the light environment within the canopy as well as plant responses to light signals (blue light, red far red ratio). Defoliation may also in some cases have a direct negative effect on leaf growth by damaging leaf meristems. Understanding the respective role of these various physiological and environmental processes requires studies where defoliation, photosynthetic active radiation and light signals are manipulated independently. Past and recent knowledge on these direct and indirect effects of defoliation on plant morphogenesis are discussed, leading to an overall integrated view of physiological and environmental processes that lead to adaptations of sward structure in response to defoliation management. Major consequences for herbage utilization efficiency are presented.

**Keywords:** sward management; defoliation; leaf growth; tillering; light; grasses

---

## 1. Introduction

### *1.1. Plasticity of Sward Structure in Relation to Defoliation and Sward Management*

Plant-herbivore relationships have been the subject of numerous earlier studies. These studies have shown that plants and swards have the capacity to adapt their structure (plasticity in size, number and spatial orientation of shoot organs), productivity and persistency, to the defoliation characteristics that result from grazing or cutting management strategies [1–3]. The plasticity of sward structure in response to successive defoliations mostly results from compensations between tiller size and tiller density. Swards that are frequently defoliated (under intensive continuous grazing management, for example) have a higher density of smaller shoot axes (tillers, branches) and conversely, swards that are defoliated infrequently (under a cutting regime, for example) have a lower density of larger shoot axes [3]. Moreover, earlier studies have also shown that the plasticity of sward structure in response to defoliation not only allows adaptation in sward productivity but also partly determines herbage utilization by animals, in conjunction with animal grazing behavior [4–9]. For a similar herbage mass, swards with a higher lamina/sheath (grasses) or leaf/stem (dicots) ratio allow a higher intake than swards with a higher sheath, pseudo-stem or stem ratio [7,10]. Maximal daily intake is related to tiller size, the vertical profile of bulk density, which depends on tiller density, and sheath length, which limits grazing in the lower sward layers [6,8,11,12]. The plasticity of sward structure in relation to defoliation and grazing management, together with the impact of sward structure on herbage utilization of grazing animals, are probably responsible for the observation that within certain limits in defoliation management, herbage productivity and herbage utilization by grazing animals are maintained at a comparable rate, despite variations in herbage mass [13].

### *1.2. Understanding Sward Structure Plasticity: The Need for Integrating Physiological and Environmental Processes*

Although it is recognized that plasticity of sward structure plays a major role in the capacity of swards to adapt to defoliation and grazing management, the underlying ecophysiological basis, *i.e.*, integration of plant physiological and environmental processes at the community level, are still poorly understood [14]. This lack of knowledge becomes even more critical when we want to consider and address the plasticity of multispecific sward communities, where specific species responses determine sward dynamics and species equilibrium. Improvement in this area of knowledge would allow a better definition of the potential and limits in the adaptability of sward structure to management [15], and would therefore help to define strategies and conditions allowing grassland sustainability. The objective of the present paper is to give an overview of the current knowledge on the processes underlying plant and sward plasticity in response to defoliation, and to highlight recent findings. For simplicity, the exercise is limited to grasses, but it could be conducted similarly on several dicots, particularly white clover, with almost similar conclusions supported by an almost similar literature background.

An earlier approach of sward structure and herbage productivity and utilization focused on the relationship between leaf area index (LAI) and growth rate. The net herbage accumulation rate is generally considered maximum for an LAI in the range of 3–5 [3,16–18]. At a lower LAI, net growth rate is limited by light interception and at a higher LAI, net growth rate is limited both by the burden of respiration, particularly respiration of shaded organs at the bottom of the canopy, and by the loss of plant material through senescence and litter fall. In addition, intake by animals of ageing and senescing organs is limited [4], and therefore at high LAI, not only net herbage productivity but also herbage utilization is restricted. However, this earlier analysis, based on the LAI-growth relationship, is derived from the study of swards under discontinuous defoliation (cutting), and needs to be adapted to conditions of continuous grazing [16], where long-term sward structure adaptations modify photosynthesis-respiration-senescence relationships.

An alternative approach has been proposed [3,18,19], in which tissue flows and tissue turnover in the sward are analyzed through several growth components at a plant axis (tiller or branch) level: leaf appearance, leaf growth and leaf senescence rates, branching pattern, and specific organ mass. By definition, this approach is closer to sward structure analysis than the previous approach relying on LAI and growth rate. Therefore, most of the current knowledge related to the previously mentioned tiller size-tiller density compensations occurring during adaptation of sward structure to defoliation management is derived in some way from the tissue turnover approach. In fact, these two approaches are complementary, since LAI is both an input of the former (growth rate model) and an output of the second (tissue turnover) approach.

In the first instance, sward LAI can be decomposed into:

$$\text{LAI} = \text{Tiller density} \times \text{Leaf Area per tiller}$$

Tiller density and leaf size are negatively correlated. The inverse relationship between tiller density and tiller size (size density compensation, SDC) has been described for many grazed swards [3,20]. Attempts have been made to relate the slope of the SDC line obtained under various grazing intensities with that of the self-thinning law, *i.e.*, $-3/2$ in log/log scale [21]. However, as demonstrated in [22], the

−3/2 slope occurs only when LAI has reached a maximum value corresponding to the environmental potential of the sward. In continuously stocked swards, LAI depends on defoliation intensity and does not often reach its maximum value. As a consequence, the SDC slope is steeper than −3/2, and is then often close to −5/2 [23–25].

Studying the tiller density-tiller size relationship clearly provides an experimental way to evaluate the impact of defoliation regimes on herbage growth and allows the delineation of situations where the compensation may or may not be effective. However, the underlying processes that drive and allow (or not) this compensation are not explicitly identified. For this purpose, a more detailed and dynamic analysis of plant axis growth and branching is necessary. Such an analysis was initiated earlier [18], but new knowledge is now available and allows a finer understanding of the involved processes.

## 2. Defoliation Patterns, Sward Structure and Light Environment

### *2.1. Defoliation Patterns in Relation to Sward Management*

Defoliation basically consists of removing part of the shoot organs of plants and is primarily characterized by its intensity (or severity) and its frequency (or its inverse, the defoliation interval). In several instances, defoliation also needs to be characterized by additional parameters, such as its spatial heterogeneity or its timing in relation to plant development, particularly floral initiation. On a tiller basis, the defoliation interval usually varies from low values, typically 7–15 days under intensive continuous grazing management and depending on the stocking density [6,12,26,27], to intermediate values of approximately 20–30 days in rotational grazing, and to high values of approximately 30–60 days in cutting systems. The tiller defoliation interval may reach higher values in extensive systems or where a long rotation is used to stockpile feed. In such situations, the impact of defoliation mostly becomes a question of biomass accumulation and its negative impacts from both plant and animal points of view. Under intensive management, defoliation intensity is generally higher under long than under short defoliation intervals. The defoliation intensity may involve over 80%–90% leaf area removal under mowing depending on the cutting height. In rotational grazing, defoliation intensity depends on the stocking density and the duration of the grazing period, which determine the residual sward height. As demonstrated in [6], a single tiller is defoliated several times during the grazing process. As a result, the intensity of defoliation in rotational grazing may vary between 50%–75% depending on grazing management. Under continuous grazing, the defoliation intensity of individual tillers, expressed on an extended tiller length basis, was found to be 35% per tiller with dairy cows [5]. It varied in the range of 40%–60% on the basis of individual leaf length [27,28]. This defoliation intensity is similar to the defoliation intensity measured in a rotational system for a single defoliation event [5,28]. From a plant physiology perspective, as well as from a sward structure point of view, it is important to note that the recently expanded leaves, located in the upper layers of the canopy, are generally defoliated more frequently and to a larger extent than older leaves, as shown under sheep grazing [27,29].

### *2.2. Plant Morphogenesis is the Driver of Sward Structure, Plasticity and Adaptation to Management*

Since leaf area is a major determinant of plant growth and since defoliation leads to the removal of part of it, adaptation of sward structure to defoliation management has to be analyzed first in terms of leaf area dynamics. Moreover, since the plasticity of sward structure in response to defoliation operates in terms of tiller size-tiller density compensation, the dynamic fluctuation of leaf area has to be analyzed on a tiller (or more generally on a plant axis) population basis.

The leaf area index of a sward is determined by 3 morphological components: (i) leaf area (often approximated in grasses by leaf length due to the linear shape of their lamina); (ii) number of mature leaves per tiller; and (iii) tiller density. A similar sward LAI can be obtained by different combinations of these components. As underlined in [18], these components result from the morphogenetic activity of tiller populations, and have to be analyzed accordingly.

In a vegetative sward, tiller morphogenesis can be described by three main variables (Figure 1): (i) leaf appearance rate (LAR); (ii) leaf elongation rate (LER); and (iii) leaf life span (LLS). These components are genetically determined and are influenced by environmental variables such as temperature, nutrient supply and plant water status, and additionally by defoliation through induced changes in light interception and light quality (see next paragraph). Leaf size or leaf length (LL) can be considered as the ratio between LER and LAR, considering that for a given genotype, the leaf elongation duration (LED) is proportional to the leaf appearance rate LAR [30]. The maximum number of green leaves per tiller (NL) can be considered as the product of LAR and leaf life span (LLS). Tiller density can be analyzed through tiller appearance and tiller death rates.

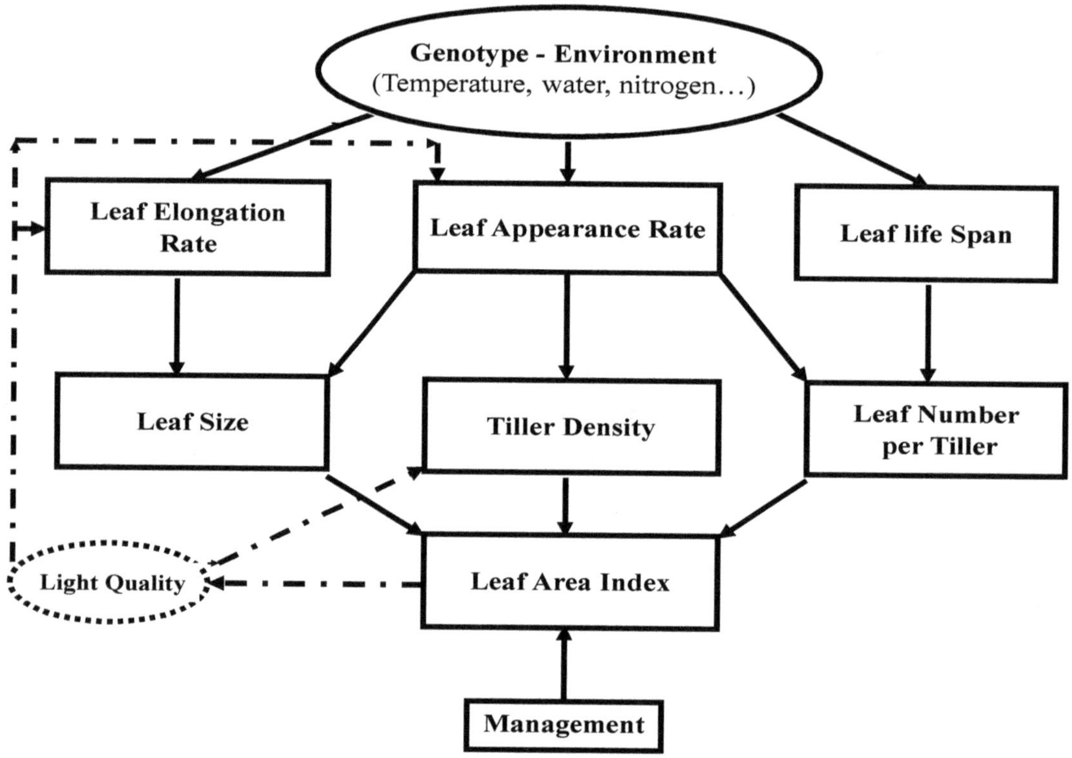


**Figure 1.** Determination of the grass sward structure in relation to components of shoot morphogenesis. Adapted from [18].

A number of sward structure parameters other than leaf area are also relevant with respect to the impact of defoliation on herbage growth and utilization. In particular, length, vertical orientation and horizontal dispersion of pseudo-stems (sheath) or stems partly determine tiller access to light under plant competition, and also partly determine herbage accessibility to mowers or grazing animals. The stage of floral induction of the apical meristem determines the short-term survival potential of grazed tillers since the probability of apex decapitation increases with the apex elevation within the sheath tube in reproductive tillers.

### *2.3. The Role of Light Signals in Plant Responses to Defoliation*

The direct role of defoliation (*i.e.*, decreased light interception due to leaf area removal and subsequent decreased rate of plant photosynthesis) in tiller morphogenesis will be discussed in detail below. However, in addition to its direct role, defoliation also has a major impact on the spectral composition of light within the sward, which in turn also affects tiller morphogenesis to a major extent and thus sward structure. Besides its role in plant photosynthesis, light also impacts plant morphogenesis as a source of light signals [31–34].

Several photoreceptor families have been identified in plants, in particular phytochrome and cryptochrome families, which are responsible for morphological responses to the red far-red ratio (R/FR) and to blue radiations (Bl), respectively. Due to a higher absorption of R than FR by leaves, foliage reflection and transmission of FR is larger than foliage reflection and transmission of R. Therefore, in the vicinity of plants and during sward development and associated leaf area index (LAI) increases, the R/FR ratio decreases within and below the canopy, and light may even be enriched in FR compared with incident light. Considering these modifications in FR light and the R/FR ratio in the plant environment, and considering that plants have the ability to morphogenetically respond to the R/FR ratio, it has been proposed [35–38] that FR (or alternatively the R/FR ratio) acts as an early signal of competition for light within canopies. This signal triggers shoot morphogenetic responses, anticipating mutual shading and reducing its consequences on plant photosynthesis. The signaling role of blue radiation operates differently than R/FR. Cryptochrome responses are observed over most of the blue light domain [39]. In addition, absorption of blue light by foliage is similar over most of the blue domain, so that gradients in blue light in the plant environment are parallel to gradients in photosynthetic active radiation (PAR). As a consequence, blue light is not generally considered as a competition anticipating light signal, in contrast with the R/FR ratio. However shoot morphogenetical responses to blue are significant and differ from responses to PAR, both in terms of physiological processes and morphological impact [40–44]. Therefore, they participate, in conjunction with responses to R/FR, in the morphological plasticity of plant swards.

Experimentally, the delineation between the direct role of defoliation (decreased plant photosynthesis due to leaf area removal) and its indirect role in the light environment (modification of both photosynthetic and morphogenetically active radiation) requires the analysis of the responses of isolated plants to various independent combinations of defoliation regimes and light quantity and quality supply. Care must be taken in the interpretation of most previous light intensity studies since in many instances, manipulation of light intensity was associated with an uncontrolled modification in light quality.

## 3. Defoliation Patterns, Light Environment and Dynamics of Sward Structural Components

### *3.1. Leaf Appearance Rate*

Defoliation sometimes appears to negatively affect LAR [45–47]. However, in several instances, the effect of defoliation on LAR is very limited or not significant [48,49], or defoliation may even increase LAR [50]. Light intensity has been found to have a positive effect on LAR in some instances [49,51], or to have a limited effect on LAR in others [52–54]. Although very few studies are devoted to the effect of light quality on LAR, in comparison to studies on the effect of visible light, it appeared for *Lolium perenne* and *Festuca arundinacea* that LAR is not significantly affected by the R/FR ratio or by blue light [49,55]. Therefore, the positive effect of light intensity on LAR that was observed in some circumstances can be interpreted as been mediated by increased photosynthetic activity and carbon supply rather than by light signals.

A number of studies show that independent of the impact of defoliation or light intensity and quality, the LAR of successive leaves of a tiller decreases during its development. This ontogenic decrease in LAR is observed similarly in plants grown from seedlings [56–61] and plants recovering from defoliation [62,63]. These studies show that the LAR of successive leaves may decrease by a factor of more than 2 during plant development, and that the decrease in LAR is systematically associated with an increase in sheath and lamina length (Figure 2A). As suggested in [64], the length of the sheath tube of older leaves may affect the LAR of the younger growing leaves. The previously mentioned ontogenic decrease in LAR during plant development may therefore be related to the ontogenic increase in the sheath length of successive leaves. The increase in sheath length over successive leaves tends to increase LAR due to the increased distance for a leaf to emerge to light. However, it was shown that this effect contributes to the decrease in LAR to only a limited extent [60]. Thus, the decrease in the LAR of successive leaves is probably more related to a profound modification of the structure of the leaf growth zone and underlying cellular division and extension processes. The shortening of the sheath by defoliation should at least partly reverse the ontogenic increase in sheath length and the decrease in LAR of successive leaves of undefoliated plants. However, in addition, the LAR of defoliated plants may also decrease due to a shortage in carbon supply to the growing leaf. These superimposed opposite effects may explain the contradictory results reported in the literature, and particularly the observation that LAR may be accelerated under a severe compared to a lax defoliation regime [48,50].

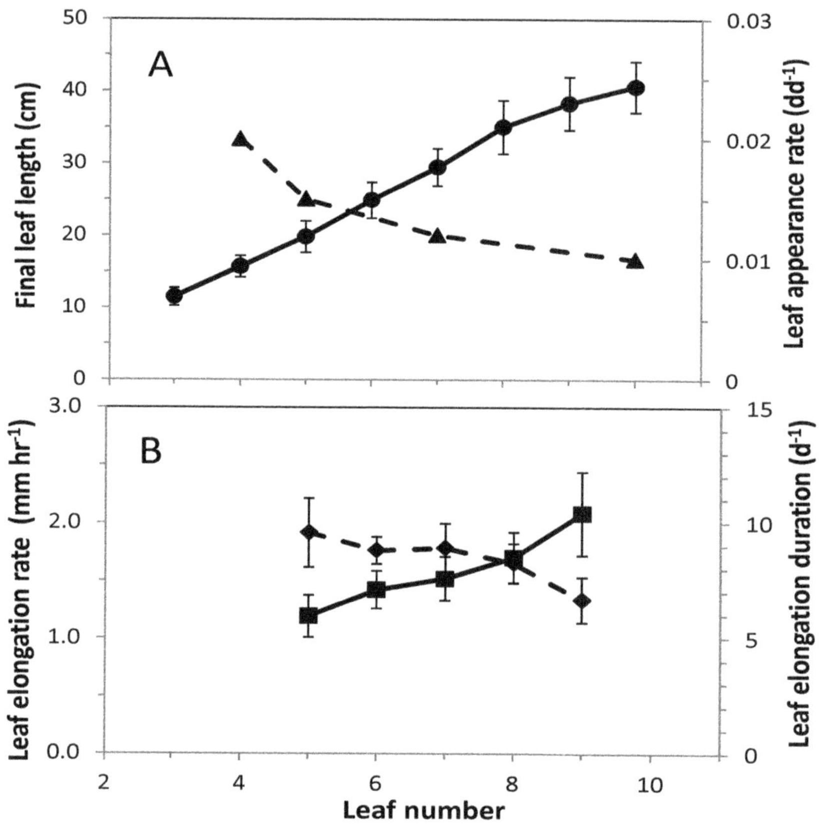


**Figure 2.** Leaf growth components (appearance and elongation rates, elongation duration, final length) of successive leaves on the ryegrass (*Lolium perenne*) main tiller of undefoliated plants. (**A**) ●: final leaf length; ▲: leaf appearance rate; (**B**) ◆: leaf elongation rate; ■: leaf elongation duration. Plants were grown under constant controlled conditions. Redrawn and completed from [61].

While LAR has long been considered to be independent of tiller ontogenic development, recent studies show that this is not the case. The ontogenic decrease in LAR is substantial and needs to be taken into account to understand the dynamics of sward structure, particularly because of its implications for tillering (see Section 3.4).

### *3.2. Leaf Growth Rate*

For many years, the physiological effect of defoliation on leaf growth and LER has been primarily considered as the result of a decrease in plant photosynthesis, induced by leaf area removal and therefore by a larger dependency on carbohydrate reserves. The numerous studies aimed at manipulating non-structural carbohydrate levels and evaluating leaf elongation or growth have shown that following defoliation, leaf growth is affected by the non-structural carbohydrate level only below a certain level [3]. However, the critical level of non-structural carbohydrates varies according to growing conditions, including air $CO_2$ concentration [65], and thus a general threshold value cannot be defined. Since these early studies on non-structural carbohydrates and LER, it has become more clearly understood that plants are able to maintain leaf elongation, following defoliation, through a number of physiological and morphological transient adaptations, reviewed in detail by [66] and presented more briefly below. These various physiological adaptations explain why leaf growth is not necessarily correlated with the non-structural carbohydrates level.

Following defoliation, the specific leaf area of new leaves or new leaf segments often increases [47], leading to a lower cost in C and N related to leaf area expansion. The increase in specific leaf area probably occurs from both a lower accumulation of soluble carbohydrates and from changes in leaf structure (width, thickness), as suggested by [67], and occurs under low irradiance [68,69]. However, the extent to which these structural adjustments occur following defoliation would need clarification. Non-structural carbohydrates (mostly fructans in $C_3$ grasses) and nitrogen compounds accumulate at relatively high concentrations in leaf intercalary meristems of grasses, and transiently participate in providing C and N to support leaf growth after defoliation [67,70–74]. Non-structural carbohydrates and nitrogen compounds from mature shoot organs left intact following defoliation generally provide the quantitatively major source of C and N mobilization to support new leaf growth [75–82]. However, increasing defoliation severity by decreasing the amount of intact shoot tissue increasingly compromises C and N mobilization potential to sustain new leaf growth. Root mobilization of carbon and more importantly nitrogen may also occur when mobilization from shoot organs is insufficient. In grasses, fructans do not accumulate to a large concentration in the mature zones of roots, and it remains unclear whether their remobilization provides substantial C supply to leaf growth or whether they are mostly used for root respiration as shown for root starch in *Medicago sativa* [83]. Root N and possibly C mobilization to shoots may be accompanied by a significant decrease, and in some cases an almost complete cessation of root growth, and may also induce root senescence ([84] and references therein). In addition to supplying nitrogen to sustain leaf growth, mobilized N compounds, essentially amino acids, also provide a substantial amount of carbon [85]. Although no definitive arguments have been made with respect to whether post defoliation leaf growth is predominantly limited by carbon or by nitrogen, recent studies suggest that C and N co-regulate leaf growth in general [85], and thus either C or N may play the predominant role depending on prevailing conditions in nitrogen supply and PAR prior to defoliation. Recent studies clearly show that the dependence of leaf growth on non-structural carbon accumulated prior to defoliation is shorter (about 2 days) than previously thought [85,86].

However, these physiological mechanisms, which allow maintenance of a significant leaf growth rate following defoliation, have limits. These limits need to be evaluated to properly understand the impact of defoliation management on leaf elongation. Under repeated defoliation, the physiological and morphological adaptive mechanisms are not necessarily sufficient to sustain the leaf growth rate. During sequences of weekly defoliation, leaf elongation of *Lolium perenne* is substantially and durably altered after the second to the third defoliation event (Figure 3A). In parallel, the non-structural carbohydrate concentration decreases and remains at low values (Figure 3B) unless defoliation pressure is suppressed, resulting in non-structural carbohydrate concentration and LER recovery. This illustrates that under repeated and frequent defoliation, carbohydrate reserves may not re-accumulate to a sufficient level between two consecutive defoliations and therefore may reduce LER, as confirmed in other studies [87–90]. These observations on isolated plants are in line with field data on tall fescue under continuous sheep grazing, showing that the LER of grazed tillers was substantially decreased compared with the LER of undefoliated tillers [91]. In grazing studies, defoliation is generally less severe in frequency and intensity than in the weekly defoliation sequence experiment previously reported in Figure 3. However, since grazing tends to preferentially remove young leaves and thus tends to leave older leaves or older leaf portions, which have a lower photosynthesis rate, the lower proportion of leaf removal occurring in the grazing experiment was probably counterbalanced by a lower photosynthetic activity of the remaining leaf segments. Thus, besides the understanding of basic physiological processes that take place following a single defoliation, it is crucial to also consider their limits and to explore various defoliation sequences.

In addition to the potential effect of defoliation on leaf elongation, in relation to carbon and nitrogen supply and mobilization, defoliation may also reduce leaf elongation by damaging part of the intercalary leaf meristem. This frequently occurs in dicots, where more meristems are located at height above ground level, and therefore are directly susceptible to suppression. Although temperate grasses are generally considered as plants adapted to grazing [14], a recent study comparing several grass species has shown significant differences in leaf meristem tolerance to defoliation [92,93]. Under severe defoliation, *Dactylis glomerata* was better able to shorten the leaf growth zone and to compensate for its shorter meristem by a higher relative growth rate of its leaf growth zone compared with *Festuca arundinacea* and *Festuca rubra*, while *Lolium perenne* had an intermediate behavior. These data are in line with other observations that *Dactylis glomerata* is more tolerant to severe defoliation than *Festuca* species [94].

Independent of carbon supply and mobilization, the effect of light quality on LER is also significant. A decreased R/FR ratio led to higher LER in several $C_3$ and $C_4$ grasses [95]. In localized illumination experiments, the response of LER was significant when the emerged portion of the growing leaf was exposed to FR [96]. Blue light decreased LER in *Lolium perenne* [97]. More recent experiments [55] have confirmed this response and have shown that blue light may reduce LER to a much larger extent than previously shown in [97]. The negative effect of blue light on LER probably explains the negative effect of visible light on LER reported previously [68]. Therefore, several light signals develop during canopy development following defoliation (decrease in the R/FR ratio and a decrease in blue) and contribute to an increased leaf extension rate.

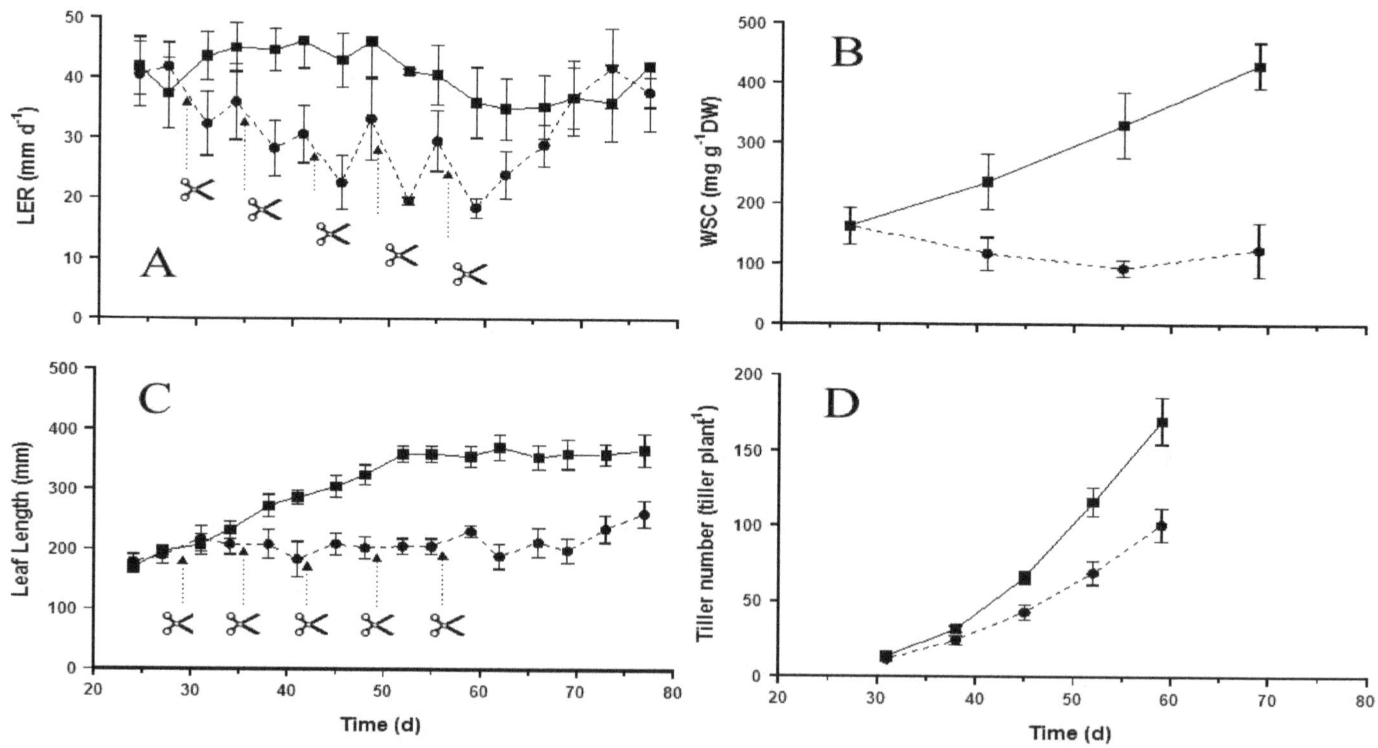


**Figure 3.** Effect of a sequence of 5 successive weekly defoliations (✂) applied to ryegrass plants, followed by a lax period allowing leaf growth recovery, on leaf elongation rate per tiller (LER, **A**),water-soluble carbohydrate concentration (WSC) in the leaf base (**B**), final leaf length (**C**) and tiller number per plant (**D**). ■ Undefoliated control plants; ● defoliated plants; d: day. Arrows indicate timing of the successive defoliations. Plants were grown under constant controlled conditions. Defoliated plants were clipped at a 5-cm stubble height [98].

### 3.3. Consequences of Defoliation on Leaf Length

During the ontogenic development of an undefoliated and isolated plant, successive leaves show a regular pattern of increase in final leaf length [47,56,61] (Figure 2A). The leaf elongation rate slightly increases in the first 2 or 3 leaves of seedlings [56], but remains approximately stable for leaves of higher rank, provided nutrient limitation or self-shading does not occur as plants get larger [61,99]. In contrast, the elongation duration of leaves of increasing rank increases very significantly [61] (Figure 2B). Therefore, the ontogenic increase in the length of successive leaves during tiller development is basically related to an increase in leaf elongation duration rather than to an increase in LER.

This ontogenic increase in leaf length with leaf number on a tiller is modulated by light quality. Blue light decreases the rate of increase of the final leaf length on a tiller to a large extent [97,100] (Figure 4). Moreover, blue light decreases sheath length to a relatively larger extent than lamina length. In contrast, FR or the R/FR ratio increases the rate of increase of the final leaf length with leaf number [96,100]. At the sward level, the implication is that as undefoliated or infrequently defoliated tillers develop, and as leaf area index increases, the ontogenic tiller development, together with the decrease in blue light and the decrease in the R/FR ratio, act simultaneously and accelerate the increase in the final length of new leaves, as shown by green shading experiments that associate changes in light quality and light intensity similarly to changes in light composition occurring within the canopy during sward development [101].

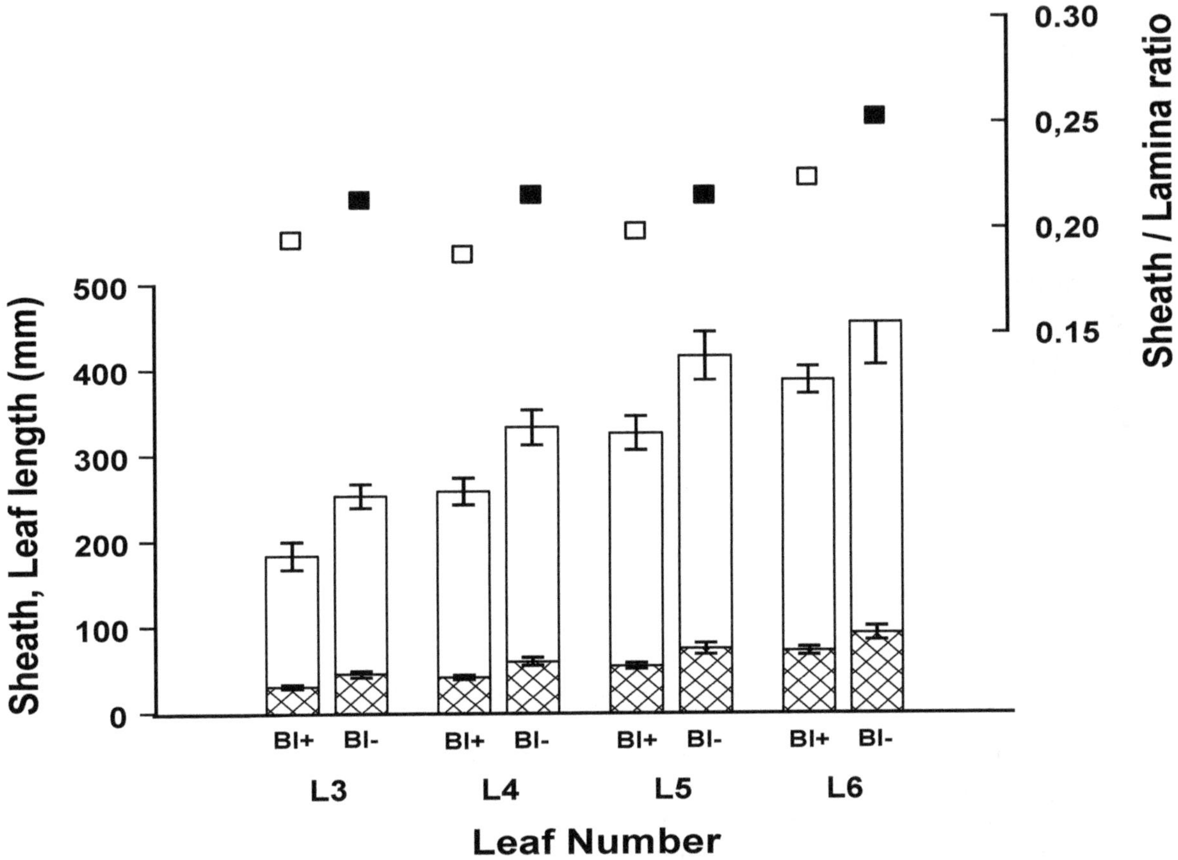


**Figure 4.** Effects of blue light on the final length of the sheath (hatched bar) and lamina (unfilled bar) of the four successive leaves of the main tiller of ryegrass (*Lolium perenne*). Light is provided by metallic iodure lamps (Blue+, □) or by sodium pressure lamps plus filters removing blue wavelengths (400–550 nm) and compensated to maintain a similar PAR (Blue−, ■). Adapted from [55].

Defoliation leads to a decrease in final leaf length in situations where leaves are cut below the ligule, but does not have a very significant effect when defoliation occurs above this point [47,48]. When significant, the negative effect of defoliation on final leaf length (Figure 3C) is simultaneously brought about by a decrease in the leaf elongation rate (see above for physiological determinants) and a decrease in leaf elongation duration. The physiological determinants are more obscure for LED than for LER. Several earlier studies have indicated that the sheath length of a mature leaf affects the final length of the following subtended leaf, and reduction in sheath tube length due to defoliation was the cause of the reduction in leaf length [47,64,102,103]. These studies suggest that the decrease in leaf length was associated with a decrease in LED. These results are confirmed by recent data showing a strong relationship between the length of the sheath tube from which a leaf emerges and the final length of this new leaf. This occurs similarly whether the source of variation in sheath length is the ontogenic increase in sheath length during development of an undefoliated tiller, whether it is the sheath length of axillary leaves and the leaves of its daughter tillers, or whether sheath length is reduced by defoliation (Figure 5). Therefore, there is good experimental evidence supporting the hypothesis that the length of the sheath tube also participates in the regulation of leaf length at the tiller level, in relation not only to its ontogenic development but also to defoliation intensity. It is currently unclear whether the impact of the sheath tube on the length of the following leaf is of the same physiological nature as the impact of blue light or R/FR, *i.e.*, whether the response to the sheath tube is determined by its effect on light quality perceived at the base of the growing leaf. Whatever the case, the fact remains that from a sward structure perspective, the influence of the sheath tube on leaf length operates in addition to the previously mentioned regulation of leaf length by light composition within the sward.

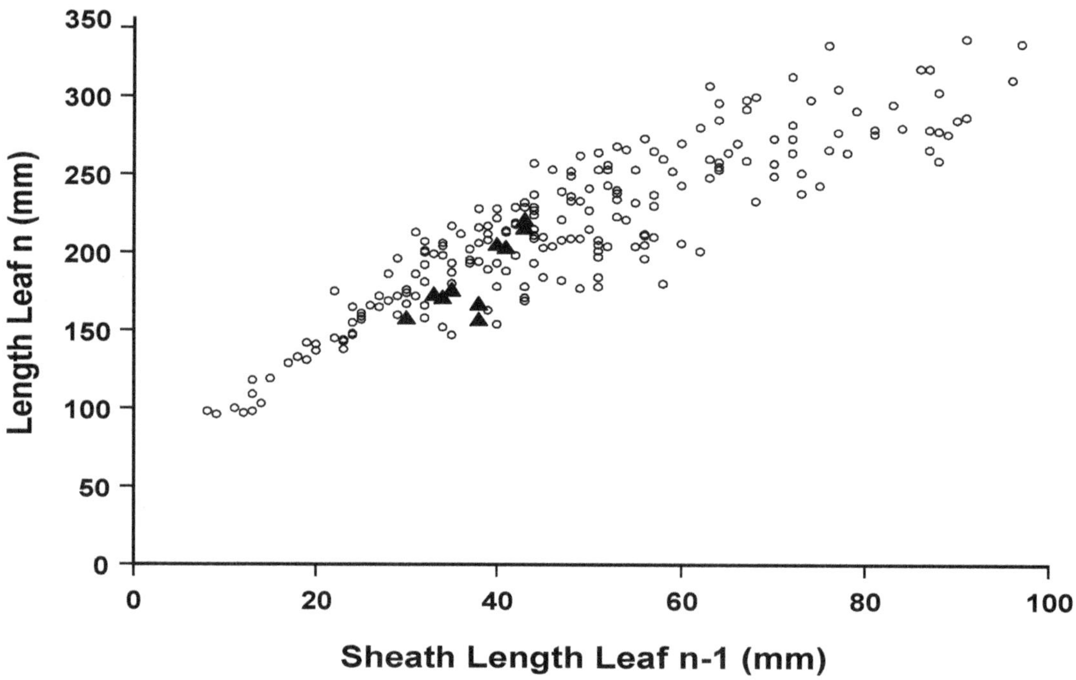


**Figure 5.** Relationship between the length of a tube sheath and the final length of the leaf that has emerged from the same tube on tillers (main and secondary tillers) of defoliated and undefoliated ryegrass (*Lolium perenne*). (○)): undefoliated plants; (▲): weekly defoliated plants. The source of variation in sheath and leaf length is the ontogenic development (successive leaves on main and secondary tillers). Redrawn from [61].

As mentioned earlier, both LAR and LED appear to be correlated with sheath length. It can be emphasized here that a correlation between LAR and LED was also observed under conditions where sheath length was not the primary source of variation [99,104]. This relationship between LAR and LED is probably derived from the principle of growth co-ordination between successive leaves (*i.e.*, coordination of the successive phases of leaf development, from initiation to emergence and achievement of final size) as proposed in [105] or more recently in [106].

This principle of growth co-ordination between successive leaves also has other major implications. It explains the observation of a constant number of leaves growing simultaneously on a tiller [3,53]. It also explains that under undisturbed conditions of growth, the onset of leaf senescence (and consequently leaf life span) is coordinated with the development of newer leaves of a tiller, thus determining a constant number of mature living (green) leaves per tiller as observed in several pasture $C_3$ grass species [3,53]. This has major implications (discussed in Section 4) with respect to herbage utilization.

### 3.4. *Defoliation and Tillering*

The appearance of new tillers occurs through the activation of axillary buds and the development of the first leaf primordium. Therefore, due to this physical link between the presence of a leaf and the presence of its subtending axillary bud (one axillary bud per leaf in many temperate grasses), tillering is limited by the number of leaves produced. Site filling [45,107] or site usage [58] indices have been used to evaluate the extent to which tillering rate is regulated by, or independently of, LAR. In addition to the physical link between leaf appearance and tillering, observations also show that the development of a tiller from an axillary bud is co-ordinated with the development of the subtending leaf, so that its probability of development is high at the end of cell division in the sheath of the subtending leaf [108], declines thereafter, and becomes low after two or three leaf initiation intervals on following phytomers [109]. Thus, the beginning of the development of a daughter tiller occurs within a narrow "window of opportunity", relative to the development of leaves on the mother tiller. This reinforces the link between the rate of appearance of leaves on a tiller and the rate of appearance of daughter tillers.

Basic principles of the regulation of tillering in swards may be illustrated in the simple case of the time trends of tiller density (the net balance between tiller birth and tiller death) in a regrowing sward following a single mechanical defoliation (Figure 6). In the first phase, soon after defoliation, tiller density rapidly and substantially increases (days 5–20, LAI below 4). In the second phase (days 25–30, LAI 4–6), the rate of increase in tiller density progressively decreases to zero, due to a very large decline, although not necessarily a complete cessation, of tiller appearance (data not shown). In the third phase (following day 35), tiller density decreases, mostly due to tiller death. As shown in [110], site filling is maximum under low LAI (phase 1 in Figure 6) and LAR is high, allowing a rapid emission of new tillers. As a sward develops and LAI increases, site filling decreases, indicating that high LAI down-regulates tillering *per se*, independently of any effect on LAR. In addition, since LAR also declines during plant ontogenesis (see Section 3.1) and sward development (Figure 6), the decline in tillering rate observed during the second phase is simultaneously determined by the decrease in LAR and by a decrease in site filling (or site usage). If site usage was limited by carbohydrate availability, the decrease in site filling would have happened in the first phase of sward development rather than in the second phase during which LAI is already restored to a large extent. There is now much evidence that tillering is regulated by light signals. Tillering is reduced under the low R/FR ratio that plants experience during sward development and at LAI values above 2–3 [49,96,111–115], mostly due to the increased delay between the appearance of a daughter tiller and the appearance of its subtending leaf. More recent data show that blue light does not play a significant regulatory role in site usage or LAR [55], and thus does not affect tillering. The absence of an effect of blue light on tillering contrasts with the large effect of the R/FR ratio.

Tiller death shown in the third phase of the above sward regrowth example is likely to occur due to the shading of young tillers that could not develop rapidly enough during sward development to maintain access to light and then C supply [3,116].

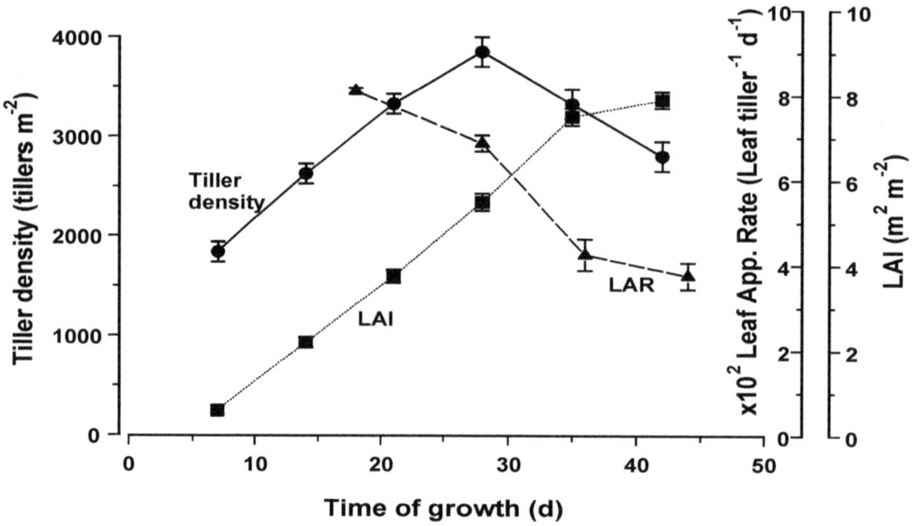


**Figure 6.** Time trends in tiller density (●), leaf area index (■, LAI), and leaf appearance rate (▲, LAR) for a tall fescue sward following a cut (at day 0). Adapted from [62].

A number of studies conducted on isolated plants show that tillering rate is reduced by defoliation, particularly when it is repeated and intense (Figure 3D; [46,47,49]). This decrease in tillering is probably due to a shortage of non-structural carbohydrates. Although the tillering rate is decreased by defoliation, it is not stopped even under frequent and intense defoliation (Figure 3D). Thus, under a sequence of repeated intense defoliation of isolated plants, the tiller number per plant continues to increase, while the leaf elongation rate progressively declines, leading to a major decrease in leaf length and therefore tiller size. This down-regulation of tillering by frequent and intense defoliation of single plants thus appears to be more limited than the large down-regulation that operates under high LAI in dense swards (during phases 2 and 3 in Figure 6). Therefore, the latter, determined simultaneously by the decrease in LAR and site filling as explained above, appears to be the basis of the size-density compensations observed at the sward level. Thus, although the impact of defoliation on leaf growth sometimes appears as a shorter-term response than morphological adaptations [17,18,84], from an ecophysiological point of view, they operate simultaneously and are part of an integrated plant-environment interplay. This probably explains why the adaptation of sward structure to defoliation, which allows the maintenance of sward productivity to a certain extent, is rapid and efficient.

## 4. Grazing Management, Herbage Production and Efficiency of Herbage Utilization

A grazed sward is a dynamic system where leaf tissues are continuously produced by tillers (growth, G) and are consumed by grazing animals (intake, I) or are senescing (S). Optimizing the quantity of herbage utilized by the grazing animal (I) requires two considerations: (i) maintaining a high rate of green herbage accumulation (G) and (ii) maximizing the efficiency of herbage utilization I/G, or equivalently minimizing herbage losses by senescence S/G. In continuous grazing systems, this highlights the conflict between maximizing G (by maintaining a high LAI) and minimizing senescence S (by maintaining a low LAI). Herbage growth increases asymptotically with sward LAI, in keeping with the quantity of intercepted radiation, while senescence increases linearly. Thus, under continuous grazing management, a compromise between herbage production and the efficiency of herbage utilization has to be found to maximize the amount of herbage intake. This optimum for pastures dominated by perennial plants in temperate regions is reached for swards maintained at LAIs of approximately 2–4 [117,118]. Under intermittent defoliation management (rotational grazing or cutting regime), the optimization of harvested herbage yield for the succession of regrowth periods requires that herbage be harvested when the maximum average growth rate is reached (Figure 1.8

in [18]). The time elapsing from the start of regrowth to optimum harvest depends greatly on the initial (or residual) LAI. The higher the initial LAI, the shorter the interval to optimum harvest [13].

In [13], a model was used to compare the relationship between average growth rate and average sward LAI under continuous and rotational sward management. It was demonstrated that the two managements are essentially part of the same continuum growth response to the average sward LAI. Thus, the key sward parameter that governs herbage production is the average LAI at which the sward is maintained. This average LAI can be maintained as constant as possible during a period under continuous stocking, or can vary from a low to a high value during the regrowth period in an intermittently defoliated system.

The efficiency of herbage utilization in a grazing system can be defined as the proportion of the gross leaf tissue production that is removed by the animals before entering the senescence state [18,28]. The calculation of the herbage utilization efficiency should consider the maintenance of a sward state that ensures the "sustainability" of herbage production, *i.e.*, that allows a more or less constant light interception by the sward. Under continuous stocking conditions, this situation corresponds to swards maintained at constant LAI or sward height [117]. For rotational grazing systems, this state corresponds to an average LAI at which the different paddocks are maintained over time [13]. The analysis of leaf tissue flux dynamics in grazed swards allows for the estimation of herbage utilization efficiency (HUE) as the proportion of the gross leaf tissue production (G) that is consumed by grazing animals (I). As demonstrated in [27] under continuous grazing conditions, the proportion of leaf length that escapes defoliation and then senesces can be estimated by the ratio between leaf lifespan and the defoliation interval. This ratio determines the average number of times an individual leaf can be defoliated before senescence [28]. For example, for an average leaf life span of 40 days and a defoliation interval of 20 days, each leaf should be defoliated 2 times before senescence. Given that the proportion of leaf length removed at each defoliation is relatively constant and averages 50% [27], it is possible to calculate the proportion of length of each leaf removed by animals. For an average leaf lifespan of about 40 days for tall fescue and an average interval of defoliation of 20 days, a theoretical herbage defoliation efficiency of 75% of leaf length can be predicted (Figure 1.7 in [18]). The defoliation interval decreases linearly with an increase in stocking density [6]. Therefore, when stocking density is increased under continuous grazing management, as a consequence of a higher herbage growth for maintaining a constant sward height (or LAI), then the leaf defoliation interval decreases, and for a given leaf life span, a greater proportion of leaf length is consumed by grazing animals, leading to an increase in herbage utilization efficiency. In particular, when herbage growth is stimulated by N supply, the increase in stocking density and the decrease in the leaf defoliation interval lead to a substantial increase in herbage use efficiency (Table 1).

**Table 1.** Comparison of herbage growth and herbage use efficiency of two tall fescue swards maintained at similar LAI in a continuous stocked system and receiving either a non-limiting N supply (N2) or a deficient N supply (N1). After [27].

| | Herbage Growth Rate ($kg \cdot OM \cdot ha^{-1} \cdot day^{-1}$) | Stocking Density ($Sheep \cdot ha^{-1}$) | Defoliation Interval (Days) | Herbage Use Efficiency (%) |
|---|---|---|---|---|
| N1 | 56 | 28 | 28.4 | 57 |
| N2 | 76 | 36 | 20.5 | 73 |

OM, organic matter.

Similar results have been obtained in natural pastures in the Pampa (Argentina) for several species where HUE increased from 50% to 67% after a supply of 100 Kg N in spring [29,119]. Thus it is possible to generalize these results and to conclude that in continuously grazed swards maintained at constant LAI, any increase in leaf tissue production, whatever the cause (temperature, nutrient or water supply, genetics) will lead to a further increase in stocking density, and accordingly to an increase in herbage use efficiency [28]. The magnitude of such an effect should be dependent on the leaf life span of the grass species in the sward. Species with a short leaf life span should be more responsible for a decrease in stocking density than species with a longer leaf life span. As a consequence, maximizing herbage use efficiency by using a high frequency of defoliation for a given leaf life span leads to a decrease in leaf litter production [16,118], thus reducing the return of C to soil. This illustrates the conflict between optimizing herbage harvested by grazing animals and accumulation of soil C, as demonstrated in [120].

In intermittently grazed swards, the leaf defoliation interval is highly disconnected from stocking density and herbage growth. The interval of defoliation is determined by the frequency of moving animals from paddock to paddock and depends mainly on farmer decisions based on animal number, paddock size and paddock number. In such a system, the instantaneous stocking density is high enough for the great majority of tillers to be defoliated several times within a short time interval of a few days. Therefore, the efficiency of herbage utilization will be determined by grazing pressure, *i.e.*, the quantity of leaf tissues offered per animal and per day. The interval of defoliation is then determined by the rest period (number of days between two successive grazing events). If the rest period remains shorter than the average leaf lifespan of the grass species considered, then the efficiency of herbage utilization will be optimized, but if the rest period becomes longer than the leaf life span, or/and to a lesser extent if the residual herbage mass after grazing is increased, then a greater proportion of leaf tissue situated within the sward horizon accessible to animals will die before the next grazing period, and the efficiency of herbage utilization will decline.

When intermittent and continuous gazing systems are compared, similar quantities of herbage intake and similar herbage utilization efficiency are obtained when the conditions for herbage growth are favorable (spring period, high level of N supply and no water deficit; [13]), and animal performance is roughly equivalent. But when herbage growth becomes limited by environmental conditions, the herbage use efficiency of continuously grazed swards declines as a result of the associated decline in stocking density and then defoliation frequency (as explained previously), while this relationship between stocking density and defoliation frequency can be disrupted in rotational grazing through the adoption by the farmer of an adequate rest period for maintaining the defoliation interval as constant as possible relative to the leaf life span of the dominant species. Thus, in extensive grazing systems, the use of an intermittent grazing system should be more efficient than continuously stocked ones.

## 5. Conclusions

Although it has been recognized from the beginning of the 1970s–1980s that swards are able to adapt their structure to defoliation, particularly in terms of tiller size and tiller density, thus allowing maintenance of productivity to a certain extent within a range of LAI, the underlying ecophysiological mechanisms were largely unclear. The present review shows that since then, our understanding of these mechanisms has progressed to a large extent, at least for several grass species. The involvement of a large number of regulating factors related to plant physiology (N and C reserves, plant morphology, structure of meristems) as well as environment (light signals) is highlighted. The contribution from responses of multiple plant morphogenetic components is also demonstrated. This complexity of regulating factors and plant morphological responses necessitates the development of an integrated comprehensive approach. Modeling methodologies have been developped during recent years. Several descriptive morphological models have been proposed [29,84,119,121,122]. Additionally, several structure-function virtual modeling approaches have been proposed more recently [106,123–127]. These modeling approaches are valuable tools to integrate current knowledge and improve our understanding of the complex nature of the regulation mechanisms involved.

Most of the present review reports data available for temperate grasses. Very similar information on the role of C and N reserves is also available for lucerne and white clover. Responses to light signals have also been described for white clover [36,124,128–130]. However the role of light signals is almost undescribed in other dicots. Therefore, dicot species deserve particular attention in future research.

**Acknowledgments:** The present work has been funded by Institut National de la Recherche Agronomique, France.

**Author Contributions:** François Gastal is the main contributor for Sections 1–3 and 5 of this review paper. Gilles Lemaire is the main contributor for Section 4.

## References

1. Jackson, D.K. The influence of patterns of defoliation on sward morphology. In *Pasture Utilization by the Grazing Animals*; Hodgson, J., Jackson, D.K., Eds.; British Grassland Society: Hurley, UK, 1976; pp. 51–60.
2. Hodgson, J.; Bircham, J.S.; Grant, S.A.; King, J. The influence of cutting and grazing management on herbage growth and utilization. In *Plant Physiology and Herbage Production*; Occasional Symposium of the British Grassland Society; Wright, C.E., Ed.; British Grassland Society: Hurley, UK, 1981; pp. 51–62.
3. Davies, A. The regrowth of grass swards. In *The Grass Crop: The Physiological Basis of Production*; Jones, M.B., Lazenby, A., Eds.; Chapman and Hall: London, UK, 1988; pp. 85–127.
4. Hodgson, J.; Capriles, J.M.; Fenlon, J.S. The influence of sward characteristics on the herbage intake of grazing calves. *J. Agric. Sci. Camb.* **1977**, *89*, 743–750. [CrossRef]
5. Wade, M.H.; Peyraud, J.L.; Lemaire, G.; Cameron, E.A. The dynamics of daily area and depth of grazing and herbage intake of cows in a five day paddock system. In Proceedings of the XVIth International Grassland Congress, Nice, France, 4–11 October 1989; Association Française pour la Production Fourragère: Versailles, France, 1989; pp. 1111–1112.
6. Wade, M.H. Factors Affecting the Availability of Vegetative *Lolium perenne* to Grazing Dairy Cows with Special Reference to Sward Characteristics, Stocking Rate and Grazing Method. Ph.D. Thesis, Rennes University, Rennes, France, March 1991.
7. Peyraud, J.L.; Gonzalez-Rodrigez, A. Relations between grass production, supplementation and intake in grazing dairy cows. In Grassland Farming. Balancing Environmental and Economic Demands, Proceedings of the 18th General Meeting of the European Grassland Federation, Aalborg, Denmark, 22–25 May 2000; Soegaard, K., Ohlsson, C., Sehested, J., Hutchings, N.J., Kristensen, T., Eds.; British Grassland Society: Reading, UK, 2000; pp. 269–282.
8. Prache, S.; Peyraud, J.L. Foraging behaviour and intake in temperate cultivated grasslands. In Proceedings of the 19th International Grassland Congress, Sao Pedro, Brazil, 11–21 February 2001; Gomide, J.A., Mattos, W.R.S., DaSilva, S.C., Eds.; FEALQ: Piracicaba, Brazil, 2001; pp. 309–319.
9. Delagarde, R.; Prache, S.; D'Hour, P.; Petit, M. Ingestion de l'herbe par les ruminants au pâturage. *Fourrages* **2001**, *166*, 189–212.
10. Peyraud, J.L.; Comeron, E.A.; Wade, M.H.; Lemaire, G. The effect of daily herbage allowance, herbage mass and animal factors upon herbage intake by grazing dairy cows. *Ann. Zootech.* **1996**, *45*, 201–217. [CrossRef]
11. Hodgson, J. Influence of sward characteristics on diet selection and herbage intake by the grazing animal. In *Nutritional Limits to Animal Production from Pasture*; Hacker, J.B., Ed.; CAB International: Wallingford, UK, 1982; pp. 153–166.
12. Barthram, G.T.; Grant, S.A. Defoliation of ryegrass-dominated swards by sheep. *Grass Forage Sci.* **1984**, *39*, 211–219. [CrossRef]
13. Parsons, A.J.; Johnson, I.R.; Harvey, A. Use of a model to optimize the interaction between frequency and severity of intermittent defoliation and to provide a fundamental comparison of the continuous and intermittent defoliation of grass. *Grass Forage Sci.* **1988**, *43*, 49–59. [CrossRef]
14. Briske, D.D. Strategies of plant survival in grazed systems: A functional interpretation. In *The Ecology and Management of Grazing Systems*; Hodgson, J., Illius, A.W., Eds.; CABI International: Wallingford, UK, 1996; pp. 37–67.
15. Hodgson, J.; Da Silva, S.C. Sustainability of grazing systems: Goals, concepts and methods. In *Grassland Ecophysiology and Grazing Ecology*; Lemaire, G., Hodgson, J., de Moraes, A., de Carvalho, P.C.F., Nabinger, C., Eds.; CABI Publishing: Wallingford, UK, 2000; pp. 1–13.
16. Parsons, A.J. The effects of season and management on the growth of grass swards. In *The Grass Crop: The Physiological Basis of Production*; Jones, M.B., Lazenby, A., Eds.; Chapman and Hall: London, UK, 1988; pp. 129–177.

17. Chapman, D.F.; Lemaire, G. Morphogenetic and structural determinants of plant regrowth after defoliation. In Proceedings of the XVII International Grassland Congress, Palmerston North, New Zealand, 8–21 February 1993; New Zealand Grassland Association: Palmerston North, New Zealand, 1993; pp. 95–104.
18. Lemaire, G.; Chapman, D. Tissue flows in grazed communities. In *The Ecology and Management of Grazing Systems*; Hodgson, J., Ed.; Centre for Agriculture and Biosciences International: Wallingford, UK, 1996; pp. 3–36.
19. Davies, A. Tissue turnover in the sward. In *Sward Measurement Handbook*, 2nd ed.; Hodgson, J., Baker, R.D., Davies, A., Laidlaw, A.S., Leaver, J.D., Eds.; The British Grassland Society Publication: Reading, UK, 1981; pp. 183–215.
20. Grant, S.A.; Barthram, G.T.; Torvell, L.; King, J.; Smith, H.K. Sward management, lamina turnover and tiller population density in continuously stocked *Lolium perenne*-dominated swards. *Grass Forage Sci.* **1983**, *38*, 333–344. [CrossRef]
21. Westoby, M. The self-thinning rule. *Adv. Ecol. Res.* **1984**, *14*, 167–225.
22. Sackville-Hamilton, N.R.; Matthew, C.; Lemaire, G. Self-thinning: A re-evaluation of concepts and status. *Ann. Bot.* **1995**, *7*, 569–577. [CrossRef]
23. Matthew, C.; Lemaire, G.; Sackville-Hamilton, N.R.; Hernandez-Garay, A. A modified self-thinning equation to describe size: Density relationships for defoliated swards. *Ann. Bot.* **1995**, *7*, 579–587. [CrossRef]
24. Hernandez-Garay, A.; Matthew, C.; Hodgson, J. Tiller size/density compensation in perennial ryegrass miniature swards subject to differing defoliation heights and a proposed productivity index. *Grass Forage Sci.* **1999**, *54*, 347–356. [CrossRef]
25. Matthew, C.; Assuero, S.G.; Black, C.K.; Sackville-Hamilton, N.R. Tiller dynamics in grazed swards. In *Grassland Ecophysiology and Grazing Ecology*; Lemaire, G., Hodgson, J., de Moraes, A., de Carvalho, P.C.F., Nabinger, C., Eds.; CABI Publishing: Wallingford, UK, 2000; pp. 127–150.
26. Hodgson, J.; Ollerenshaw, J.H. The frequency and severity of defoliation of individual tillers in set-stocked swards. *Grass Forage Sci.* **1969**, *24*, 226–234. [CrossRef]
27. Mazzanti, A.; Lemaire, G. Effect of nitrogen fertilization on herbage production of tall fescue swards continuously grazed by sheep. 2. Consumption and efficiency of herbage utilization. *Grass Forage Sci.* **1994**, *49*, 352–359. [CrossRef]
28. Lemaire, G.; Da Silva, S.C.; Agnusdei, M.; Wade, M.; Hodgson, J. Interactions between leaf lifespan and defoliation frequency in temperate and tropical pastures: A review. *Grass Forage Sci.* **2009**, *64*, 341–353. [CrossRef]
29. Lemaire, G.; Agnusdei, M. Leaf tissue turnover and efficiency of herbage utilization. In *Grassland Ecophysiology and Grazing Ecology*; Lemaire, G., Hodgson, J., de Moraes, A., de Carvalho, P.C.F., Nabinger, C., Eds.; CABI Publishing: Wallingford, UK, 2000; pp. 265–287.
30. Dale, J.E. Some effects of temperature and irradiance on growth of the first four leaves of wheat *Triticum aestivum*. *Ann. Bot.* **1982**, *50*, 851–858.
31. Kasperbauer, M.J. Spectral distribution of light in a tobacco canopy and effects of end-of-day light quality on growth and development. *Plant Physiol.* **1971**, *47*, 775–778. [CrossRef] [PubMed]
32. Smith, H. The ecological functions of the phytochrome family. Clues to a transgenic program of crop improvement. *Photochem. Photobiol.* **1992**, *56*, 815–822. [CrossRef]
33. Smith, H. Physiological and ecological function within the phytochrome family. *Ann. Rev. Plant Biol.* **1995**, *46*, 289–315. [CrossRef]
34. Smith, H. Phytochromes and light signal perception by plants—An emerging synthesis. *Nature* **2000**, *407*, 585–591. [CrossRef] [PubMed]
35. Ballaré, C.L.; Sanchez, R.A.; Scopel, A.L.; Casal, J.J.; Ghersa, C.M. Early detection of neighbour plants by phytochrome perception of spectral changes in reflected sunlight. *Plant Cell Environ.* **1987**, *10*, 551–557.
36. Robin, C.; Varlet-Grancher, C.; Gastal, F.; Flenet, F.; Guckert, A. Photomorphogenesis of white clover (*Trifolium repens* L.): Phytochrome mediated effects on 14C-assimilate partitioning. *Eur. J. Agron.* **1992**, *1*, 235–240. [CrossRef]
37. Ballaré, C.; Casal, J. Light signals perceived by crop and weed plants. *Field Crops Res.* **2000**, *67*, 149–160. [CrossRef]
38. Kebrom, T.; Brutnell, T. The molecular analysis of the shade avoidance syndrome in the grasses has begun. *J. Exp. Bot.* **2007**, *58*, 3079–3089. [CrossRef] [PubMed]

39. Taiz, L.; Zeiger, E. Growth and development. In *Plant Physiology*; Sinauer Associates Inc.: Sunderland, MA, USA, 1998; pp. 483–517.
40. Ballaré, C.; Scopel, A.; Sanchez, R. Photocontrol of stem elongation in plant neighborhoods—Effects of photon fluence rate under natural conditions of radiation. *Plant Cell Environ.* **1991**, *14*, 57–65. [CrossRef]
41. Gautier, H.; Varlet-Grancher, C.; Baudry, N. Effects of blue light on the vertical colonization of space by white clover and their consequences for dry matter distribution. *Ann. Bot.* **1997**, *80*, 665–671. [CrossRef]
42. Aphalo, P.; Ballaré, C.; Scopel, A. Plant-plant signalling, the shade-avoidance response and competition. *J. Exp. Bot.* **1999**, *50*, 1629–1634. [CrossRef]
43. Christophe, A.; Moulia, B.; Varlet-Grancher, C. Quantitative contributions of blue light and PAR to the photocontrol of plant morphogenesis in *Trifolium repens* (L.). *J. Exp. Bot.* **2006**, *57*, 2379–2390. [CrossRef] [PubMed]
44. Barillot, R.; Frak, E.; Combes, D.; Durand, J.L.; Escobar-Gutierrez, A.J. What determines the complex kinetics of stomatal conductance under blueless PAR in *Festuca arundinacea*? Subsequent effects on leaf transpiration. *J. Exp. Bot.* **2010**, *61*, 2795–2806. [CrossRef] [PubMed]
45. Davies, A. Leaf tissue remaining after cutting and regrowth in perennial ryegrass. *J. Agric. Sci.* **1974**, *82*, 165–172. [CrossRef]
46. Hume, D.E. Effect of cutting on production and tillering in prairie grass (*Bromus willdenowii* Kunth) compared with two ryegrass (*Lolium*) species. 1. Vegetative plants. *Ann. Bot.* **1991**, *67*, 533–541.
47. Van Loo, E.N. On the Relation between Tillering, Leaf Area Dynamics and Growth of Perennial Ryegrass (*Lolium perenne* L.). Ph.D. Thesis, Wageningen University, Wageningen, The Netherlands, May 1993.
48. Grant, S.A.; Barthram, G.T.; Torvell, L. Components of regrowth in grazed and cut *Lolium perenne* swards. *Grass Forage Sci.* **1981**, *36*, 155–168. [CrossRef]
49. Gautier, H.; Varlet-Grancher, C.; Hazard, L. Tillering responses to the light environment and to defoliation in populations of perennial ryegrass (*Lolium perenne* L.) selected for contrasting leaf length. *Ann. Bot.* **1999**, *83*, 423–429. [CrossRef]
50. Hernandez-Garay, A.; Matthew, C.; Hodgson, J. The influence of defoliation height on dry-matter partitioning and $CO_2$ exchange of perennial ryegrass miniature swards. *Grass Forage Sci.* **2000**, *55*, 372–376. [CrossRef]
51. Bos, H.J.; Neuteboom, J.H. Morphological analysis of leaf and tiller number dynamics of wheat (*Triticum aestivum* L.): Responses to temperature and light intensity. *Ann. Bot.* **1998**, *81*, 131–139. [CrossRef]
52. Thomas, H.; Norris, I.A. The influence of light and temperature during winter on growth and death in simulated swards of *Lolium perenne*. *Grass Forage Sci.* **1981**, *36*, 107–116. [CrossRef]
53. Robson, M.J.; Ryle, G.J.A.; Woledge, J. The grass plant—Its form and function. In *The Grass Crop: The Physiological Basis of Production*; Jones, M.B., Lazenby, A., Eds.; Chapman and Hall: London, UK, 1988; pp. 25–83.
54. Berone, G.D.; Lattanzi, F.A.; Agnusdei, M.G.; Bertolotti, N. Growth of individual tillers and tillering rate of Lolium perenne and Bromus stamineus subjected to two defoliation frequencies in winter in Argentina. *Grass Forage Sci.* **2008**, *63*, 504–512. [CrossRef]
55. Gastal, F.; Verdenal, A.; Barre, P. The effect of blue light on leaf growth and plant development in two morphologically contrastd perennial ryegrass genotypes: Cellular basis and ecological implications. In XXth International Grassland Congress, Dublin, Ireland, 26 June–1 July 2005; O'Mara, F.P., Wilkins, R.J., 't Mannetje, L., Lovett, D.K., Rogers, D.A.M., Boland, T.M., Eds.; Wageningen Academic Publishers: Wageningen, Netherlands, 2005; p. 202.
56. Robson, M.J. The effect of temperature on the growth of S.170 tall fescue (*Festuca arundinacea*). I. Constant temperature. *J. Appl. Ecol.* **1972**, *9*, 643–653. [CrossRef]
57. Miglietta, F. Effect of photoperiod and temperature on leaf initiation rates in wheat (*Triticum* spp.). *Field Crops Res.* **1989**, *21*, 121–130. [CrossRef]
58. Skinner, R.H.; Nelson, C.J. Estimation of potential tiller production and site usage during tall fescue canopy development. *Ann. Bot.* **1992**, *70*, 493–499.
59. Skinner, R.H.; Nelson, C.J. Role of leaf appearance rate and the coleoptile tiller in regulating tiller production. *Crop Sci.* **1994**, *34*, 71–75. [CrossRef]
60. Billiard, D. Rythme de développement des feuilles sur une talle de graminée tropicale (*Hyparrhenia diplandra*): 1. Résultats expérimentaux. *Acta Oecol.* **1994**, *15*, 577–591.

61. Lestienne, F.; Gastal, F.; Moulia, B.; Thornton, B. Pattern of leaf and tiller development of perennial ryegrass plants. In Multi-Function Grasslands: Quality Forages, Animal Products and Landscapes, Proceedings of the 19th General Meeting of the European Grassland Federation, La Rochelle, France, 27–30 May 2002; Durand, J.L., Emile, J.C., Huyghe, C., Lemaire, G., Eds.; Association Française pour la Production Fourragère: Versailles, France, 2002; pp. 332–333.
62. Bélanger, G. Incidence de la Fertilisation Azotée et de la Saison sur la Croissance, L'assimilation et la Répartition du Carbone dans un Couvert de Fétuque Elevée en Conditions Naturelles. Ph.D. Thesis, Paris XI University, Orsay, France, October 1990.
63. Duru, M.; Ducrocq, H. Growth and senescence of the successive grass leaves on a tiller. Ontogenic development and effect of temperature. *Ann. Bot.* **2000**, *85*, 635–643. [CrossRef]
64. Davies, A.; Evans, M.E.; Exley, J.K. Regrowth of perennial ryegrass as affected by simulated leaf sheaths. *J. Agric. Sci.* **1983**, *101*, 131–137. [CrossRef]
65. Skinner, R.H.; Morgan, J.A.; Hanson, J.D. Carbon and nitrogen reserve remobilization following defoliation: Nitrogen and elevated $CO_2$ effects. *Crop Sci.* **1999**, *39*, 1749–1756. [CrossRef]
66. Schnyder, H.; Schaüfele, R.; de Visser, R.; Nelson, C.J. An integrated view of C and N uses in leaf growth zones of defoliated grasses. In *Grassland Ecophysiology and Grazing Ecology*; Lemaire, G., Hodgson, J., de Moraes, A., de Carvalho, P.C.F., Nabinger, C., Eds.; CABI Publishing: Wallingford, UK, 2000; pp. 41–60.
67. De Visser, R.; Vianden, H.; Schnyder, H. Kinetics and relative significance of remobilized and current C and N incorporation in leaf and root growth zones of *Lolium perenne* after defoliation: Assessment by $^{13}C$ and $^{15}N$ steady-state labelling. *Plant Cell Environ.* **1997**, *20*, 37–46. [CrossRef]
68. Schnyder, H.; Nelson, C.J. Growth rates and assimilate partitioning in the elongation zone of tall fescue leaf blades at high and low irradiance. *Plant Physiol.* **1989**, *90*, 1201–1206. [CrossRef] [PubMed]
69. Allard, G.; Nelson, C.J.; Pallardy, S.G. Shade effects on growth of tall fescue: I. Leaf anatomy and dry matter partitioning. *Crop Sci.* **1991**, *31*, 163–171. [CrossRef]
70. Volenec, J.J.; Nelson, C.J. Carbohydrate metabolism in leaf meristems of tall fescue. 2. Relationship to leaf elongation rates modified by nitrogen fertilization. *Plant Physiol.* **1984**, *74*, 595–600. [CrossRef] [PubMed]
71. Schnyder, H.; Nelson, C.J. Growth rates and carbohydrate fluxes within the elongation zone of tall fescue leaf blades. *Plant Physiol.* **1987**, *85*, 548–553. [CrossRef] [PubMed]
72. Gastal, F.; Nelson, C.J. Nitrogen use within the growing leaf blade of tall fescue. *Plant Physiol.* **1994**, *105*, 191–197. [PubMed]
73. Schaüfele, R.; Schnyder, H. Cell growth analysis during steady and non-steady growth in leaves of perennial ryegrass (*Lolium perenne* L.) subject to defoliation. *Plant Cell Environ.* **2000**, *23*, 185–194. [CrossRef]
74. Morvan-Bertrand, A.; Boucaud, J.; Le Saos, J.; Prud'homme, M.P. Roles of the fructans from leaf sheaths and from the elongating leaf bases in the regrowth following defoliation of *Lolium perenne* L. *Planta* **2001**, *213*, 109–120. [CrossRef] [PubMed]
75. Davidson, J.L.; Milthorpe, F.L. The effect of defoliation on the carbon balance in *Dactylis glomerata*. *Ann. Bot.* **1966**, *30*, 185–198.
76. Ryle, G.J.A.; Powell, C.E. Defoliation and regrowth in the gramineous plant: The role of current assimilates. *Ann. Bot.* **1975**, *39*, 297–310.
77. Pollock, C.J.; Cairns, A.J. Fructan metabolism in grasses and cereals. *Ann. Rev. Plant Biol.* **1991**, *42*, 77–101. [CrossRef]
78. Ourry, A.; Boucaud, J.; Duyme, M. Sink control of nitrogen uptake and assimilation during regrowth after-cutting of ryegrass (*Lolium perenne* L.). *Plant Cell Environ.* **1990**, *13*, 185–189. [CrossRef]
79. Thornton, B.; Millard, P. Increased defoliation frequency depletes remobilization of nitrogen for leaf growth in grasses. *Ann. Bot.* **1997**, *80*, 89–95. [CrossRef]
80. Morvan-Bertrand, A.; Boucaud, J.; Prud'homme, M.P. Influence of initial levels of carbohydrates, fructans, nitrogen, and soluble proteins on regrowth of *Lolium perenne* L.cv. Bravo following defoliation. *J. Exp. Bot.* **1999**, *50*, 1817–1826. [CrossRef]
81. Thornton, B.; Millard, P.; Bausenwein, U. Reserve formation and recycling of carbon and nitrogen during regrowth of defoliated plants. In *Grassland Ecophysiology and Grazing Ecology*; Lemaire, G., Hodgson, J., de Moraes, A., de Carvalho, P.C.F., Nabinger, C., Eds.; CABI Publishing: Wallingford, UK, 2000; pp. 85–99.
82. Lestienne, F.; Thornton, B.; Gastal, F. Impact of defoliation intensity and frequency on N uptake and mobilization in Lolium perenne. *J. Exp. Bot.* **2006**, *57*, 997–1006. [CrossRef] [PubMed]

83. Avice, J.C.; Ourry, A.; Lemaire, G.; Boucaud, J. Nitrogen and carbon flows estimated by $^{15}N$ and $^{13}C$ pulse-chase labelling during regrowth of alfalfa. *Plant Physiol.* **1996**, *112*, 282–290.
84. Richards, J.H. Physiology of plants recovering from defoliation. In Proceedings of the XVII International Grassland Congress, Palmerston North, New Zealand, 8–21 February 1993; New Zealand Grassland Association: Palmerston North, New Zealand, 1993; pp. 85–94.
85. Schnyder, H.; de Visser, R. Fluxes of reserve-derived and currently assimilated carbon and nitrogen in perennial ryegrass recovering from defoliation. The regrowing tiller and its component functionally distinct zones. *Plant Physiol.* **1999**, *119*, 1423–1435. [CrossRef] [PubMed]
86. Morvan-Bertrand, A.; Pavis, N.; Boucaud, J.; Prud'homme, M.P. Partitioning of reserve and newly assimilated carbon in roots and leaf tissues of *Lolium perenne* during regrowth after defoliation: Assessment by $^{13}C$ steady-state labelling and carbohydrate analysis. *Plant Cell Environ.* **1999**, *22*, 1097–1108. [CrossRef]
87. Lee, J.M.; Donaghy, D.J.; Sathish, P.; Roche, J.R. Interaction between water-soluble carbohydrate reserves and defoliation severity on the regrowth of perennial ryegrass (*Lolium perenne* L.)-dominant swards. *Grass Forage Sci.* **2009**, *64*, 266–275. [CrossRef]
88. Donaghy, D.J.; Turner, L.R.; Adamczewski, K.A. Effect of defoliation management on water-soluble carbohydrate energy reserves, dry matter yields, and herbage quality of tall fescue. *Agron. J.* **2008**, *100*, 122–127. [CrossRef]
89. Turner, L.R.; Donaghy, D.J.; Lane, P.A.; Rawnsley, R.P. Effect of defoliation management, based on leaf stage, on perennial ryegrass (*Lolium perenne* L.), prairie grass (*Bromus willdenowii* Kunth.) and cocksfoot (*Dactylis glomerata* L.) under dryland conditions. 1. Regrowth, tillering and water-soluble carbohydrate concentration. *Grass Forage Sci.* **2006**, *61*, 164–174.
90. Turner, L.R.; Donaghy, D.J.; Lane, P.A.; Rawnsley, R.P. Patterns of leaf and root regrowth, and allocation of water-soluble carbohydrate reserves following defoliation of plants of prairie grass (*Bromus willdenowii* Kunth.). *Grass Forage Sci.* **2007**, *62*, 497–506. [CrossRef]
91. Mazzanti, A.; Lemaire, G.; Gastal, F. The effect of nitrogen fertilisation upon the herbage production of tall fescue swards continuously grazed with sheep. 1. Herbage growth dynamics. *Grass Forage Sci.* **1994**, *49*, 111–120. [CrossRef]
92. Gastal, F.; Dawson, L.; Thornton, B. N use efficiency of perennial grasses is affected by their tolerance to defoliation. In Proceedings of the 19th General Meeting of the European Grassland Federation, La Rochelle, France, 27–30 May 2002; Durand, J.L., Emile, J.C., Huyghe, C., Lemaire, G., Eds.; Association Française pour la Production Fourragère: Versailles, France, 2002; pp. 296–297.
93. Gastal, F.; Dawson, L.A.; Thornton, B. Responses of plant traits of four grass species from contrasting habitats to defoliation and N supply. *Nut. Cycl. Agroecosyst.* **2010**, *88*, 245–258. [CrossRef]
94. Cullen, B.R.; Chapman, D.F.; Quigley, P.E. Comparative defoliation tolerance of temperate perennial grasses. *Grass Forage Sci.* **2006**, *61*, 405–412. [CrossRef]
95. Casal, J.J.; Sanchez, R.A.; Deregibus, V.A. The effect of light quality on shoot extension growth in three species of grasses. *Ann. Bot.* **1987**, *59*, 1–7.
96. Skinner, R.H.; Simmons, S.R. Modulation of leaf elongation, tiller appearance and tiller senescence in spring barley by far-red light. *Plant Cell Environ.* **1993**, *16*, 555–562. [CrossRef]
97. Gautier, H.; Varlet-Grancher, C. Regulation of leaf growth of grass by blue light. *Physiol. Plant.* **1996**, *98*, 424–430. [CrossRef]
98. Gastal, F.; INRA, Lusignan, France. Unpublished work. 2003.
99. Skinner, R.H.; Nelson, C.J. Effect of tiller trimming on phyllochron and tillering regulation during tall fescue development. *Crop Sci.* **1994**, *34*, 1267–1273. [CrossRef]
100. Casal, J.J.; Alvarez, M.A. Blue light effects on the growth of *Lolium multiflorum* Lam. leaves under natural radiation. *New Phytol.* **1988**, *109*, 41–45. [CrossRef]
101. Bahmani, I.; Hazard, L.; Varlet-Grancher, C.; Betin, M.; Lemaire, G.; Matthew, C.; Thom, E.R. Differences in tillering of long- and short-leaved perennial ryegrass genetic lines under full light and shade treatments. *Crop Sci.* **2000**, *40*, 1095–1102. [CrossRef]
102. Wilson, R.E.; Laidlaw, A.S. The role of the sheath tube in the development of expanding leaves in perennial ryegrass. *Ann. Appl. Biol.* **1985**, *106*, 385–391. [CrossRef]
103. Casey, I.A.; Brereton, A.J.; Laidlaw, A.S.; McGilloway, D.A. Effects of sheath tube length on leaf development in perennial ryegrass (*Lolium perenne* L.). *Ann. Appl. Biol.* **1999**, *134*, 251–257. [CrossRef]

104. Bos, H.J.; Neuteboom, J.H. Growth of individual leaves of spring wheat (*Triticum aestivum* L.) as influenced by temperature and light intensity. *Ann. Bot.* **1998**, *81*, 141–149. [CrossRef]
105. Skinner, R.H.; Nelson, C.J. Elongation of the grass leaf and its relationship to the phyllochrone. *Crop Sci.* **1995**, *35*, 4–10. [CrossRef]
106. Verdenal, A.; Combes, D.; Escobar-Gutierrez, A.J. A study of ryegrass architecture as a self-regulated system, using functional-structural plant modelling. *Funct. Plant Biol.* **2008**, *35*, 911–924. [CrossRef]
107. Neuteboom, J.H.; Lantiga, E.A. Tillering potential and relationship between leaf and tiller production in perennial ryegrass. *Ann. Bot.* **1989**, *63*, 265–270.
108. Skinner, R.H.; Nelson, C.J. Epidermal cell division and the coordination of leaf and tiller development. *Ann. Bot.* **1994**, *74*, 9–15. [CrossRef] [PubMed]
109. Nelson, C.J. Shoot morphological plasticity of grasses: Leaf growth *vs.* Tillering. In *Grassland Ecophysiology and Grazing Ecology*; Lemaire, G., Hodgson, J., de Moraes, A., de Carvalho, P.C.F., Nabinger, C., Eds.; CABI Publishing: Wallingford, UK, 2000; pp. 101–126.
110. Simon, J.C.; Lemaire, G. Tillering and leaf area index in grasses in vegetative phase. *Grass Forage Sci.* **1987**, *42*, 373–380. [CrossRef]
111. Deregibus, V.A.; Sanchez, R.A.; Casal, J.J. Effects of light quality on tiller production in *Lolium* sp. *Plant Physiol.* **1983**, *72*, 900–912. [CrossRef] [PubMed]
112. Casal, J.J.; Deregibus, V.A.; Sanchez, R.A. Variations in tiller dynamics and morphology in *Lolium multiflorum* Lam. vegetative and reproductive plants as affected by differences in red/far-red irradiation. *Ann. Bot.* **1985**, *56*, 553–559.
113. Skalova, H. Morphological plasticity of *Festuca rubra* clones from three neighbouring communities in response to red: Far-red levels. *Folia Geobot.* **2005**, *40*, 77–90. [CrossRef]
114. Evers, J.; Vos, J.; Andrieu, B.; Struik, P. Cessation of tillering in spring wheat in relation to light interception and red: Far-red ratio. *Ann. Bot.* **2006**, *97*, 649–658. [CrossRef] [PubMed]
115. Williamson, M.; Wilson, G.; Hartnett, D. Controls on bud activation and tiller initiation in $C_3$ and $C_4$ tallgrass prairie grasses: The role of light and nitrogen. *Botany* **2012**, *90*, 1221–1228. [CrossRef]
116. Ong, C.J. The physiology of tiller death in grasses. 1. The influence of tiller age, size and position. *J. Br. Grass. Soc.* **1978**, *33*, 197–203. [CrossRef]
117. Bircham, J.S.; Hodgson, J. The influence of sward conditions on rates of herbage growth and senescence in mixed swards under continuous stocking management. *Grass Forage Sci.* **1983**, *38*, 323–331. [CrossRef]
118. Parsons, A.J.; Leafe, E.L.; Colett, B.; Penning, P.D.; Lewis, J. The physiology of grass production under grazing. 2. Photsynthesis, crop growth and animal intake of continuously grazed swards. *J. App. Ecol.* **1983**, *20*, 127–139. [CrossRef]
119. Rodriguez Palma, R. Fertilizacion Nitrogenada de un Pastizal de la Pampa Defirimada: Cecimiento y Utilization del Forraje Bajo Pastoreo de Vacunos. Master's Thesis, Faculty of Agricultural Sciences, University of Mar del Plata, Mar del Plata, Argentina, 1998.
120. Soussana, J.F.; Lemaire, G. Coupling carbon and nitrogen cycles for environmentally sustainable intensification of grasslands and crop-livestock systems. *Agric. Ecosyst. Environ.* **2014**, *190*, 9–17. [CrossRef]
121. Billiard, D. Development rate of leaves along the tiller of a tropical grass (*Hyparrhenia Diplandra*). 2. Modelling. *Acta Oecol.* **1994**, *15*, 715–726.
122. Hirata, M. Linking management, environment and morphogenetic and structural components of a sward for simulating tiller density dynamics in Bahiagrass (*Paspalum notatum*). *Agriculture* **2015**, *5*, 330–343. [CrossRef]
123. Durand, J.L.; Schaufele, R.; Gastal, F. Grass leaf elongation rate as a function of developmental stage and temperature: Morphological analysis and modelling. *Ann. Bot.* **1999**, *83*, 577–588. [CrossRef]
124. Gautier, H.; Mech, R.; Prusinkiewicz, P.; Varlet-Grancher, C. 3D architectural modelling of aerial photomorphogenesis in white clover (*Trifolium repens* L.) using L-systems. *Ann. Bot.* **2000**, *85*, 359–370. [CrossRef]
125. Fournier, C.; Durand, J.L.; Ljutovac, S.; Schaufele, R.; Gastal, F.; Andrieu, B. A functional-structural model of elongation of the grass leaf and its relationships with the phyllochron. *New Phytol.* **2005**, *166*, 881–894. [CrossRef] [PubMed]

126. Fournier, C.; Andrieu, B.; Buck-Sorlinz, G.; Evers, J.B.; Drouet, J.L.; Escobar-Gutierrez, A.J.; Vos, J. Functional-structural modelling of Gramineae. In *Functional-Structural Plant Modelling in Crop Production*; Vos, J., Marcelis, L.F.M., DeVisser, P.H.B., Struik, P.C., Evers, J.B., Eds.; Springer: Dordrecht, The Netherlands, 2007; Volume 2, pp. 175–186.
127. Evers, J.B.; Vos, J.; Chelle, M.; Andrieu, B.; Fournier, C.; Struik, P. Simulating the effects of localized red:far-red ratio on tillering in spring wheat (*Triticum aestivum*) using a three-dimensional virtual plant model. *New Phytol.* **2007**, *176*, 325–336. [CrossRef] [PubMed]
128. Robin, C.; Hay, M.J.M.; Newton, P.C.D.; Greer, D.H. Effect of light quality (red/far-red ratio) at the apical bud of the main stolon on morphogenesis of *Trifolium repens* L. *Ann. Bot.* **1994**, *74*, 119–123. [CrossRef]
129. Robin, C.; Hay, M.J.M.; Newton, P.C.D. Effect of light quality (red/far-red ratio) and defoliation treatments applied at a single phytomer on axillary bud outgrowth in *Trifolium repens* L. *Oecologia* **1994**, *100*, 236–242. [CrossRef]
130. Christophe, A.; Moulia, B.; Varlet-Grancher, C.; Durand, J.L. Red:far red ratio responses of spacer length in two contrasted clones of white clover. In Multi-Function Grasslands: Quality Forages, Animal Products and Landscapes, Proceedings of the 19th General Meeting of the European Grassland Federation, La Rochelle, France, 27–30 May 2002; Durand, J.L., Emile, J.C., Huyghe, C., Lemaire, G., Eds.; Association Française pour la Production Fourragère: Versailles, France, 2002; pp. 288–289.

# 8

# Nutrient Composition, Forage Parameters and Antioxidant Capacity of Alfalfa (*Medicago sativa*, L.) in Response to Saline Irrigation Water

**Jorge F. S. Ferreira *, Monica V. Cornacchione †,‡, Xuan Liu ‡ and Donald L. Suarez ‡**

US Salinity Laboratory, 450 W. Big Springs Rd., Riverside, CA 92507, USA; Xuan.Liu@ars.usda.gov (X.L.); Donald.Suarez@ars.usda.gov (D.L.S.)

* Author to whom correspondence should be addressed; Jorge.Ferreira@ars.usda.gov

† Currently at INTA- Estación Experimental Agropecuaria Santiago del Estero, Jujuy 850, Santiago del Estero 4200, Argentina; cornacchione.monica@inta.gob.ar.

‡ These authors contributed equally to this work.

Academic Editor: Cory Matthew

**Abstract:** Although alfalfa is moderately tolerant of salinity, the effects of salinity on nutrient composition and forage parameters are poorly understood. In addition, there are no data on the effect of salinity on the antioxidant capacity of alfalfa. We evaluated four non-dormant, salinity-tolerant commercial cultivars, irrigated with saline water with electrical conductivities of 3.1, 7.2, 12.7, 18.4, 24.0, and 30.0 $dS \cdot m^{-1}$, designed to simulate drainage waters from the California Central Valley. Alfalfa shoots were evaluated for nutrient composition, forage parameters, and antioxidant capacity. Salinity significantly increased shoot N, P, Mg, and S, but decreased Ca and K. Alfalfa micronutrients were also affected by salinity, but to a lesser extent. Na and Cl increased significantly with increasing salinity. Salinity slightly improved forage parameters by significantly increasing crude protein, the net energy of lactation, and the relative feed value. All cultivars maintained their antioxidant capacity regardless of salinity level. The results indicate that alfalfa can tolerate moderate to high salinity while maintaining nutrient composition, antioxidant capacity, and slightly improved forage parameters, thus meeting the standards required for dairy cattle feed.

**Keywords:** alfalfa; salinity; forage quality; nutrient composition; antioxidant capacity; total phenolics

---

## 1. Introduction

Alfalfa (*Medicago sativa*, L.) is the most cultivated legume worldwide and the fourth most cultivated crop in the United States. Alfalfa is cultivated in most continents and in more than 80 countries occupying more than 35 million ha [1]. In the USA, it is among the top three field crops cultivated in 26 states, thus contributing more than US $10 billion a year to the farm economy, primarily as an animal feed [2]. Alfalfa is considered to be the most important forage crop for providing protein to dairy and beef cattle, sheep, horses, birds, and other livestock [1]. Feeding of alfalfa hay to lactating dairy cows has decreased sharply in the past 10 years, primarily as a result of economic issues associated with high water use, the costs of multiple harvests, and storage [3]. These authors also mentioned the increased use of corn and cereal silages in animal diets to replace alfalfa. However, dry matter intake is significantly higher for cows fed alfalfa and barley silages than for cows fed oat and triticale silages [4]. According to these authors, alfalfa silage contains higher concentrations of all minerals analyzed compared with cereal silages, except for Na. Moreover, the cows also absorbed K better from alfalfa silage (89%) than from cereal silages (74% to 83%). Alfalfa is highly important to livestock considering its fast canopy recovery after each harvest, its relative tolerance of salinity, its capacity to endure temperature extremes (e.g., hot days and cold nights), its nutritional value, and palatability to livestock.

In arid lands, irrigation is necessary for high forage mass production. However, this irrigation is often associated with salinization. Among the approximately 270 million hectares of irrigated land worldwide, about 40% is located in arid/semiarid zones [5] where soil salinization generally occurs. Some of the typical agronomic parameters used to evaluate the salinity tolerance of crops include yield, survival, plant height, and relative growth rate or reduction [6–8]. Few researchers have evaluated alfalfa forage mass production, nutrient composition, and forage parameters for livestock under high salinity stress [9–12]. Further, we found no published reports on the effects of salinity on the antioxidant capacity of alfalfa. It has been reported that salinity stress imposed on a model legume (*Lotus japonicus*) increased antioxidant enzyme levels in leaves [13], and that the expression of genes associated with antioxidant enzymes increased in response to excessive levels of reactive oxygen species (ROS) generated by salinity stress [14]. These authors postulated that these enzymes protect plant tissues from ROS damage triggered by salinity stress, but there are no reports on the biosynthesis of non-enzymatic antioxidants, such as flavonoids and phenolic compounds, by alfalfa in response to salinity. Alfalfa shoots are a rich source of antioxidant flavonoids, mainly apigenin, tricin, luteolin, and chrysoeriol glycosides [15], and of phenolic compounds reported to have anti-inflammatory [16], antioxidant, and neuroprotective activity in mice [17]. The ratio of alfalfa antioxidant flavones acylated with hydroxycinnamic acid to non-acylated (lower antioxidant capacity) flavones increases in summer when plants are exposed to a higher amount of UV-B radiation [15]. Antioxidant flavonoids in *Ligustrum vulgare* were reported to increase under both UV-B and NaCl salinity stress [18]. Thus, although alfalfa is fed to livestock for its high protein content, digestibility, and palatability, there is a scarcity of information on the effects of salinity on alfalfa mineral composition and forage quality, while there is no information on its antioxidant capacity under salinity stress.

In this work, we evaluated four commercial alfalfa cultivars, tolerant to salinity, for their response to salinity when cultivated in outdoor sand tanks and irrigated at six salinity levels with water high in sodium, chloride, and sulfate. The goal of our work was to evaluate the effects of increasing salinity on the mineral nutritional composition, forage quality, and antioxidant capacity of alfalfa shoots.

## 2. Experimental Section

### 2.1. Plant Material and Growth Conditions

Four commercial non-dormant, salinity-tolerant, *Medicago sativa* L. cultivars "Salado", "SW8421S", "SW9215", and "SW9720" (S&W, Fresno, CA, USA, www.swseedco.com) were grown from seeds in 24 outdoor sand tanks from 23 June 2011 to 17 April 2012 at the Salinity Laboratory (USDA-ARS) in Riverside, California. Irrigation water at different levels of electrical conductivity (EC) was applied to four cultivars in a split-plot design. The irrigation water EC (measured in deciSiemens per meter) levels consisted of a control using Riverside tap water (EC = 0.6 $dS{\cdot}m^{-1}$) plus fertilizers (EC = 3.1 $dS{\cdot}m^{-1}$), and treatments of 7.2, 12.7, 18.4, 24.0 and 30.0 $dS{\cdot}m^{-1}$, with four tanks (replicates) per treatment. The tanks measured 82 cm wide by 202 cm long by 85 cm deep. Further details on sowing density per cultivar and irrigation frequency are described elsewhere [19]. Salinity treatments and the irrigation water control (EC of 3.1 $dS{\cdot}m^{-1}$) were designed to simulate the drainage water composition of the Central Valley, CA, with subsequent concentration of salts considering mineral precipitation (calcite and/or gypsum) using the UNSATCHEM model [20], which simulates typical soil water interactions. All reservoirs had modified Hoagland's solution, and added $Na^+$, $SO_4^{2-}$, and $Cl^-$ (including control water) to reach the target EC; the detailed composition is described elsewhere [19]. The composition of Riverside tap water (EC = 0.6 $dS{\cdot}m^{-1}$) in $mmolc{\cdot}L^{-1}$ was: 3.4 $Ca^{2+}$, 0.8 $Mg^{2+}$, 1.6 $Na^+$, 0.1 $K^+$, 1.3 $SO_4^{2-}$, 0.8 $Cl^-$, and 0.49 $NO_3^-$. The water composition of all the treatment waters is shown in Table 1.

### 2.2. Plant Growth and Nutrient Composition

Growth and forage mass measurements were collected at seven harvest dates except for the plants that were irrigated with water with an EC = 24.0 dS·m$^{-1}$, which were harvested three times (4th, 6th, and 7th harvests) during the 299 days of cultivation and are presented elsewhere [19]. For this work, we present data on ionic and nutrient composition at 84 days after seeding (DAS) (2nd harvest, on 15 September 2011) and at 299 DAS (7th harvest, on 17 April 2012). The second harvest was conducted when the control plants were at the early flowering stage, corresponding to morphological stage 5 [21]. The seventh harvest was conducted when the control plants were at a late vegetative stage (due to the absence of flowering). The shoot fresh and dry weights (dried at 60 °C for 48 h) were recorded at each harvest and all plants were cut back to 5–8 cm above the sand surface.

**Table 1.** Chemical composition of the water used in the six salinity treatments in this study. EC, electrical conductivity of irrigation water that defines each salinity level (in deciSiemen per meter); mmolc·L$^{-1}$, millimole of charge of each cation or anion listed.

| Treatment<br>EC (dS·m$^{-1}$) | 1<br>3.1 | 2<br>7.2 | 3<br>12.7 | 4<br>18.4 | 5<br>24.0 |
|---|---|---|---|---|---|
| Ion Concentration in mmolc·L$^{-1}$ | | | | | |
| $Ca^{2+}$ | 6.4 | 19.2 | 25.0 | 29.4 | 28.4 |
| $Mg^{2+}$ | 4.0 | 14.3 | 24.1 | 40.7 | 58.5 |
| $Na^{+}$ | 15.5 | 54.2 | 101 | 169 | 229 |
| $K^{+}$ | 6.4 | 6.4 | 6.2 | 6.4 | 6.6 |
| $SO_4^{2-}$ | 15.3 | 53.3 | 85.0 | 132 | 182 |
| $Cl^{-}$ | 8.0 | 31.8 | 62.9 | 104 | 133 |
| $PO_4^{3-}$ | 0.3 | 0.3 | 0.3 | 0.4 | 0.5 |
| $NO_3^{-}$ | 5.5 | 5.6 | 5.5 | 6.0 | 6.0 |

All salinity levels had the following added nutrients, (in mmolc·L$^{-1}$): 0.3 $KH_2PO_4$, 5.0 $KNO_3$, 3.1 $MgSO_4.7H_2O$, 3.0 $CaCl_2$, and 1.0 KCl. Table modified from [19]. Highest salinity level (30 dS·m$^{-1}$) not shown as all plants died at this level.

The levels of the macronutrients N, P, K, Ca, Mg, and total S, and of the micronutrients Fe, Cu, Mn, Zn, and Mo were determined from nitric acid digestions of the dried and ground plant material using Inductively Coupled Plasma Optical Emission Spectrometry (ICP-OES, 3300DV, Perkin-Elmer Corp., Waltham, MA, USA). There was insufficient plant material to analyze samples from the EC = 24 dS·m$^{-1}$ treatment at 84 DAS, and there are no data from the EC = 30 dS·m$^{-1}$ treatment as all plants died at this salinity level.

### 2.3. Oxygen Radical Absorbance Capacity (ORAC) and Total Phenolics (TP) Analyses

Ground dried samples (0.5 g) of alfalfa tops were mixed with 5 g of sand. Each mixture was then extracted in a pressurized stainless steel cell (ASE 350, Thermo Scientific/Dionex, Sunnyvale, CA, USA) using hexane to extract the lipophilic fraction and acetone:water:acetic acid (70:29.5:0.5 by volume) for the hydrophilic fraction. The extraction time was 5 min, followed by a 100% flush, a 60-s purge with 2 cycles, at 80 °C and 1500 psi. The hexane extract was evaporated to dryness with nitrogen in an evaporator (N-EVAP, Organomation, Berlin, MA, USA) at 37 °C and then redissolved in 10 mL of pure acetone; a 50-µL aliquot was collected for dilution and lipophilic ORAC analysis. After extraction with aqueous acetone by the ASE 350, the samples were made up to a volume of 25 mL in the acetone-water-acetic acid solution. A 150-µL aliquot of the aqueous acetone extracts was diluted for hydrophilic ORAC analysis. The ORAC assay is based on the inhibition of the peroxyl-radical-induced oxidation initiated by thermal decomposition of azo-compounds such as [2,2′-azobis(2-amidino-propane) dihydrochloride (AAPH)] [22]. Samples were analyzed for their antioxidant capacity (ORAC) in triplicate. The same ASE 350 aqueous acetone extracts were used for quantification of TP according to the Folin-Ciocalteu method [23,24] using gallic acid (cat. No. 398225,

Sigma-Aldrich, Saint Louis, MO, USA) as the standard. A 20-μL aliquot of the extracts or a gallic acid standard solution was pipetted into a cell of a 96-cell microplate, followed by the addition of 100 μL of 0.4 N Folin Ciocalteu phenol reagent (stock solution F9252, Sigma-Aldrich, Saint Louis, MO, USA) and the addition of 80 μL of 0.94 M $Na_2CO_3$. The plate was covered with a plastic plate cover and allowed to develop color for 5 min at 50 °C. The absorbance was read at 765 nm using a microplate spectrophotometer (xMark™, BIO-RAD, Hercules, CA, USA).

### *2.4. Forage Quality*

Shoots were dried at 60 °C for 48 h. Samples were ground to a size of 1.0 mm and analyzed for acid detergent fiber (ADF), neutral detergent fiber (NDF), and moisture by an independent laboratory (Analytical Feed & Food Laboratory, Visalia, CA, USA), according to AOAC International Methodology [25]. The parameters and analytical methods used were AOAC 973.18 for ADF, AOAC 2002.04 for NDF, and AOAC 930.15 for moisture. The parameters calculated according to ADF, NDF, and/or moisture include the net energy for lactation (NEL), calculated as NEL = 0.8611 − (0.00835 × ADF); relative feed value (RFV), calculated as RFV = (DMD × DMI)/1.29; dry matter intake (DMI), calculated as DMI = 120/NDF; and dry matter digestibility (DMD), calculated as DMD = 88.9 – (0.779 × ADF), according to National Forage Testing Association [26]. Crude protein (CP) was estimated as N% × 6.25 [27]. Nitrogen was determined by sample combustion in pure oxygen and measured by thermal conductivity detection (AOAC, 2000; ID 990.03) using a Vario Pyro Cube® (Elementar Americas, Inc., Mt. Laurel, NJ, USA).

### *2.5. Statistical Analysis*

The nutrient composition data for each harvest were analyzed using a split-plot procedure, with the following statistical model:

$$Y_{ijk} = \mu + S_j + R_i + C_k + (SC)_{jk} + \varepsilon_{ijk}$$

where R, S and C represent the replicates (i = 1, . . . 4), salinity level (j = 1, . . . 5), and cultivars (k = 1, . . . 4) respectively. All effects were considered as fixed. Thus, $Y_{ijk}$ is the response to replicate i in $S_j$ and $C_k$, μ is the overall mean; and $\varepsilon_{ijk}$ represents the random error. The significance in the split-plot design was calculated by deriving the mean squares in the analysis of variance using the InfoStat program [28] with a completely randomized design (CRD). The significance of the main plot (salinity, S) was tested by S > R (salinity inside replicate) as an experimental error of the main plot, and the mean square error was used to test significance of the subplot (C) and the interaction S × C (salinity per cultivar). The mean differences were determined using the Fisher LSD test at $p \leq 0.05$. Chemical analyses for forage parameters were performed on two samples per cultivar, which were combined to represent each salinity level ($n$ = 8) per harvest. These data (Figure 1) were subjected to a one-way (salinity) ANOVA with means compared by the Fisher LSD test. For total phenolics (TP) and antioxidant capacity (ORAC) analyses, samples were analyzed in triplicate, where total phenolics were quantified from a gallic acid standard curve. The effects of salt as a main plot, cultivar as a subplot, and the interaction between salt and cultivar (salt × cultivar) for ORAC and TP concentrations were analyzed at $p \leq 0.05$ using the GLM procedure with a standard split-plot test format in SAS (version 9.3; SAS Institute, Cary, NC, USA). The differences in ORAC and TP between the two harvests were analyzed at $p \leq 0.05$ using the *T*-test procedure in SAS (version 9.3; SAS Institute, Cary, NC, USA).

## 3. Results

### *3.1. Forage Quality*

The impact of salinity on forage quality, expressed as the mean of the four cultivars at each salinity level per harvest, is presented in Figure 1. The parameters used to evaluate forage quality include acid detergent fiber (ADF), neutral detergent fiber (NDF), net energy for lactation (NEL), crude protein (CP), and relative feed value (RFV).

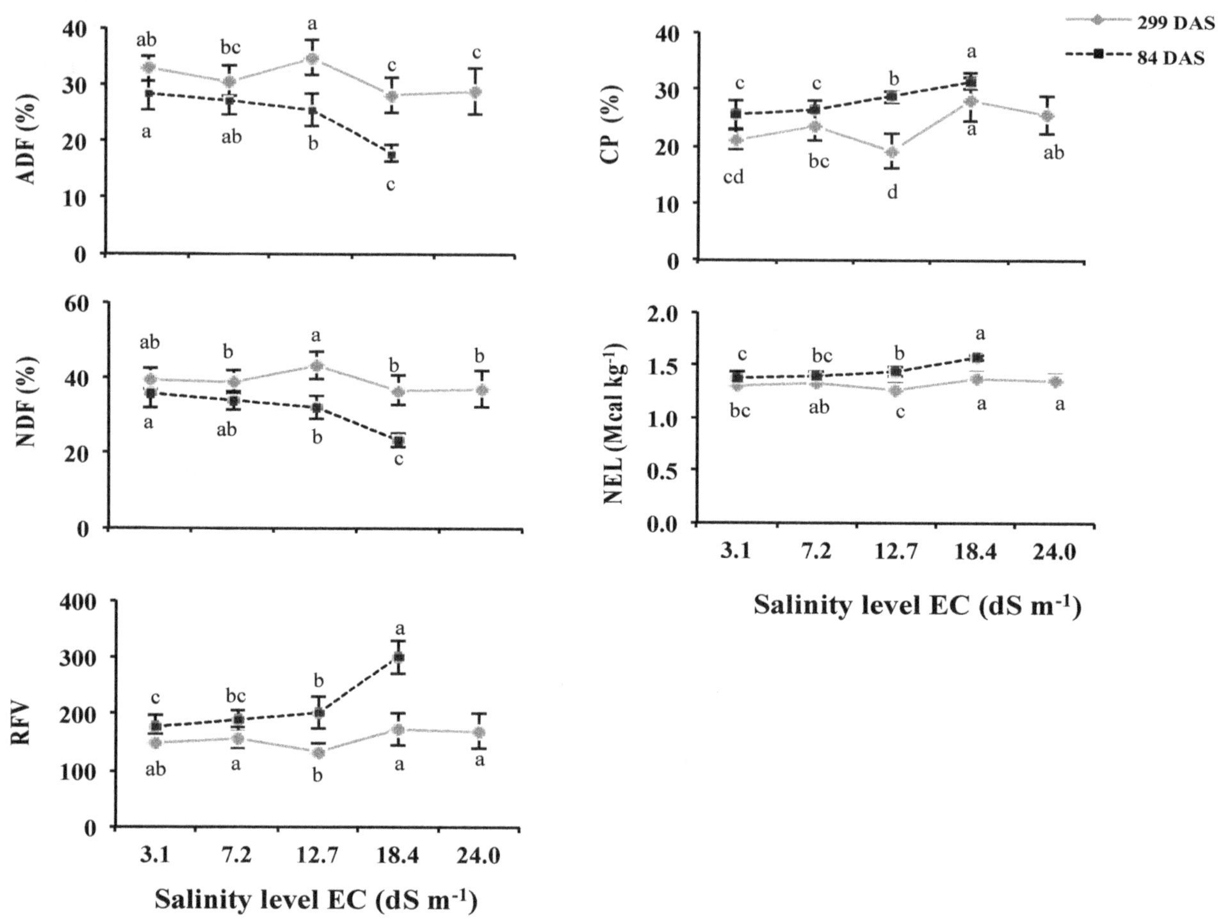


**Figure 1.** Impact of salinity increase on acid detergent fiber (ADF), neutral detergent fiber (NFD), net energy of lactation (NEL), crude protein (CP), and relative feed value (RFV) of salt-tolerant alfalfa. Data points represent the means (±SD) of the salinity-tolerant cultivars ($n = 8$). Means with the same letter are not significantly different according to a Fisher LSD test ($p \leq 0.05$). For the harvest at 84 DAS, the lack of data at 24 dS·m$^{-1}$ was due to there being insufficient plant material for analysis because of growth limitations.

Salinity had a significant effect on the forage quality for both harvests ($p \leq 0.001$). At 84 DAS, there were no differences up to EC = 7.2 dS·m$^{-1}$ for all parameters evaluated. Above that level, ADF and NDF decreased by approximately 8% and 9%, respectively, from 12.7 to 18.4 dS·m$^{-1}$. Consequently, the RFV (related to the ADF and NDF contents) increased sharply between those levels. CP increased by 5.2% from 7.2 to 18.4 dS·m$^{-1}$ (Figure 1). In addition, the mean NEL increased as salinity increased. At 299 DAS, salinity also affected all forage parameters ($p \leq 0.05$). In contrast to 84 DAS, at 299 DAS significant differences between the control and salinity treatments generally were first observed at 12.7 dS·m$^{-1}$ instead of at 7.2 dS·m$^{-1}$ (Figure 1).

### 3.2. *Nutrient Composition of Alfalfa*

#### 3.2.1. Macronutrients

The macronutrient (modified from [19]) data, including N and P, are expressed on a dry matter (DM) basis (Table 2). The main macronutrients found in alfalfa shoots (g·kg$^{-1}$ DM) at both harvests were N, K, and Ca, while total S, Mg, and P were present at much lower levels (Table 2). Salinity had a significant effect on all macronutrients for both harvests, except for total S at 299 DAS. Nitrogen increased with salinity for both harvests, reaching levels that were significantly higher than those of the control at and above 12.7 dS·m$^{-1}$ (84 DAS), and at and above 18.4 dS·m$^{-1}$ (299 DAS). Shoot K decreased significantly ($p \leq 0.01$) for all cultivars and harvests as salinity increased. The calcium content remained constant up to 7.2 dS·m$^{-1}$ (84 DAS) or up to 12 dS·m$^{-1}$ (299 DAS), but decreased significantly for both harvests (more drastically at 299 DAS) as salinity increased. The Mg levels significantly increased for both harvests, with salinity, from the control to the highest level of salinity (84% and 48% increases for 84 DAS and 299 DAS, respectively). Sulfur concentrations increased with salinity, being significant ($p \leq 0.01$) at 84 DAS, but not at 299 DAS. Concentrations of P remained constant up to 12.7 dS·m$^{-1}$, but increased significantly ($p \leq 0.01$) above that salinity level for both harvests (Table 2). There was a significant ($p \leq 0.01$) cultivar effect for all macronutrients (except for N) at 84 DAS, while at 299 DAS, there was a significant cultivar effect only for Ca and Mg (both at $p \leq 0.05$). Both Na and Cl increased significantly ($p \leq 0.01$) in shoots with increasing salinity, but these and detailed data by cultivar and salinity are presented in a companion paper [19].

**Table 2.** Average macronutrients (±SE) in alfalfa shoot dry matter (DM) according to salinity levels. EC, electrical conductivity of irrigation water in deciSiemens per meter. ND, not determined (insufficient biomass). Modified from [19].

| | N | P | K | Ca | Mg | Total S |
|---|---|---|---|---|---|---|
| | | | **DM (g·kg$^{-1}$) EC dS·m$^{-1}$ Second Harvest (84 DAS)** | | | |
| **3.1** | 40.8 $^{c}$ ± 1.43 | 2.6 $^{b}$ ± 0.09 | 46.4 $^{a}$ ± 1.05 | 14.1 $^{a}$ ± 0.4 | 2.6 $^{c}$ ± 0.14 | 3.5 $^{d}$ ± 0.08 |
| **7.2** | 42.1 $^{c}$ ± 1.04 | 2.7 $^{b}$ ± 0.09 | 41.4 $^{b}$ ± 0.94 | 13.5 $^{a}$ ± 0.5 | 2.7 $^{c}$ ± 0.16 | 3.9 $^{c}$ ± 0.10 |
| **12.7** | 46.0 $^{b}$ ± 0.56 | 2.9 $^{b}$ ± 0.08 | 38.6 $^{c}$ ± 0.62 | 13.0 $^{c}$ ± 0.69 | 3.4 $^{b}$ ± 0.22 | 4.8 $^{b}$ ± 0.20 |
| **18.4** | 50.5 $^{a}$ ± 0.80 | 3.8 $^{a}$ ± 0.13 | 34.3 $^{d}$ ± 0.88 | 12.1 $^{b}$ ± 0.24 | 4.8 $^{a}$ ± 0.07 | 7.4 $^{a}$ ± 0.17 |
| **24** | ND | ND | ND | ND | ND | ND |
| | | | **Seventh Harvest (299 DAS)** | | | |
| **3.1** | 34.1 $^{d}$ ± 1.07 | 3.4 $^{b}$ ± 0.17 | 40.3 $^{a}$ ± 1.12 | 18.0 $^{a}$ ± 0.51 | 2.5 $^{c}$ ± 0.08 | 3.8 $^{a}$ ± 0.12 |
| **7.2** | 37.6 $^{bc}$ ±1.37 | 3.1 $^{b}$ ± 0.06 | 30.4 $^{bc}$ ± 0.74 | 18.3 $^{a}$ ± 0.61 | 2.8 $^{bc}$ ± 0.12 | 4.6 $^{a}$ ± 0.20 |
| **12.7** | 30.8 $^{d}$ ± 1.77 | 2.8 $^{b}$ ± 0.14 | 31.0 $^{b}$ ± 0.68 | 16.7 $^{a}$ ± 0.51 | 3.2 $^{ab}$ ± 0.12 | 4.8 $^{a}$ ± 0.15 |
| **18.4** | 45.3 $^{a}$ ± 2.11 | 4.1 $^{a}$ ± 0.12 | 27.3 $^{cd}$ ± 0.56 | 12.1 $^{b}$ ± 0.45 | 3.0 $^{bc}$ ± 0.10 | 4.8 $^{a}$ ± 0.15 |
| **24** | 40.8 $^{a}$ ±1.92 | 4.3 $^{a}$ ±0.16 | 26.7 $^{d}$ ± 0.61 | 11.0 $^{b}$ ± 0.83 | 3.6 $^{a}$ ± 0.20 | 5.3 $^{a}$ ± 0.39 |

Different small letters within each column, and between EC levels, represent significantly different means according to Fisher's LSD test ($p \leq 0.05$), where $n$ = 16 (except for N, $n$ = 8) for EC levels.

### 3.2.2. Micronutrients

Shoot micronutrients analyzed for the four alfalfa cultivars were iron (Fe), copper (Cu), manganese (Mn), zinc (Zn), and molybdenum (Mo) (Table 3). At 84 DAS, there were no differences in mean Fe concentrations (ranging from 99.1 to 109.6 mg·kg$^{-1}$ DM) or Cu (2.07–3.11 mg·kg$^{-1}$ DM) as a function of increasing salinity (EC). Mean concentrations of Mn and Mo tended to increase with increasing salinity with significant ($p \leq 0.05$ and $p \leq 0.01$, respectively) differences between the control and the highest salinity level (18.4 dS·m$^{-1}$) at 84 DAS. There was a significant ($p \leq 0.01$) increase in Zn concentration at each level of salinity increase at 84 DAS. At 299 DAS, the Fe, Cu, Mn, and Zn levels remained mostly unchanged, but there was a small but significant ($p \leq 0.05$) decline (16%–28%) in the Fe levels between the 3.1 dS·m$^{-1}$ control (116 mg·kg$^{-1}$ DM) and the other saline treatments. Mn showed a transient increase of 42% (17.3 to 24.6 mg·kg$^{-1}$ DM) as salinity increased from 3.1 to 7.2 dS·m$^{-1}$, and then declined to the salinity control levels. In general, the shoot Mo concentrations for all levels of salinity were significantly ($p \leq 0.05$) higher than those of the control (Table 3).

**Table 3.** Average micronutrient concentrations (±SE) in alfalfa shoot dry matter (DM), according to salinity levels. EC, electrical conductivity of irrigation water in deciSiemens per meter. ND, not determined (insufficient biomass).

| | Fe | Cu | Mn | Zn | Mo |
|---|---|---|---|---|---|
| | | | DM ($mg \cdot kg^{-1}$) | | |
| | | | EC $dS \cdot m^{-1}$ Second Harvest (84 DAS) | | |
| **3.1** | 104.0 $^{a}$ ± 6.29 | 2.1 $^{a}$ ± 0.27 | 25.5 $^{b}$ ± 3.38 | 40.9 $^{d}$ ± 1.32 | 2.0 $^{c}$ ± 0.09 |
| **7.2** | 99.1 $^{a}$ ± 4.90 | 2.3 $^{a}$ ± 0.10 | 31.7 $^{ab}$ ± 4.8 | 45.9 $^{c}$ ± 1.00 | 3.1 $^{b}$ ± 0.11 |
| **12.7** | 106.5 $^{a}$ ± 5.89 | 3.1 $^{a}$ ± 0.16 | 34.8 $^{a}$ ± 4.10 | 54.9 $^{b}$ ± 1.11 | 3.2 $^{b}$ ± 0.14 |
| **18.4** | 109.6 $^{a}$ ± 5.0 | 3.1 $^{a}$ ± 0.19 | 34.8 $^{a}$ ± 1.10 | 60.5 $^{a}$ ± 1.25 | 4.1 $^{a}$ ± 0.11 |
| **24** | ND | ND | ND | ND | ND |
| | | | Seventh Harvest (299 DAS) | | |
| **3.1** | 116.1 $^{a}$ ± 6.35 | 5.8 $^{a}$ ±0.83 | 17.2 $^{b}$ ± 0.91 | 97.6 $^{a}$ ± 3.36 | 2.7$^{c}$ ± 0.19 |
| **7.2** | 97.7 $^{b}$ ± 7.35 | 6.1 $^{a}$ ± 0.64 | 24.6 $^{a}$ ± 1.44 | 89.9 $^{a}$ ± 3.26 | 6.4 $^{a}$ ± 0.43 |
| **12.7** | 89.9 $^{b}$ ± 7.35 | 6.5 $^{a}$ ± 0.41 | 18.9 $^{b}$ ± 0.99 | 105.6 $^{a}$ ± 3.18 | 6.3 $^{a}$ ± 0.44 |
| **18.4** | 83.5 $^{b}$ ± 3.17 | 5.3 $^{a}$ ± 0.26 | 17.4 $^{b}$ ± 1.05 | 101.3 $^{a}$ ± 3.26 | 4.7 $^{c}$ ± 0.36 |
| **24** | 92.3 $^{b}$ ± 7.69 | 5.7 $^{a}$ ± 0.49 | 14.8 $^{b}$ ± 1.04 | 98.3 $^{a}$ ± 3.85 | 4.2 $^{c}$ ± 0.21 |

Different lower case letters within each column, and between EC levels, represent significantly different means according to Fisher's LSD test ($p \leq 0.05$), where $n = 16$.

### 3.3. *Antioxidant Capacity of Alfalfa*

Salinity had no effect ($p > 0.05$) on either the oxygen radical absorbance capacity (ORAC) or the total phenolic levels of the four alfalfa cultivars. The hydrophilic fractions of shoots had most (68%–99%) of the shoot total antioxidant capacity (Table 4). At early plant development (84 DAS), alfalfa shoots had hydrophilic ORAC ($ORAC_{Hydro}$) levels that ranged from 190–230 $\mu moles \cdot TE \cdot g^{-1}$ DM (Figure 2), while at 299 DAS, $ORAC_{Hydro}$ ranged from 229–274 $\mu moles \cdot TE \cdot g^{-1}$ DM, and the shoot total antioxidant capacity ranged from 244–287 $\mu moles \cdot TE \cdot g^{-1}$ DM (Figure 2, Table 4). Total phenolic (TP) concentrations ranged from 5.0–5.6 $mg \cdot GAE \cdot g^{-1}$ DM for both harvests (Figure 2).

**Table 4.** Oxygen radical absorbance capacity of the lipophilic ($ORAC_{Lipo}$) and hydrophilic ($ORAC_{Hydro}$) fractions, and total antioxidant capacity ($ORAC_{Hydro}$ + $ORAC_{Lipo}$), in micromoles of trolox equivalents per gram of dry matter ($\mu moles \cdot TE \cdot g^{-1}$ DM) of alfalfa irrigated with water of different electrical conductivities (EC). Plants were sampled on 17 April 2012 (299 DAS). Data are means ± SE combined for the four cultivars with two replicated analyses per sample ($n = 8$).

| EC | $ORAC_{Lipo}$ | $ORAC_{Hydro}$ | $ORAC_{Total}$ |
|---|---|---|---|
| ($dS \cdot m^{-1}$) | | ($\mu moles \cdot TE \cdot g^{-1}$ DM) | |
| 3.1 | 15.0 ± 2.4 | 239.5 ± 12.8 | 254.5 ± 13.6 |
| 7.2 | 11.2 ± 1.5 | 252.1 ± 11.6 | 263.3 ± 11.0 |
| 12.7 | 13.4 ± 2.5 | 273.6 ± 14.3 | 286.9 ± 14.2 |
| 18.4 | 16.4 ± 1.7 | 268.4 ± 14.0 | 284.8 ± 15.5 |
| 24.0 | 15.3 ± 1.0 | 228.8 ± 18.3 | 244.1 ± 18.2 |

There was no effect of salinity (expressed as EC), cultivar, or the salt × cultivar interaction.

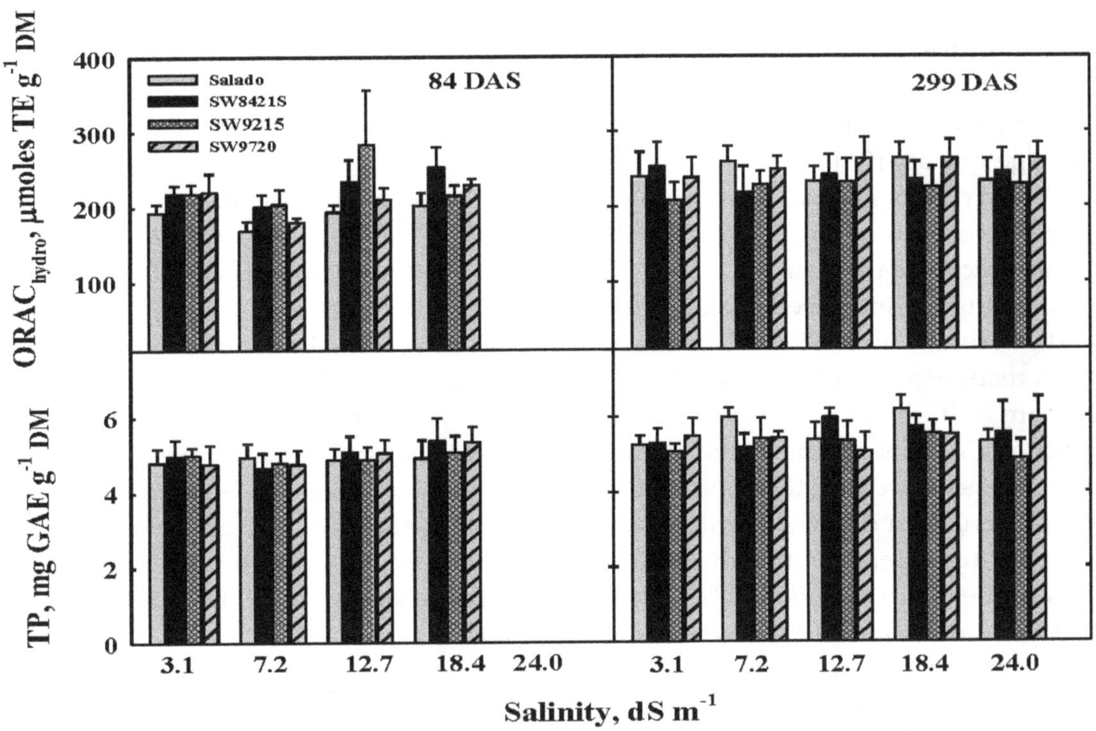


**Figure 2.** Total phenolics (TP) and hydrophilic shoot oxygen radical absorbance capacity (ORAC) of four salinity-tolerant alfalfa cultivars irrigated with saline water with different electrical conductivity levels. ORAC was measured in micromoles of trolox equivalents per gram of dry matter (µmoles·TE·g$^{-1}$ DM). TP was measured as mg of gallic acid equivalents per gram of dry matter (mg·GAE·g$^{-1}$ DM). Bars represent means (±SD), where $n = 4$. Plants were sampled at 84 and 299 days after sowing. For the harvest at 84 DAS, the lack of data at 24 dS·m$^{-1}$ was due to growth limitations.

## 4. Discussion

### 4.1. Forage Quality

Forage quality was based on laboratory analyses of shoot biomass and evaluated in relation to recommended forage standards for livestock production output (e.g., milk, body weight gain) for animals consuming alfalfa of similar nutritional value and energy content [3,29]. Lower NDF translates into both increased DMI and milk yield within a forage family [3]. Regarding alfalfa protein, approximately 80% is degraded in the rumen of polygastric animals, but addition of tannins to alfalfa feed decreases rumen protein degradability and increases protein absorption [30].

Plant maturity is the main factor affecting forage quality [31], but the interaction between environmental and agronomic factors with maturity will influence the quality of alfalfa, even if harvested at the same stage of development [32]. Similarly, approaching harvest time, any stress that delays or accelerates alfalfa maturation affects the leaf-to-stem ratio and consequently, forage quality. The stems contain mostly structural components and are low in N, while the leaves contain mainly photosynthetic components and are richer in N than the stems. As a result, leaves have two to three times more CP than stems [33]. Increased leaf N leads to increased leaf area, thus increasing the leaf/stem ratio [34,35], but this could also be accounted for by the reduced stem height caused by salinity. The leaf-to-stem ratio increase leads to decreases in both ADF and NDF. Decreased ADF and NDF and increased shoot N lead to higher shoot CP levels in alfalfa irrigated with saline water. As reported in a previous study [19], plant height was significantly reduced by salinity only at 84 DAS, with the average difference in plant height between the control and EC = 18.4 dS·m$^{-1}$ being 23 cm. Thus, we hypothesize that the decrease in height (shorter internodes) in salt-affected plants may have increased the leaf-to-stem ratio, shoot N, and CP by 61 $g \cdot kg^{-1}$ DM (6%). This decreased height of salt-affected plants also led to decreases in ADF and NDF of 107 and 122 $g \cdot kg^{-1}$ DM (10.7% and 12.2%) at 84 DAS and of 2.5% and 4% at 299 DAS, respectively, improving forage potential quality (Figure 1). This is in agreement with a previous report that salinity increased alfalfa leaf-to-stem ratio, slightly improving forage quality [36].

Al-Khatib and collaborators [7] reported that the leaf-to-stem ratio of alfalfa increased while forage mass decreased in response to increasing NaCl until 20 $dS \cdot m^{-1}$ (200 mM NaCl). At 299 DAS, there was also a significant increase in CP of 42.1 $g \cdot kg^{-1}$ DM (4.2%) between the control plants and those under 24 $dS \cdot m^{-1}$ (reflecting the increased accumulation of leaf N with increased salinity). This increase in CP was observed at both 84 and 299 DAS because N accumulation in shoots increased by 23% and 33%, respectively, in response to increased salinity (Table 2, Figure 1). Although plants had a fairly constant supply of N from $NO_3^-$ in all irrigation treatments (Table 1), shoots significantly accumulated $NO_3^-$-N, leading to higher CP. This could be due to morphological changes (e.g., increased leaf-to-stem ratio) under salinity stress or because the roots in the sand tanks were found to be associated with rhizobia. Despite the differences in developmental stages between the second and seventh harvest, there was a tendency for CP to increase with salinity levels up to 18.4 $dS \cdot m^{-1}$. Although plants irrigated with salinity levels higher than the control had different stages of maturity, plant height has been used to predict forage parameters under field conditions [33,37].

Differences in forage parameters changed more sharply at 84 DAS (late summer) with salinity than at 299 DAS (early spring). These changes were likely caused by differences in climatic conditions combined with salinity [19]. Both climate and intervals between harvests (24 days before the second and 54 days before the seventh harvest) have a direct impact on maturity [33,38]. The RFV of alfalfa shoots in this experiment were similar to the values reported for alfalfa cultivars grown under field conditions with EC values ranging from 4–16 $dS \cdot m^{-1}$, although RFV did not change with salinity [39].

According to the classification of alfalfa hay [40], and judging from the parameters evaluated in this study, alfalfa herbage grown at the highest tolerated salinity fell within the "supreme" category. In comparison, forage grown at control salinity levels would be classified as "good" and "premium". Hence, our results indicated that forage quality improved with increasing salinity (despite some variation), independently of the changes between harvest seasons. Similar increases in CP and decreases in ADF in the salinity-tolerant cultivars Salado and SW9720 under salinity stress have been reported [9,11]. An increase in CP of alfalfa cultivars less tolerant to salinity was also reported when salinity increased from 2.1 to 7.8 $dS \cdot m^{-1}$ [41] or when salinity ranged from 0.3–4.5 $dS \cdot m^{-1}$ in one out of three years of cultivation [42]. Both drought and salinity restrict the growth of alfalfa, and mild drought also improves the forage quality of alfalfa [43]. These authors explained that the increase in quality with drought was due to a delay in plant maturation and an increase in the leaf-to-stem ratio; the latter is related to a reduction in stem length. However, the results of a 90-day pot experiment indicated that there were no differences in CP or N concentrations in alfalfa shoots when an EC of 15 $dS \cdot m^{-1}$ was applied using only NaCl [44].

The NEL values of alfalfa irrigated with increasing salinity, and ranging from 1.38–1.58 $Mcal \cdot kg^{-1}$ for the second harvest (84 DAS) and from 1.3 to 1.37 $Mcal \cdot kg^{-1}$ for the seventh harvest (299 DAS), were within the average (1.47 $Mcal \cdot kg^{-1}$) required for lactating cows [29], although some supplementation may be required to maintain the required energy levels.

### *4.2. Mineral Nutrient Composition*

When irrigated with non-saline water, the predominant macronutrients in alfalfa are N, K, Ca, Mg, P, and S [45]. In our plants, which were fertilized to achieve the desired macro and micronutrients concentrations for ideal crop growth, and irrigated with saline water, the three main shoot macronutrients were also N, K, and Ca, followed by Cl and Na (data presented in [19]) and S, as these were added to the irrigation water to achieve high salinity, then followed by Mg and P at similar concentrations (Table 2). This suggests that alfalfa plants were provided adequate nutrients for growth, and our results express mostly the effects of salinity in a properly fertilized crop. The discussion on macro- and micronutrient requirements is based on the specifications for lactating dairy cattle provided by the Nutrient Requirements of Dairy Cattle [29]. The NRC requirement level for animals producing 35 kg milk·day$^{-1}$ (Holstein or Jersey) was used, based on the average milk production for 2012 in California [46].

*Macronutrients and sodium*—Although adequate mineral nutrition alone will not prevent animal diseases, susceptibility to infectious diseases in response to malnourishment has been recognized for several centuries [47]. Thus, it is important to know if crop stress induced by salinity alters the nutrient composition of alfalfa.

The lowest Ca concentration in shoots in response to salinity (11 g·kg$^{-1}$) was still above the daily dietary requirement (6.1 g·kg$^{-1}$) for dairy cattle [29], while the highest Ca concentrations (18 g·kg$^{-1}$) were observed at ECs of 3.1 and 7.2 dS·m$^{-1}$ at 299 DAS (Table 2). While dietary Ca concentrations above 10 g·kg$^{-1}$ have been associated with reduced dry matter intake (Miller, 1983, in [29]), diets as high as 18 g·kg$^{-1}$ have been fed to non-lactating dairy cows without problems (Beede *et al.*, 1991, in [29]). Feeding Ca in excess of daily dietary requirements is suggested to improve performance, mainly when cows are fed corn silage diets [29]. Potassium is the third most abundant element in mammals and is important for cellular osmotic balance. The cellular homeostasis of Na and K is maintained by $Na^+/K^+$ pumps located inside the cell membrane. These two cations play an important role in electrical activity of nerve and muscle cells, in the acid-base balance, and in water retention. Potassium is a cofactor for the activation of enzymes, including those involved in protein synthesis and carbohydrate metabolism [48]. Because of increasing levels of $Cl^-$ in irrigation water, shoot absorption of potassium decreased significantly ($p \leq 0.01$) for both harvests (by 26%–33%). Sodium significantly increased (by 60%), both with salinity and harvest date (presented elsewhere [19]), which was expected due to its elevated concentration in the saline treatment water. The levels of K across harvests and salinity (2.6%–4.6%) were well above the required levels (1.04%) for average lactating cows [29]. However, diets supplemented with potassium carbonate increased K from 1.6% to 4.6% (w/w) and decreased milk yield and feed intake [49]. Thus, K levels in alfalfa shoots irrigated with saline water containing 6 to 6.5 mmolc·L$^{-1}$ could be of concern, depending on forage intake.

A continuous supply of Mg from feed is desirable because a high K level in forage decreases Mg absorption from the rumen and can lead to tetany [50]. The frequency of tetany in cows, triggered by low Mg and/or Ca, and high K in forage, increases when the ratio of K: (Ca + Mg) exceeds 2.2 [51]. In our results, the ratio of K: (Ca + Mg) was higher than 2.2 at 84 DAS, but lower than 2.2 at 299 DAS, suggesting that Mg levels should be monitored in alfalfa irrigated with saline water. Thus, although our results indicate that salinity can lead to a small, but significant accumulation of Mg by alfalfa shoots, Mg supplementation is still a must due to its poor absorption (13% to 16% from ration) by cows [52].

Sulfur (S) is an important component of cysteine and methionine, of many enzymes, and of antioxidants such as glutathione and thioredoxin, but elevated concentrations of S in alfalfa shoots can be detrimental to animal feed intake and function. Although we discuss the concentrations of S in shoots of different ages, the saline water used here was sulfate-dominant to mimic the drainage waters of California's Central Valley. Thus, levels of S might not be of concern where waters are $Cl^-$ dominant. However, the S levels in our experiment remained similar at 299 DAS across salinity treatments. The lack of significant S uptake at 299 DAS may be explained by cooler temperatures and lower evapotranspiration before that harvest. The S concentration in shoots ranged from 0.38%, at the lowest EC, to 0.54% at the highest EC observed at 299 DAS. Regardless of season, a decrease in S in a later harvest (as seen here) was reported previously for alfalfa irrigated with sulfate-dominant water at both 15 and 25 dS·m$^{-1}$ [53]. The authors reported an S range in alfalfa of 0.5%–0.9% at 25 dS·m$^{-1}$. In the S range recorded at 299 DAS for this study, and considering that the average consumption of alfalfa is 4.26 kg·cow$^{-1}$ [3], the S consumption would be 16.2 to 23.0 g·day$^{-1}$, well below the 32 g S·day$^{-1}$ upper limit recommended for a mature grazing beef cow [54], but 1.9 to 2.7 times above the 8.52 g

S·day$^{-1}$ (0.2% S/day) required for dairy cows [29]. Although no S toxicity has been reported [29], it is important to balance the diet in order to maintain S intake at a safe level (below 0.4% of DM daily), as levels of S of 0.4% in bailed alfalfa can lead to molybdenosis and reduced uptake of Cu and Se in beef cattle if alfalfa is the only source of feed [45].

The P requirement in the daily diet of average-producing dairy cows is 0.35% [29], but P levels regarded as adequate in alfalfa shoots are 0.08% to 0.15% [45]. P deficiency will lead to osteomalacia (softening of the bones) and fragile bones. The average levels of P in our alfalfa shoots at 299 DAS (0.28% to 0.44% DW) are considered to be high for shoot levels, relative to alfalfa grown in soils of the Mediterranean and desert zones [45]. In addition, according to nutrient tables presented by these authors, our Mg levels (0.25%–0.37%) were adequate, while shoot K and S were high.

Salinity significantly increased Na and Cl levels for both harvest dates by 40%–60%, as presented in a companion paper [19], resulting in shoot Na levels two to five times higher than the level required (0.23%) for average-producing lactating dairy cows [29]. Our data showed that alfalfa accumulates more Na and $Cl^-$ over time, even at the same irrigation salinity level. As previously reported [19], shoot Na ranged from 3.5–10 g·kg$^{-1}$, and Cl from 7–14 g·kg$^{-1}$, across salinity levels and harvest times. We found no reference reporting Na toxicity to livestock, but increasing Na in the diet from 5.5–8.8 g·kg$^{-1}$ caused no reduction of feed intake, milk yield, or toxicity (Schneider *et al.* 1986, in [29]). NaCl, often added to feed mixes, can be tolerated up to 3% (lactating cows) or 4.5% (growing animals) of dietary dry matter. Thus, Na and Cl levels in alfalfa irrigated with saline water present no safety concern.

*Micronutrients*—Micronutrients and some vitamins are essential for animals to achieve optimal immune function, growth, and reproduction. Cattle can have sufficient amounts of these minerals for growth and reproduction, but not have enough for optimal immune function [47]. Examples are Cu and Zn, which are required for the activity of the antioxidant enzymes Cu-Zn superoxide dismutase (SOD) [55].

The average iron concentration was not affected by salinity and ranged from 83.5–116 mg·kg$^{-1}$ across harvests, regardless of salinity treatment. Concentrations of 50 to 100 mg·kg$^{-1}$ of Fe in a basal ration are within the requirements for the growth of grazing cattle [47,56] and concentrations of 15 mg·kg$^{-1}$ in daily feed are recommended for average lactating cows [29]. Iron is essential for the formation of new red blood cells and only levels ≥4000 mg·kg$^{-1}$ affect weight gain and cause diarrhea in young calves [47].

Copper (Cu) and zinc (Zn) are important micronutrients for immune function, and levels of 20 mg·kg$^{-1}$ Cu and 40–60 mg·kg$^{-1}$ Zn were suggested as optimal for feeding in the total diet of dairy cattle [57], while levels of 11 mg·kg$^{-1}$ Cu and 48 mg·kg$^{-1}$ Zn are recommended for average lactating dairy cows [29]. The Cu levels found in shoots for both harvests were below 7.0 mg·kg$^{-1}$, indicating the need for supplementation. In addition, the ratio of Cu to Mo in shoots was always approximately 1:1, well below the ratio of 10:1 that is considered a threshold for potential Cu toxicity [58].

Salinity significantly increased the Zn concentration in young plants (84 DAS) but not in established alfalfa plants (299 DAS), with concentrations ranging from 90–106 mg·kg$^{-1}$. Considering that a minimum Zn concentration of 48 mg·kg$^{-1}$ is required for average lactating cows [29], our plants contained levels more than adequate to support a healthy immune function in livestock [57]. Manganese levels in alfalfa shoots were the third highest, after Fe and Zn. Manganese is important for its role in enzymatic systems but it is poorly absorbed (14%–18%) and if deficient, can reduce fertility and delay estrous [56]. This author mentions that Mn deficiency can lead to abortion and deformed calves at birth, but elevated Mn in the diet is generally not toxic. Levels of Mn in our alfalfa cultivars were at least 14 mg·kg$^{-1}$, as recommended for average lactating cows (NRC 2001). However, considering the poor absorption of Mn, mineral supplementation would be recommended.

### 4.3. Antioxidant Capacity of Alfalfa

Antioxidant flavonoids in the diet are believed to have health-promoting benefits to both humans and animals. In addition to protein, alfalfa is a rich source of flavonoid antioxidants and phytoestrogens including luteolin, coumestrol, and apigenin [59]. Phenolic compounds (including flavonoids) protect plants against the damaging effects of excessive reactive oxygen species (ROS) triggered by abiotic stresses, including salinity [60,61]. Although oxygen radical absorbance capacity (ORAC) has been widely accepted by industry to gauge the total antioxidant capacity of fruits, vegetables, spices, and other items consumed by humans, ORAC has only recently been used to estimate the antioxidant capacity of plants destined for livestock consumption [62–64]. The total antioxidant capacity is the sum of the lipophilic ($ORAC_{Lipo}$) and hydrophilic ($ORAC_{Hydro}$) fractions extracted from plants by hexane (lipophilic) and 70:30 acetone:aqueous buffer (hydrophilic). Our ORAC data (Table 4) confirmed those of others [63,64] who reported that the hydrophilic fractions of plant extracts contain most (68%–99%) of the total antioxidant capacity of shoots. Alfalfa shoots grown with saline water had 94%–96% of the total antioxidant capacity in the hydrophilic fraction with only 4%–6% in the lipophilic fraction, indicating that alfalfa shoots are low in lipophilic antioxidants such as tocopherols, carotenes, and fatty acids. The oven-dried alfalfa plants in our study had $ORAC_{Hydro}$ values that ranged from 229–274 $\mu$moles·TE·$g^{-1}$ DM (Table 4, Figure 2). Although these values may seem small compared with those of other leguminous forages, such as *Lespedeza cuneata* ($ORAC_{Hydro}$ = 530 $\mu$moles·TE·$g^{-1}$ DM), previously reported [63] alfalfa flavonoids and isoflavonoids present in hydrophilic (aqueous) extracts reduced oxidative stress and exerted hepatoprotective activity in rats treated with the liver-damaging compound carbon tetrachloride [65]. These results indicate that when animals consume alfalfa on a regular basis, it can provide benefits other than nutritional value.

The values for both ORAC and total phenolics (TP) remained unaltered by increased salinity, without differences for either ORAC or TP among cultivars (Figure 2). Our results agree with a previous report where there were no differences in antioxidant compounds among different cultivars of alfalfa in the absence of salt stress [15]. These authors also reported that the major antioxidants in alfalfa shoots, determined by HPLC, were tricin and apigenin glycosides (each approximately 40% of the total HPLC peaks), and luteolin and chrysoeriol glycosides (10% or less of the total HPLC peaks). Our results suggest that the salinity levels tested did not highly stress these salt-tolerant alfalfa cultivars. Previously, mostly the aglycons (flavonoids stripped of sugar moieties by acidic or enzymatic hydrolysis) have been determined, but the determination of full glycosidic forms (flavonoid plus sugar moieties) has also been conducted [59]. Flavonoids from alfalfa have the typical structure of several other flavonoids reported as beneficial to human diets and found in fruits and vegetables. Although sun drying (used to produce alfalfa hay) drastically decreased the antioxidant capacity of the antioxidant herb *Artemisia annua*, oven drying at 45 °C only slightly reduced the antioxidant capacity compared with freeze drying [66]. Thus, we consider that our oven-dried alfalfa shoots had an antioxidant capacity close to that of freeze-dried (or fresh) shoots. We could not find any published work on the antioxidant capacity of alfalfa shoots determined by ORAC or TP, except that the total ORAC ($ORAC_{Hydro+Lipo}$) of alfalfa hay was 171 $\mu$moles·TE·$g^{-1}$, and the $ORAC_{Lipo}$ was only 3% of the total ORAC [63]. The antioxidant capacity of all cultivars used here was not affected by salinity, thus expanding the value of alfalfa beyond its contents of CP and minerals.

Although the value of antioxidants in animal and human nutrition is still debated by some, several benefits (e.g., anti-cancer, anti-inflammatory, *etc.*) of antioxidant-rich diets have been proposed. Dairy cows supplemented daily with 500 g of oregano (2082 $\mu$moles·TE·$g^{-1}$ DM) increased their milk fat concentration, feed and milk NEL efficiencies, and fat-corrected milk yield by 3.5% [67]. Although oregano has an ORAC value 8–9 fold higher than our oven-dried alfalfa shoots (225 to 256 $\mu$moles·TE·$g^{-1}$), the average consumption of alfalfa shoots by cows is 5.4 kg·$day^{-1}$, which is 10-fold higher than the 500 g·$day^{-1}$ oregano supplement from the above-mentioned study. Thus, daily alfalfa consumption can provide as much antioxidant flavonoid intake as oregano, thus adding to the forage value of alfalfa.

## 5. Conclusions

The effect of salinity in irrigation water on the suitability of alfalfa as a forage was based on shoot levels of macro- and micronutrients, and the forage quality estimated from ADF, NDF, and CP. Additional forage value was based on the antioxidant capacity and total phenolics in response to salinity. The nutrient composition of alfalfa can vary with salinity. Although our saline irrigation waters provided 27%–87% more $SO_4$ than Cl and 60%–94% more Na than Cl, alfalfa shoots contained 20%–190% more Cl than total S and 20%–120% more Cl than Na. Although Na and Cl in shoots increased with salinity, reducing the K concentration by 26%–32% and Ca by 15%–32% in shoots, shoot K and Ca were considered high and adequate [1,45], respectively, at all salinity levels. Increased salinity also increased shoot N (23%–33%), P (21%–46%), Mg (20%–84%), and total S (100%–110%) for both harvests. In general, the levels of macro- and micronutrients were adequate or high for alfalfa forage [1,29,45] regardless of salinity. However, when irrigation water was sulfate-dominant, the S concentrations in alfalfa were close to the upper limits recommended for safe animal consumption and require monitoring for water EC higher than 12.7 $dS \cdot m^{-1}$. Regarding forage potential quality, shoots from plants irrigated with salinity levels higher than the control remained unaltered, or slightly improved compared with the salinity control levels, with NDF and CP at levels recommended for various classes of milking cows, but below the NDF values required for bulls and dry cows [39]. The antioxidant capacity was 15–23 fold higher for hydrophilic than for lipophilic fractions, but remained mostly unaltered by salinity, indicating that total antioxidant compounds, including phenolics and flavonoids (postulated to neutralize reactive oxygen species triggered by salinity stress), may remain fairly constant in alfalfa cultivars that are tolerant to salinity. These constant antioxidant levels, regardless of salinity stress, may play an extra beneficial role in helping to maintain animal health, as accepted for antioxidants in humans. Except for numeric values (such as reduced K and increased S), salinity levels up to 24 $dS \cdot m^{-1}$ did not alter the potential nutritional value and antioxidant capacity of alfalfa for livestock. The nutrient composition and antioxidant capacity of alfalfa are expected to play a dual role in the maintenance of health, body index, and milk production in dairy cows. This is the first report we are aware of, which has determined the total antioxidant capacity of alfalfa in response to salinity. Further studies involving animal performance are required to confirm the potential feed value of salt-stressed alfalfa under field conditions.

**Acknowledgments:** We acknowledge Nedda Saremi for help with the macro- and micronutrient analyses and Nahid Vishteh for determining the chemical composition of the saline water used in this study.

**Author Contributions:** Jorge Ferreira was responsible for the antioxidant method (ORAC), the data interpretation and discussion of forage nutritional value and antioxidant capacity, and the writing of the manuscript with Monica Cornacchione and Donald Suarez. Monica Cornacchione conducted the experiments, analyzed the data, and helped write the manuscript. Xuan Liu performed the tests for antioxidant activity (ORAC) and total phenolics (TP) and helped write the experimental section. Donald Suarez developed the experimental design, including the composition of the saline water, and assisted with the writing of the manuscript.

## Abbreviations

| | |
|---|---|
| ADF | acid detergent fiber |
| NDF | neutral detergent fiber |
| NEL | net energy for lactation |
| CP | crude protein |
| RFV | relative feed value |
| ORAC | oxygen radical absorbance capacity |
| TP | total phenolics |

## References


1. Radović, J.; Sokolović, D.; Marković, J. Alfalfa—Most important perenial forage legume in animal husbandry. *Biotechnol. Anim. Husb.* **2009**, *25*, 465–475. [CrossRef]
2. USDA-ARS. Roadmap for Alfalfa Research. Available online: http://ars.usda.gov/SP2UserFiles/Place/54281000/alfalfaroadmap2.pdf (accessed on 11 March 2014).
3. DePeters, E. Forage Quality: Important Attributes & Changes on the Horizon. In the Proceedings of California Alfalfa and Grains Symposium, Sacramento, CA, USA, 10–12 December 2012; UC Cooperative Extension, Plant Sciences Department, University of California, Davis: Davis, CA, USA, 2012.
4. Khorasani, G.R.; Janzen, R.A.; McGill, W.B.; Kennelly, J.J. Site and extent of mineral absorption in lactating cows fed whole-crop cereal grain silage of alfalfa silage. *J. Anim.Sci.* **1997**, *75*, 239–248. [PubMed]
5. Smedema, L.K.; Shiati, K. Irrigation and salinity: A perspective review of the salinity hazards of irrigation development in the arid zone. *Irrig. Drain. Syst.* **2002**, *16*, 161–174. [CrossRef]
6. Ashraf, M.; Harris, P.J.C. Potential biochemical indicators of salinity tolerance in plants. *Plant Sci.* **2004**, *166*, 3–16. [CrossRef]
7. Al-Khatib, M.; McNeilly, T.; Collins, J.C. The potential of selection and breeding for improved salt tolerance in lucerne (*Medicago sativa* L.). *Euphytica* **1992**, *65*, 43–51. [CrossRef]
8. Mass, E.V.; Grattan, S.R. Crop yields as affected by salinity. In *Agricultural Drainage*; Agron. Monograph 38; Skaggs, R.W., van Schilfgaarde, J., Eds.; ASA, CSSA, SSA: Madison, WI, USA, 1999; pp. 55–108.
9. Robinson, P.H.; Grattan, S.R.; Getachew, G.; Grieve, C.M.; Poss, J.A.; Suarez, D.L.; Benes, S.E. Biomass accumulation and potential nutritive value of some forages irrigated with saline-sodic drainage water. *Anim. Feed Sci. Technol.* **2004**, *111*, 175–189. [CrossRef]
10. Grattan, S.R.; Grieve, C.M.; Poss, J.A.; Robinson, P.H.; Suarez, D.L.; Benes, S.E. Evaluation of salt-tolerant forages for sequential water reuse systems: I. Biomass production. *Agric. Water Manag.* **2004**, *70*, 109–120. [CrossRef]
11. Suyama, H.; Benes, S.E.; Robinson, P.H.; Grattan, S.R.; Grieve, C.M.; Getachew, G. Forage yield and quality under irrigation with saline-sodic drainage water: Greenhouse evaluation. *Agric. Water Manage.* **2007**, *88*, 159–172. [CrossRef]
12. Steppuhn, H.; Acharya, S.N.; Iwaasa, A.D.; Gruber, M.; Miller, D.R. Inherent responses to root-zone salinity in nine alfalfa populations. *Can. J. Plant Sci.* **2012**, *92*, 235–248. [CrossRef]
13. Rubio, M.C.; Bustos-Sanmamed, P.; Clemente, M.R.; Becana, M. Effects of salt stress on the expression of antioxidant genes and proteins in the model legume *Lotus japonicus*. *New Phytol.* **2009**, *181*, 851–859. [CrossRef] [PubMed]
14. Mhadhbi, H.; Fotopoulos, V.; Mylona, P.V.; Jebara, M.; Elarbi Aouani, M.; Polidoros, A.N. Antioxidant gene–enzyme responses in *Medicago truncatula* genotypes with different degree of sensitivity to salinity. *Physiol. Plant.* **2011**, *141*, 201–214. [CrossRef] [PubMed]
15. Stochmal, A.; Oleszek, W. Seasonal and structural changes in flavones in alfalfa (*Medicago sativa*) aerial parts. *Int. J. Food Agric. Environ.* **2007**, *5*, 170–174.
16. Choi, K.C.; Hwang, J.M.; Bang, S.J.; Kim, B.T.; Kim, D.H.; Chae, M.; Lee, S.A.; Choi, G.J.; Kim, D.H.; Lee, J.C. Chloroform extract of alfalfa (*Medicago sativa*) inhibits lipopolysaccharide-induced inflammation by downregulating ERK/NF-κB signaling and cytokine production. *J. Medic. Food* **2013**, *16*, 410–420. [CrossRef] [PubMed]
17. Bora, K.S.; Sharma, A. Phytochemical and pharmacological potential of *Medicago sativa*: A review. *Pharm. Biol.* **2011**, *49*, 211–220. [CrossRef] [PubMed]
18. Agati, G.; Biricolti, S.; Guidi, L.; Ferrini, F.; Fini, A.; Tattini, M. The biosynthesis of flavonoids is enhanced similarly by UV radiation and root zone salinity in *L. vulgare* leaves. *J. Plant Physiol.* **2011**, *168*, 204–212. [CrossRef] [PubMed]
19. Cornacchione, M.V.; Suarez, D.L. Emergence, forage production, and ion relations of alfalfa in response to saline waters. *Crop Sci.* **2015**, *55*, 444–457. [CrossRef]
20. Suarez, D.L.; Simunek, J. Unsatchem: Unsaturated water and solute transport model with equilibrium and kinetic chemistry. *Soil Sci. Soc. Am. J.* **1997**, *61*, 1633–1646. [CrossRef]
21. Kalu, B.A.; Fick, G. Quantifying morphological development of alfalfa for studies of herbage quality. *Crop Sci.* **1981**, *21*, 267–271. [CrossRef]

22. Prior, R.L.; Hoang, H.; Gu, L.; Wu, X.; Bacchiocca, M.; Howard, L.; Hampsch-Woodill, M.; Huang, D.; Ou, B.; Jacob, R. Assays for hydrophilic and lipophilic antioxidant capacity [oxygen radical absorbance capacity (ORAC)] of plasma and other biological and food samples. *J. Agric. Food Chem.* **2003**, *51*, 3273–3279. [CrossRef] [PubMed]
23. Singleton, V.L.; Rossi, J.A. Colorimetry of total phenolics with phosphomolybdic-phosphotungstic acid reagents. *Am. J. Enol. Vitic.* **1965**, *16*, 144–158.
24. Slinkard, K.; Singleton, V.L. Total phenol analysis: Automation and comparison with manual methods. *Am. J. Enol. Vitic.* **1977**, *28*, 49–55.
25. AOAC. *Official Methods of Analysis of AOAC International*, 17th ed.; Association of Official Analytical Chemists: Gaithersburg, MD, USA, 2000; p. 2000.
26. National Forage Testing Association. Forage Analysis Procedures. Available online: http://www.foragetesting.org/files/LaboratoryProcedures.pdf (accessed on 27 May 2013).
27. Atwater, W.O.; Bryant, A.P. *The Chemical Composition of American Food Materials*; USDA Office of Experiment Stations, Ed.; US Government Printing Office: Washington, DC, USA, 1906; p. 87.
28. Di Rienzo, J.A.; Casanoves, F.; Balzarini, M.G.; González, L.; Tablada, M.; Robledo, C.W. Infostat. Grupo Infostat; FCA Universidad Nacional de Córdoba, Argentina. Available online: Http://www.Infostat.Com.Ar (accessed on 30 August 2013).
29. NRC. *Nutrient Requirements of Dairy Cattle*, 7th ed.; National Academy Press: Washington, DC, USA, 2001.
30. Getachew, G.; Pittroff, W.; DePeters, E.J.; Putnam, D.H.; Dandekar, A.; Goyal, S. Influence of tannic acid application on alfalfa hay: *In vitro* rumen fermentation, serum metabolites and nitrogen balance in sheep. *Animal* **2008**, *2*, 381–390. [CrossRef] [PubMed]
31. Minson, D.J. *Forage in Ruminant Nutrition*; Academic Press: San Diego, CA, USA, 1990; p. 463.
32. Buxton, D.R. Quality-related characteristics of forages as influenced by plant environment and agronomic factors. *Anim. Feed Sci. Technol.* **1996**, *59*, 37–49. [CrossRef]
33. Putnam, D.H.; Robinson, P.; DePeters, E. Forage quality and testing. In *Irrigated Alfalfa Management for Mediterranean and Desert Zones*; Publication 3512; Summers, C.G., Putnam, D.H., Eds.; University of California/Agricultural and Natural Resources: Davis, CA, USA, 2008; pp. 241–264.
34. Lemaire, G.; Avice, J.C.; Kim, T.H.; Ourry, A. Developmental changes in shoot N dynamics of lucerne (*Medicago sativa* L.) in relation to leaf growth dynamics as a function of plant density and hierarchical position within the canopy. *J. Exp. Bot.* **2005**, *56*, 935–943. [CrossRef] [PubMed]
35. Lemaire, G.; Khaity, M.; Onillon, B.; Allirand, J.M.; Chartier, M.; Gosse, G. Dynamics of accumulation and partitioning of N in leaves, stems and roots of lucerne (*Medicago sativa* L.) in a dense canopy. *Ann. Bot.* **1992**, *70*, 429–435.
36. Hoffman, G.J.; Maas, E.V.; Rawlins, S.L. Salinity-ozone interactive effects on alfalfa yield and water relations. *J. Environ. Qual.* **1975**, *4*, 326–331. [CrossRef]
37. Mueller, S.C.; Teuber, L.R. Alfalfa growth and development. In *Irrigated Alfalfa Management for Mediterranean and Desert Zones*; Publication 3512; Summers, C.G., Putnam, D.H., Eds.; University of California/Agricultural and Natural Resources: Davis, CA, USA, 2008; pp. 31–38.
38. Orloff, S.B.; Putnam, D.H. Harvest strategies for alfalfa. In *Irrigated Alfalfa Management for Mediterranean and Desert Zones*; Publication 3512; Summers, C.G., Putnam, D.H., Eds.; University of California/Agricultural and Natural Resources: Davis, CA, USA, 2008; pp. 197–207.
39. Yurtseven, S. The nutrient and energy contents of medicago varieties growth in salt-affected soils of the harran plain. *Hayvansal Üretim* **2011**, *52*, 39–45.
40. USDA-CO, D.O.A.M.N.S. California hay report. Available online: http://www.ams.usda.gov/mnreports/ml_gr311.txt (accessed on 9 May 2013).
41. Hussain, G.; Al-Jaloud, A.A.; Ai-Shammary, S.F.; Karimulla, S. Effect of saline irrigation on the biomass yield, and the protein, nitrogen, phosphorus, and potassium composition of alfalfa in a pot experiment. *J. Plant Nutr.* **1995**, *18*, 2389–2408. [CrossRef]
42. Isla, R.; Aragüés, R. Response of alfalfa (*Medicago sativa* L.) to diurnal and nocturnal saline sprinkler irrigations. I: Total dry matter and hay quality. *Irrig. Sci.* **2009**, *27*, 497–505. [CrossRef]
43. Halim, R.A.; Buxton, D.R.; Hattendorf, M.J.; Carlson, R.E. Water-stress effects on alfalfa forage quality after adjustment for maturity differences. *Agron. J.* **1989**, *81*, 189–194. [CrossRef]

44. Pessarakli, M.; Huber, J.T. Biomass production and protein synthesis by alfalfa under salt stress. *J. Plant Nutr.* **1991**, *14*, 283–293. [CrossRef]
45. Meyer, R.D.; Marcum, D.B.; Orloff, S.B.; Schmierer, J.L. Alfalfa fertilization strategies. In *Irrigated Alfalfa Management for Mediterranean and Desert Zones*; Publication 3512; Summers, C.G., Putnam, D.H., Eds.; University of California/Agricultural and Natural Resources: Davis, CA, USA, 2008; pp. 73–87.
46. te Velde, G. *Milking Jersey's vs. Holstein's on a Commercial Dairy in California: Milk Production, Feed Efficiency, Intake, Costs, and Advantages*; BS, California Politechnic State University: San Luis Obispo, CA, USA, 2013.
47. Koong, L.-J.; Wise, M.B.; Barrick, E.R. Effect of elevated dietary levels of iron on the performance and blood constituents of calves. *J. Anim. Sci.* **1970**, *31*, 422–427. [PubMed]
48. Ammerman, C.B.; Goodrich, R.D. Advances in mineral nutrition in ruminants. *J. Anim. Sci.* **1983**, *57*, 519–533. [PubMed]
49. Fisher, L.J.; Dinn, N.; Tait, R.M.; Shelford, J.A. Effect of level of dietary potassium on the absorption and excretion of calcium and magnesium by lactating cows. *Can. J. Anim. Sci.* **1994**, *74*, 503–509. [CrossRef]
50. Grattan, S.R.; Grieve, C.M.; Poss, J.A.; Robinson, P.H.; Suarez, D.L.; Benes, S.E. Evaluation of salt-tolerant forages for sequential water reuse systems: III. Potential implications for ruminant mineral nutrition. *Agric. Water Manage.* **2004**, *70*, 137–150. [CrossRef]
51. Grunes, D.L.; Stout, P.R.; Brownell, J.R. Grass tetany of ruminants. In *Advances in Agronomy*; Brady, N.C., Ed.; Academic Press: London, UK, 1970; Volume 22, pp. 331–374.
52. Jittakhot, S.; Schonewille, J.T.; Wouterse, H.; Focker, E.J.; Yuangklang, C.; Beynen, A.C. Effect of high magnesium intake on apparent magnesium absorption in lactating cows. *Anim. Feed Sci. Technol.* **2004**, *113*, 53–60. [CrossRef]
53. Grieve, C.M.; Poss, J.A.; Grattan, S.R.; Suarez, D.L.; Benes, S.E.; Robinson, P.H. Evaluation of salt-tolerant forages for sequential water reuse systems: II. Plant–ion relations. *Agric. Water Manage.* **2004**, *70*, 121–135. [CrossRef]
54. Arthington, J. Know the Sulfur Content of Your Forage—Test It. Available online: http://rcrec-ona.ifas.ufl.edu/pdf/publications/ona-reports/2013/5%202013/or5-13.html (accessed on 9 May 2013).
55. Spears, J.W.; Weiss, W.P. Role of antioxidants and trace elements in health and immunity of transition dairy cows. *Vet. J.* **2008**, *176*, 70–76. [CrossRef] [PubMed]
56. Corah, L. Trace mineral requirements of grazing cattle. *Anim. Feed Sci. Technol.* **1996**, *59*, 61–70. [CrossRef]
57. Scaletti, R.W.; Amaral-Phillips, D.M.; Harmon, R.J. *Using Nutrition to Improve Immunity Against Disease in Dairy Cattle: Copper, Zinc, Selenium, and Vitamin E*; University of Kentucky: Lexington, KY, USA, 1999; pp. 1–4.
58. Jones, M.; van der Merwe, D. *Copper Toxicity in Sheep is on the Rise in Kansas and Nebraska*; Kansas State University/Veterinary Medical Teaching Hospital: Manhattan, KS, USA, 2008; p. 5.
59. Stochmal, A.; Piacente, S.; Pizza, C.; De Riccardis, F.; Leitz, R.; Oleszek, W. Alfalfa (*Medicago sativa* L.) flavonoids. 1. Apigenin and luteolin glycosides from aerial parts. *J. Agric. Food Chem.* **2001**, *49*, 753–758. [CrossRef] [PubMed]
60. Petridis, A.; Therios, I.; Samouris, G.; Tananaki, C. Salinity-induced changes in phenolic compounds in leaves and roots of four olive cultivars (*Olea europaea* L.) and their relationship to antioxidant activity. *Environ. Experim. Bot.* **2012**, *79*, 37–43. [CrossRef]
61. Tattini, M.; Remorini, D.; Pinelli, P.; Agati, G.; Saracini, E.; Traversi, M.L.; Massai, R. Morpho-anatomical, physiological and biochemical adjustments in response to root zone salinity stress and high solar radiation in two mediterranean evergreen shrubs, *Myrtus communis* and *Pistacia lentiscus*. *New Phytol.* **2006**, *170*, 779–794. [CrossRef] [PubMed]
62. Brisibe, E.A.; Umoren, U.E.; Brisibe, F.; Magalhäes, P.M.; Ferreira, J.F.S.; Luthria, D.; Wu, X.; Prior, R.L. Nutritional characterisation and antioxidant capacity of different tissues of *Artemisia annua* L. *Food Chem.* **2009**, *115*, 1240–1246. [CrossRef]
63. Ferreira, J.F.S. Artemisia Species in Small Ruminant Production: Their Potential Antioxidant and Anthelmintic Effects. In *Appalachian Workshop and Research Update: Improving Small Ruminant Grazing Practices*; Morales, M., Ed.; Mountain State University/USDA: Beaver, WV, USA, 2009; pp. 53–70.
64. Katiki, L.M.; Ferreira, J.F.S.; Gonzalez, J.M.; Zajac, A.M.; Lindsay, D.S.; Chagas, A.C.S.; Amarante, A.F.T. Anthelmintic effect of plant extracts containing condensed and hydrolyzable tannins on *Caenorhabditis elegans*, and their antioxidant capacity. *Vet. Parasitol.* **2013**, *192*, 218–227. [CrossRef] [PubMed]

65. Al-Dosari, M.S. *In vitro* and *in vivo* antioxidant activity of alfalfa (*Medicago sativa* L.) on carbon tetrachloride intoxicated rats. *Am. J. Chin. Med.* **2012**, *40*, 779. [CrossRef] [PubMed]
66. Ferreira, J.F.S.; Luthria, D.L. Drying affects artemisinin, dihydroartemisinic acid, artemisinic acid, and the antioxidant capacity of *Artemisia annua* L. Leaves. *J. Agric. Food Chem.* **2010**, *58*, 1691–1698. [CrossRef] [PubMed]
67. Tekippe, J.A.; Hristov, A.N.; Heyler, K.S.; Cassidy, T.W.; Zheljazkov, V.D.; Ferreira, J.F.S.; Karnati, S.K.; Varga, G.A. Rumen fermentation and production effects of *Origanum vulgare* L. Leaves in lactating dairy cows. *J. Dairy Sci.* **2011**, *94*, 5065–5079. [CrossRef] [PubMed]

# 9

# Leaf Length Variation in Perennial Forage Grasses

**Philippe Barre [1,*], Lesley B. Turner [2] and Abraham J. Escobar-Gutiérrez [1]**

1 INRA, UR4, Le Chêne RD 150, Lusignan 86600, France; abraham.escobar@lusignan.inra.fr
2 IBERS, Aberystwyth University, Gogerddan Campus, Aberystwyth SY23 3EE, UK; lbt@aber.ac.uk
* Author to whom correspondence should be addressed; philippe.barre@lusignan.inra.fr

Academic Editor: Cory Matthew

**Abstract:** Leaf length is a key factor in the economic value of different grass species and cultivars in forage production. It is also important for the survival of individual plants within a sward. The objective of this paper is to discuss the basis of within-species variation in leaf length. Selection for leaf length has been highly efficient, with moderate to high narrow sense heritability. Nevertheless, the genetic regulation of leaf length is complex because it involves many genes with small individual effects. This could explain the low stability of QTL found in different studies. Leaf length has a strong response to environmental conditions. However, when significant genotype × environment interactions have been identified, their effects have been smaller than the main effects. Recent modelling-based research suggests that many of the reported environmental effects on leaf length and genotype × environment interactions could be biased. Indeed, it has been shown that leaf length is an emergent property strongly affected by the architectural state of the plant during significant periods prior to leaf emergence. This approach could lead to improved understanding of the factors affecting leaf length, as well as better estimates of the main genetic effects.

**Keywords:** leaf length; forage; turf; grass; plant modelling

---

## 1. Introduction

The leaf length of forage grasses shows high variability between species, ranging from a few centimeters to more than a meter. The choice of species for sowing in a sward depends on the use of the sward (e.g., grazing, silage or hay production, biomass production, and permanent *versus* short term grassland) and on environmental conditions (climate and soil). Once the choice of species has been made, there is still great within-species variation which can be optimized. In this paper we focus on the origin of this within-species variation in leaf length.

Leaf length is a key factor determining the vegetative yield of forage grasses, and has therefore become one of the main breeding objectives [1,2]. Many studies have been conducted to determine the morphological and physiological traits which could explain vegetative yield variation in swards. Leaf length, leaf elongation rate and yield per tiller (which are generally highly positively correlated) seem to be most important, ahead of tiller density which tends to become stabilized in dense canopies [3–7]. Apparently in contradiction with this, a study on perennial ryegrass [8] showed that the rate of tiller production, rather than leaf length, explained the difference in vegetative production after three cycles of divergent selection for dry matter yield. However, the yield data were obtained from spaced plants [9]. This confirms the hypothesis that the yields measured from spaced plants and dense canopies are not entirely explained by the same morphological characteristics, and that the yield per plant of spaced plants is not a good criterion for selection aiming to increase yield in swards [10]. The leaf length that maximizes vegetative yield depends on the cutting frequency [5,6,11]. Under infrequent cutting, long-leaved genotypes yield more than short-leaved genotypes. Conversely, under frequent cutting, short-leaved genotypes tend to yield more than long-leaved genotypes. Nevertheless, long-leaved genotypes show higher plasticity than short-leaved genotypes with regard to the cutting regime. Long-leaved genotypes can decrease their leaf length when cut frequently, whereas short-leaved genotypes cannot increase their leaf length when cut infrequently [12]. The consequence of this is that long-leaved genotypes seem better for vegetative yield in swards than short-leaved genotypes, irrespective of the cutting regime. However, cutting height is also important in this context. Indeed, there is genotype-dependent variation in the ability to adapt to severe defoliation by decreasing the height of the leaf growth zone, thus protecting caulinary meristems [13]. In contrast to dry matter yield, leaf length and related parameters evaluated on spaced plants, even on seedlings, can be good selection criteria for improving the vegetative yield of swards [14–17]. Moreover, leaf length is positively correlated with short-term intake when grazed by dairy cows [18,19].

Leaf length in grasses plays an essential role in shaping the physical structure of the canopy and consequently on competition for light within the sward. One of the major adaptive responses to light competition in plants is an increase of plant height, *i.e.*, leaf length during the vegetative period in grasses [20–24]. This increase in plant height is affected by phenotypic plasticity. Nevertheless, in a sward composed of different genotypes and/or species, phenotypic plasticity cannot always compensate for genetic differences between plants, ultimately leading to the death of some genotypes. For example, in a sward with long- and short-leaved genotypes of perennial ryegrass under infrequent cutting, the proportion of short-leaved genotypes decreases due to competition for light, as in Figure 1 [25]. Like other phototrophic organisms, light acquisition is essential for the survival of perennial forage grasses. In sown grasslands consisting of many genotypes and often of several species, the plant height of the different constituents should be optimized in order to avoid the fade-out of one of the species [26]. There are two possible strategies to optimize mixture composition. The first is to include constituents with similar patterns of seasonal growth which are therefore in competition for light acquisition. In this case, constituents with similar competitive ability, including plant height and tillering, must be chosen in order to avoid exclusion. The second strategy is to include constituents which grow at different periods of the year, *i.e.* asynchronous growth, which would provide more stable production over the growing season [27].

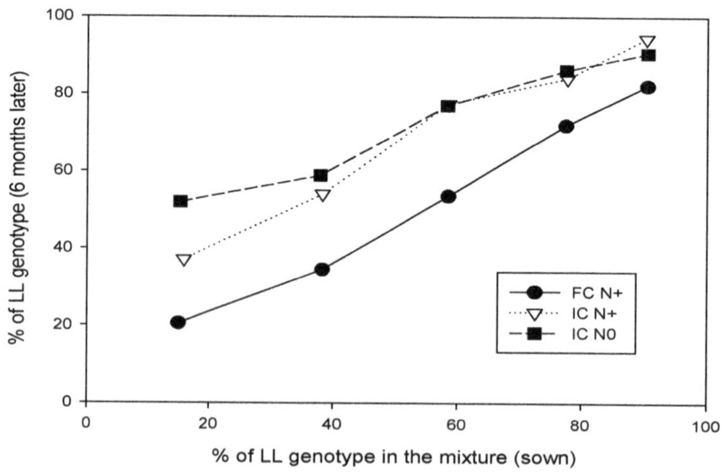


**Figure 1.** Genotypic responses of mixtures under three management regimes showing a decrease in the percentage of short-leaved genotypes under infrequent cutting and no change under frequent cutting. (FC N+, frequent cutting with nitrogen; IC N+, infrequent cutting with nitrogen; IC N0, infrequent cutting with no nitrogen). LL: long-leaved genotype. Figure from [25].

It is important to remember that leaf length in grasses is greatly influenced by the developmental stage of the plant: reproductive or vegetative [1]. Growth rate increases markedly following flower induction and before any visible stem elongation (Figure 2) [28–30]. This change in leaf growth rate seems to be due to an increase in cell division which could be related to environmental regulation of the gibberellins pathway [31–33]. Consequently, for a given genotype, leaf length varies greatly depending on whether the leaf grows on an axis which has been induced for flowering or not. Moreover, leaf elongation rate during the two growth phases seems to be genetically independent to a large extent [34]. This implies that a genotype × growth season (reproductive *versus* vegetative, *i.e.*, spring *versus* fall) interaction is expected. In this paper we will focus on variation within a growing phase and the trait of interest will be leaf length (sheath and lamina) and not stem elongation.

The objective of this paper is to discuss the origin of variation in leaf length within perennial forage grasses, *i.e.*, genetics (heritability and genetic architecture), environment (temperature, nitrogen, light) and genetic × environment interaction, and to produce new insights into this variation by including recent advances in plant morphogenesis modelling.

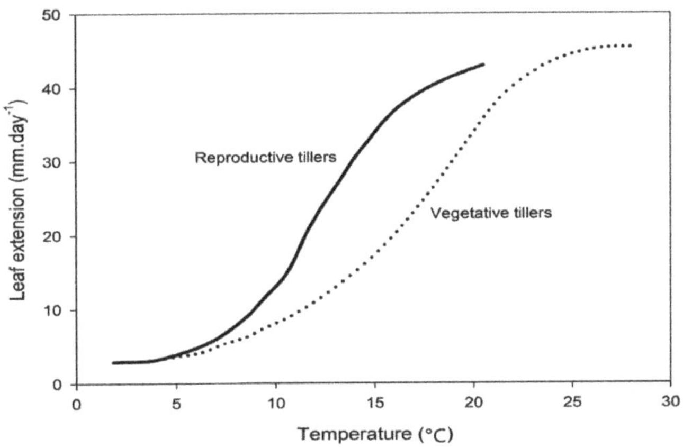


**Figure 2.** Contrasted responses of leaf extension to temperature before and after flowering induction in perennial ryegrass. Adapted from [30] with data from [35,36].

## 2. Genetic Variation in Leaf Length

### 2.1. Heritability

Broad-sense heritability ($H^2$) reflects all the genetic contributions to a population's phenotypic variance. It is calculated as the genetic variance divided by the sum of the genetic and the environmental variances: $H^2 = \sigma^2_G/(\sigma^2_G + \sigma^2_E)$. By definition it depends on the genotypes included in the population and on the experimental design (field heterogeneity, one or several locations and years). Another estimate of broad-sense heritability, taking into account the number of replicates, is commonly used: $H^2_{average} = \sigma^2_G/(\sigma^2_G + \sigma^2_E/n)$ with n being the number of replicates. The $H^2_{average}$ is useful to assess the accuracy in the prediction of genotypic values, but since it depends on the experimental design it should not be used to compare studies. Generally comparison of heritabilities should be done with caution. In one location, during one growing phase, on spaced-plants and on equivalent leaves (same rank), leaf length broad-sense heritability ($H^2$) is high: above 0.65 [37–39]. It decreases when several environments and/or years are taken into account and also, as expected, when reproductive and vegetative growing stages are included: 0.3–0.6 [38,40–43]. Differences in vernalization requirements between genotypes exist in perennial grasses [44–46] and could lead to differences in the date of flower induction, which in turn could lead to differences in leaf length. These differences in leaf length between genotypes do not reflect true differences in leaf length potential but rather differences in earliness of flower induction. This phenomenon could lead to false genotype × environment interactions with regard to leaf length.

Narrow sense heritability ($h^2$) is defined as the additive variance divided by the phenotypic variance: $h^2 = \sigma^2_A/\sigma^2_P$, with the additive variance being the variance of the average effects of the alleles, representing the genetic component of variance responsible for parent-offspring resemblance.

$h^2$ is directly linked to the expected genetic gain. The deviation from this expectation is due to interaction between alleles of the same locus (dominance) or of different loci (epistasis). Narrow sense heritability of leaf length is high, above 0.65, which reveals large additive effects and small dominance and epistatic effects [17,47].

Leaf length has been demonstrated to respond to selection. Several divergent selections for leaf length or leaf growth parameters evaluated on spaced plants showed strong responses with realized heritabilities from 0.2 to 0.6 depending on the plant material, as in Figure 3 [15,48,49]. Moreover, selection for leaf length on spaced plants had an effect on the vegetative yield in swards [6,14,16].

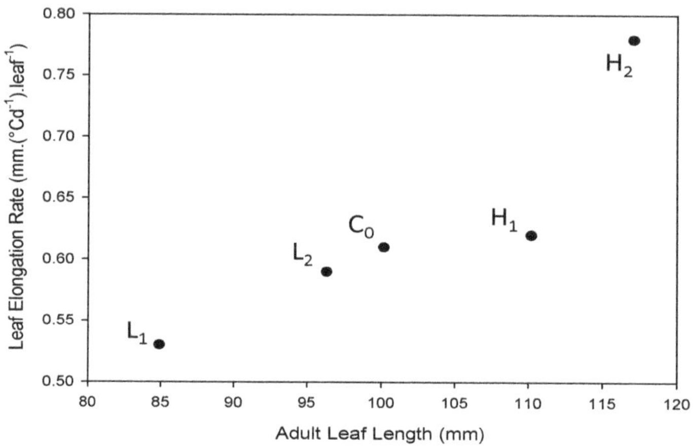


**Figure 3.** Response of leaf elongation rate and adult leaf length to divergent selection for lamina length on spaced plants (adapted from [15]). Means of long-leaved (H) and short-leaved (L) populations after 1 or 2 years of selection and mean of the initial ($C_0$) population. Examples of a turf and a forage variety are also presented.

In conclusion, considerable variation in leaf length exists within grass species, and, when measured properly (same leaf rank, same growing stage: reproductive or vegetative, no stem elongation), leaf length appears to be highly heritable and to respond to selection even when genotype × environment interactions are present.

### *2.2. Genetic Architecture*

Several QTL studies on leaf length or related parameters (leaf elongation rate LER, plant height at vegetative stage) have been performed on forage grasses (mainly on perennial ryegrass) showing the complex genetic architecture of these traits Table 1 [50]. QTL with small effects, *i.e.*, explaining less than 15% of the phenotypic variance, have been detected on all seven chromosomes. Moreover, the QTL together commonly explained less than half of the phenotypic variance. This seems to be the case even in crosses between forage and turf genotypes [51]. QTL often have inconsistencies between cuts within a year, between years and between locations.

These results could be seen to contradict the high heritability of this trait, but there are several possible explanations. It is not surprising to find different QTL for leaf length parameters in the reproductive stage in spring (even very early in floral development) and in the vegetative stage in autumn. Indeed, the limitations to leaf growth in the two stages are not the same [32]. Another source of variability arises from the way in which the measurements of leaf length are taken. Ideally, to be comparable between genotypes, the same leaf rank must be measured. Often this is not possible in the field and the youngest fully emerged leaf or plant/leaf height is measured. Regrowth after cutting to a particular plant height has also been used. The most appropriate measurement will depend on the circumstances of the experiment. Other than these physiological considerations, if leaf length genetic variation is based on many genes with small effects, it is impossible to detect them all with the population sizes historically used in QTL studies on forage grasses (from 100 to 400 genotypes).

Randomly different QTL can be obtained in the same environment with two small (100–500) sets of plants from the same population; this is called the Beavis effect [52]. Different QTL could be detected in different environments if an environment has a high error variance that prevents the detection of a QTL [53]. In addition, QTL × environment interactions may well impact on the inconsistency of QTL.

**Table 1.** QTL for leaf length and related parameters in perennial ryegrass. The percentages of phenotypic variance are given for each linkage group (LG).

| Ref. | Parents | Traits | LG1 | LG2 | LG3 | LG4 | LG5 | LG6 | LG7 |
|---|---|---|---|---|---|---|---|---|---|
| [39] | Pop8490 | Leaf length | | | | 13 | | | |
| [41] | WSC F2 Perma × Aurora | | 12 | 11 | 9 | 10 | 25 | 31–38 | 15 |
| [54] | ILGI p152/112 mapping family | | | | | | 6 | | |
| [42] | North African × Aurora F1 (NAx × AU6) | Leaf area | | | 12 | | | 6 | |
| [39] | Pop8490 (FL42 × FC61) | Lamina length in spring | | 9 | | 9 | | | 16 |
| [39] | Pop8490 (FL42 × FC61) | Lamina length in autumn | | 8–18 | | 10 | | 12 | 9 |
| [40] | Grasslands Impact × Grasslands Samson | Lamina length in autumn | 13 | 5 | 14–10 | 14–14 | | | 6 |
| [43] | WSC F2 Perma × Aurora | Leaf extension rate (LER) | | 14 | 11 | | | | |
| [40] | Grasslands Impact × Grasslands Samson | LER in spring | | | | | | 13 | |
| [39] | Pop8490 | LER in autumn | | | | 11 | | | |
| [40] | Grasslands Impact × Grasslands Samson | LER in autumn | 9 | | 26–27 | | | | 5–8 |
| [55] | Three connected populations (elite material) | Vegetative plant height in spring | | 6–5 | 5 | | 4–4 | 4–4 | 4–9 |
| [55] | Three connected populations (elite material) | Vegetative plant height in autumn | 3 | | | 7–6 | 4 | | 4 |
| [56] | WSC F2 Perma × Aurora | Flag lamina length * | | | | | | | 10 |
| [56] | ILGI p152/112 mapping family | | | | | | | | 20 |
| [57] | Italian Veyo × Danish Falster | | | 11 | | 13–12 | | 17 | |

* Flag lamina lengths have been added even though not directly related to leaf length before stem elongation.

In conclusion, leaf length in forage grasses has a complex genetic architecture which seems to impede the detection of consistent QTL. The consequence for plant breeding is that, unless some strong QTL are identified (alleles leading to a dwarf or giant phenotype), it would be better to use molecular markers for predicting genetic values than for pyramiding favorable alleles. Phenotypic selection seems to have accumulated favorable alleles at different loci and often in a heterozygous state.

## 3. Environmental and Genetic × Environmental Interaction Effects on Leaf Length

Abundant empirical evidence demonstrates that leaf length exhibits a very high plasticity to environmental factors. Indeed, both theoreticians and experimentalists recognize that leaf length responds to sward management and to various environmental factors such as, but not limited to, temperature, nitrogen and water supply, defoliation frequency and intensity, light quantity and quality [29,58–61].

The length of a leaf is determined by its constituent cells and their length. The number of cells and the length of these cells result from cell division and elongation processes. Cell division plays a major role in the variation of leaf length within and between species [62,63]. It appears that these cellular processes are under the influence of the length of the enclosing sheaths both directly and indirectly. Experimental modifications to incise or artificially increase the pseudostem tube length directly affected the length of the leaf elongation zone and the final length of the cells [63,64]. Furthermore, modification of the pseudostem tube length with, for example, an opaque plastic tube, could also indirectly affect cell dynamics *via* control of the timing of leaf tip emergence [65]. Delaying or anticipating this event could modify both the placement of the sheath-blade boundary and the total cell number, as cessation of cell division at the base of the leaf could be triggered at the moment of leaf tip emergence from the previous sheath [63,64]. These effects may be light mediated. In order to determine (i) if physical factors other than light are involved in these responses, (ii) if this putative light effect is changed by qualitative or quantitative spectral modification, and (iii) if sheath elongation is also dynamically affected by pseudostem length, [66,67] tested the effect of pseudostem extension with plastic tubes on the leaf growth of uncut tall fescue plants. Tubes with contrasting optical properties were used: red-colored tubes which affect the "blue" domain of the spectrum, green-colored tubes which affect the Red: Far Red ratio, transparent tubes and opaque foil tubes. It appeared that reducing the passage of light through the tubes increased leaf elongation, and the length of leaves and sheaths. The effects of red and green tubes were not significantly different. These results support the hypothesis that light mediates the pseudostem morphogenetic effect. Furthermore, in this context, leaf elongation does not react to a qualitative modification of a unique domain in the light spectrum, but rather to a quantitative general decrease in irradiance [67]. Consequently, the pseudostem seems to play an essential morphogenetic role in the control of leaf elongation, mainly due to its impact on the length of the leaf growth zone and the timing of leaf tip emergence.

Genetic × environmental interactions on leaf length have been observed in multi-site trials and in trials in semi-controlled environments varying for environmental factors such as temperature, nitrogen and water supply, defoliation frequency and intensity, and light quantity and quality. Of course, the level of interaction depends on both the genetic and the environmental variation, but in general the effect of the interaction is smaller than the principal effects. For example, a study on the response of leaf length to light quality and quantity in several perennial ryegrass genotypes showed a significant genotype × environment interaction but with a smaller effect than the principal effects [20]. Furthermore, in perennial ryegrass, a divergent selection for LER in response to light (green filter *versus* transparent filter) did not create progeny with significantly different LER in response to light, as seen in Figure 4 [51].

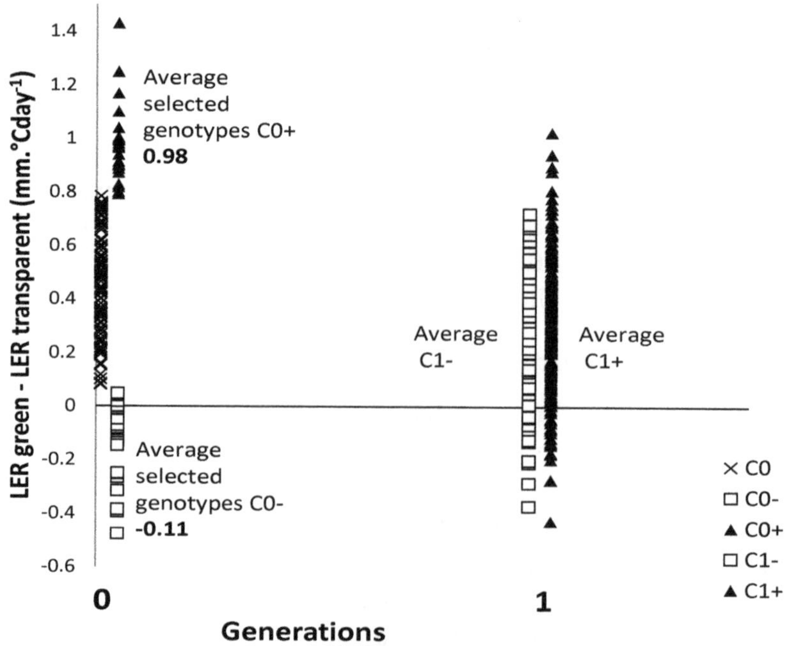


**Figure 4.** Divergent mass selection for LER in response to light (LER under green filter minus LER under transparent filter) in perennial ryegrass. Initial population: C0 and the next generation after intercrossing the selected genotypes: C1 (EU project GRASP).

## 4. The Added Value of Plant Modelling

Recent modelling-based research suggests that many of the reported environmental effects on leaf length and genotype × environment interactions could be biased; they have included both the effect of environment and the effect of self-regulatory processes during plant development [66]. Indeed, it has been shown that leaf length is an emergent property strongly affected by the architectural state of the plant during significant periods prior to leaf emergence.

As discussed above, leaf length is under genetic control, is highly heritable, and shows significant genetic variation. Nevertheless, this trait displays high plasticity that could be mediated by self-regulatory processes. Thus, leaf length is directly affected by the sheath length of the preceding leaf on the same tiller [61,63,64] in a sort of recursive loop (Figure 5). These concepts were integrated into a cybernetic framework [66]. Briefly, leaf growth follows a Beta integral function [68]. The relationship between the length of the pseudostem *i.e.*, series of sheaths from which a leaf emerge and the final length of the leaves, has been conceptualized and used to generate a growth potential (created by cell division and by the length of the leaf elongation zone) that is integrated while the leaf grows inside the pseudostem, before its tip emerges. The first phase of the growth, inside the pseudostem tube, is generic for all leaves. When the leaf tip emerges, the value of the growth potential is carried forward as the final length parameter of the growth function. This integration is taken as a synthesis-degradation process. Therefore, the longer the time elapsed from the beginning of leaf growth, the longer the final length. At this moment the proportion of sheath and of blade will also be determined according to a function described in [61]. It was found that the ratio between leaf length and sheath length of the preceding leaf is quasi-constant for a given genotype under a given environment. Model behavior and emergent properties were highly consistent with observations regarding plant morphological development, genetic variability and plasticity.

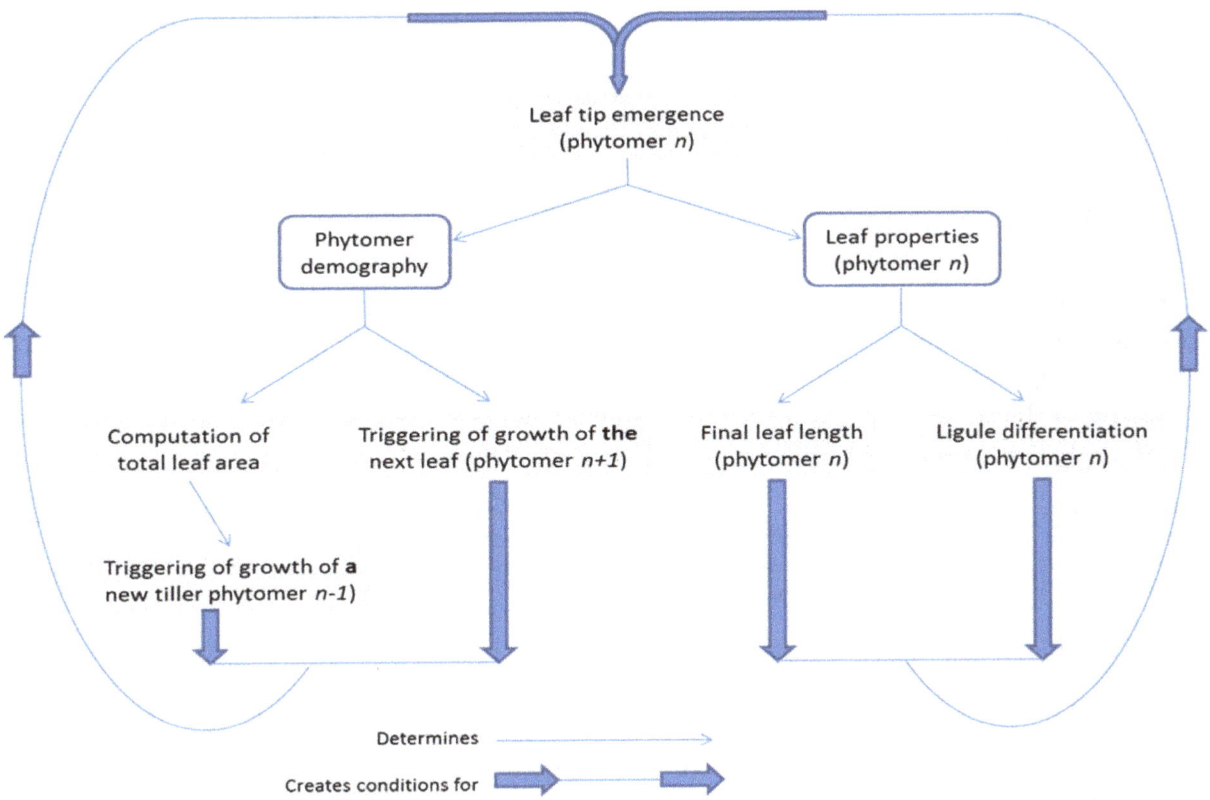


**Figure 5.** Diagram showing the cybernetic scheme implemented in the model. The recursive call to these rules automatically generates the morphology of the virtual plant by determining the number of leaves and their lengths.

A practical application of the model is that it is always possible to invert the process and to estimate the parameters of leaf growth from a range of measurements of leaf and sheath length for the preceding leaf of a given genotype in a given environment. All the measured plants have to be either vegetative or reproductive but not a mix of both. Thus, it is possible to compare genotypes and environments either by comparing the parameters of the model (not always explicit) or by comparing the simulated length of leaves of the same rank or of the same sheath length for the preceding leaf. Finally, it allows the architectural structure of the tiller and plant to be taken into account. Thus, genetic and environment effects can be properly separated from architectural effects (Figure 6).

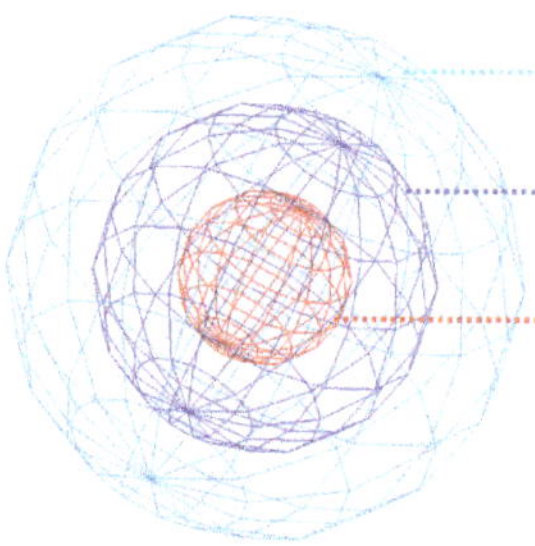


**Figure 6.** Putative nested levels of control determining ryegrass leaf length. Genetic factors regulate upstream processes that are modulated by lower contingent factors (adapted from [69]).

## 5. Conclusions

Leaf length of forage grasses is a key agronomic trait showing high intra-specific variation and plasticity in response to environmental factors. Part of the genotype × environment interaction could be explained by the methods used to measure leaf length. Since leaf length is strongly influenced by micro-environment, *i.e.*, the status of the leaf in the plant including interactions with other organs, this micro-environment should be taken into account during the estimation of genetic values. Plant-morphogenesis models could help in extracting the genetic component of leaf length variation from variability due to uncontrolled micro-environments.

The identification of the respective biological scales, or levels in the regulatory network, at which genetic and/or environmental controls on leaf length are important is challenging.

Leaf length genetic architecture seems to involve many genes with small effects but with relatively high additive effects compared with dominance effects. This complex genetic architecture suggests larger populations should be used for QTL identification and for genomic selection. A better understanding of the sources of variation in leaf length should allow a better estimation of its genetic, components which should lead to the discovery of more consistent QTL.

**Acknowledgments:** The authors thank the European Union for founding the GRASP project, Thomas Lübberstedt for its coordination, and Isabelle Cameleyre, who participated actively in the experiment on LER in response to light quality and quantity. The authors thank Laurent Hazard, Anthony Parsons, and Oxford University Press for Figures 1–3. Alban Verdenal kindly provided Figure 6.

**Author Contributions:** All authors contributed equally to the conception and writing of the paper.

## References

1. Wilkins, P.W. Breeding perennial ryegrass for agriculture. *Euphytica* **1991**, *52*, 201–204. [CrossRef]
2. Humphreys, M.O. Genetic improvement of forage crops—Past, present and future. *Can. J. Agr. Sci.* **2005**, *143*, 441–448. [CrossRef]
3. Zarrough, K.M.; Nelson, C.J.; Coutts, J.H. Relationship between tillering and forage yield of tall fescue. I. Yield. *Crop Sci.* **1983**, *23*, 333–337. [CrossRef]

4. Horst, G.L.; Nelson, C.J.; Asay, K.H. Relationship of leaf elongation to forage yield of tall fescue genotypes. *Crop Sci.* **1978**, *18*, 715–719. [CrossRef]
5. Rhodes, I. The relationship between productivity and some components of canopy structure in ryegrass (*Lolium* spp.). I. Leaf length. *Can. J. Agr. Sci.* **1969**, *73*, 315–319. [CrossRef]
6. Hazard, L.; Ghesquière, M. Productivity under contrasting cutting regimes of perennial ryegrass selected for short and long leaves. *Euphytica* **1997**, *95*, 295–299. [CrossRef]
7. Zarrough, K.M.; Nelson, C.J.; Coutts, J.H. Relationship between tillering and forage yield of tall fescue. II. Pattern of tillering. *Crop Sci.* **1983**, *23*, 338–342. [CrossRef]
8. Ceccarelli, S.; Falcinelli, M.; Damiani, F. Selection for dry matter yield in *Lolium perenne* L. II. Correlated responses under two cutting regimes. *Can. J. Plant. Sci.* **1980**, *60*, 501–508. [CrossRef]
9. Ceccarelli, S.; Falcinelli, M.; Damiani, F. Selection for dry matter yield in *Lolium perenne* L. I. Direct response to selection. *Can. J. Plant. Sci.* **1980**, *60*, 491–500. [CrossRef]
10. Hayward, M.D.; Vivero, J.L. Selection for yield in *Lolium perenne*. II. Performance of spaced plant selections under competitive conditions. *Euphytica* **1984**, *33*, 787–800. [CrossRef]
11. Hazard, L.; Ghesquière, M.; Betin, M. Breeding for management adaptation in perennial ryegrass (*Lolium perenne* L.). I. Assessment of yield under contrasting cutting frequencies and relationships with leaf morphogenesis components. *Agronomie* **1994**, *14*, 259–266. [CrossRef]
12. Hazard, L. Plasticity gives a greater flexibility to forage grass use. *Fourrages* **1996**, *147*, 293–302.
13. Gastal, F.; Dawson, L.A.; Thornton, B. Responses of plant traits of four grasses from contrasting habitats to defoliation and n supply. *Nutr. Cycling Agroecosyst.* **2010**, *88*, 245–258. [CrossRef]
14. Jones, R.J.; Nelson, C.J.; Sleper, D.A. Seedling selection for morphological characters associated with yield of tall fescue. *Crop Sci.* **1979**, *19*, 631–634. [CrossRef]
15. Hazard, L.; Ghesquière, M.; Barraux, C. Genetic variability for leaf development in perennial ryegrass populations. *Can. J. Plant. Sci.* **1996**, *76*, 113–118. [CrossRef]
16. Rhodes, I.; Mee, S. Changes in dry matter yield associated with selection for canopy characters in ryegrass. *Grass Forage Sci.* **1980**, *35*, 35–39. [CrossRef]
17. Rhodes, I. The relationship between productivity and some components of canopy structure in ryegrass (*Lolium* spp.). III. Spaced plant characters, their heritabilities and relationship to sward yield. *Can. J. Agr. Sci.* **1973**, *80*, 171–176. [CrossRef]
18. Barrett, P.D.; Laidlaw, A.S.; Mayne, C.S.; Christie, H. Pattern of herbage intake rate and bite dimensions of rotationally grazed dairy cows as sward height declines. *Grass Forage Sci.* **2001**, *56*, 362–373. [CrossRef]
19. Barre, P.; Emile, J.C.; Betin, M.; Surault, F.; Ghesquière, M.; Hazard, L. Morphological characteristics of perennial ryegrass leaves that influence short-term intake in dairy cows. *Agron. J.* **2006**, *98*, 978–985. [CrossRef]
20. Barre, P.; Gueye, B.; Gastal, F. Effect of light quality and quantity on leaf growth in *Lolium perenne* L. In Sustainable Use of Genetic Diversity in Forage and Turf breeding, Proceedings of the EUCARPIA 2009, La Rochelle, France, 11–14 May 2009; Huyghe, C., Ed.; Springer: Berlin, Germany, 2010; pp. 61–65.
21. Kephart, K.D.; Buxton, D.R.; Taylor, S.E. Growth of C3 and C4 perennial grasses under reduced irradiance. *Crop Sci.* **1992**, *32*, 1033–1038. [CrossRef]
22. Buxton, D.R.; Lentz, E.M. Performance of morphologically diverse orchardgrass clones in spaced and sward plantings. *Grass Forage Sci.* **1993**, *48*, 336–346. [CrossRef]
23. Bahmani, I.; Hazard, L.; Varlet-Granchet, C.; Betin, M.; Lemaire, G.; Matthew, C.; Thom, E.R. Differences in tillering of long- and short-leaved perennial ryegrass genetic lines under full light and shade treatments. *Crop Sci.* **2000**, *40*, 1095–1102. [CrossRef]
24. Gautier, H.; Varlet-Grancher, C. Regulation of leaf growth of grass by blue light. *Physiol. Plant.* **1996**, *98*, 424–430. [CrossRef]
25. Hazard, L.; Ghesquière, M. Evidence from the use of isozyme markers of competition in swards between short-leaved and long-leaved perennial ryegrass. *Grass Forage Sci.* **1995**, *50*, 241–248. [CrossRef]
26. Rhodes, I.; Stern, W.R. Competition for light. In *Plant Relations in Pastures*; CSIRO: East Melborne, Australia, 1978; pp. 175–189.
27. Prieto, I.; Violle, C.; Barre, P.; Durand, J.L.; Ghesquière, M.; Litrico, I. Complementary effects of species and genetic diversity on productivity and stability of sown grasslands. *Nat. Plant.* **2015**, *1*, 1–5. [CrossRef]
28. Kemp, D.R.; Eagles, C.F.; Humphreys, M.O. Leaf growth and apex development of perennial ryegrass during winter and spring. *Ann. Bot.* **1989**, *63*, 349–355.

29. Gastal, F.; Belanger, G.; Lemaire, G. A model of the leaf extension rate of tall fescue in response to nitrogen and temperature. *Ann. Bot.* **1992**, *70*, 437–442.
30. Parsons, A.J.; Robson, M.J. Seasonal changes in the physiology of S24 perennial ryegrass (*Lolium perenne* L.). I. Response of leaf extension to temperature during the transition from vegetative to reproductive growth. *Ann. Bot.* **1980**, *46*, 435–444.
31. Ball, C.C.; Parsons, A.J.; Rasmussen, S.; Shaw, C.; Rowarth, J.S. Seasonal differences in the capacity of perennial ryegrass to respond to gibberellin explained. *Proc. N. Z. Grassl. Assoc.* **2012**, *74*, 183–188.
32. Parsons, A.J.; Rasmussen, S.; Liu, Q.; Xue, H.; Ball, C.; Shaw, C. Plant growth—Resource or strategy limited: Insights from responses to gibberellin. *Grass Forage Sci.* **2013**, *68*, 577–588. [CrossRef]
33. Bernier, G. The control of floral evocation and morphogenesis. *Ann. Rev. Plant Phys. Plant Mol. Biol.* **1988**, *39*, 175–219. [CrossRef]
34. Wilkins, P.W. Genotype harvesting frequency and genotype nitrogen level interactions for annual dry-matter yield in *Lolium perenne* in relation to breeding. *Euphytica* **1989**, *41*, 207–214.
35. Peacock, M. Temperature and leaf growth in *Lolium perenne*. II the site of temperature perception. *J. Appl. Ecol.* **1975**, *12*, 115–123. [CrossRef]
36. Peacock, J.M. Temperature and leaf growth in *Lolium perenne*. III. Factors affecting seasonal differences. *J. Appl. Ecol.* **1975**, *12*, 685–697. [CrossRef]
37. Cooper, J.P.; Edwards, D. The genetic control of leaf development in *Lolium*. I. Assessment of genetic variation. *Heredity* **1961**, *16*, 63–82. [CrossRef]
38. Auzanneau, J.; Huyghe, C.; Escobar-Gutierrez, A.J.; Julier, B.; Gastal, F.; Barre, P. Association study between the gibberellic acid insensitive gene and leaf length in a *Lolium perenne* L. synthetic variety. *BMC Plant Biol.* **2011**, *11*, 183–196. [CrossRef] [PubMed]
39. Barre, P.; Moreau, L.; Mi, F.; Turner, L.; Gastal, F.; Julier, B.; Ghesquiere, M. Quantitative trait loci for leaf length in perennial ryegrass (*Lolium perenne* L.). *Grass Forage Sci.* **2009**, *64*, 310–321. [CrossRef]
40. Sartie, A.M.; Matthew, C.; Easton, H.S.; Faville, M.J. Phenotypic and QTL analyses of herbage production-related traits in perennial ryegrass (*Lolium perenne* L.). *Euphytica* **2011**, *182*, 295–315. [CrossRef]
41. Kobayashi, S.; Humphreys, M.O.; Tase, K.; Sanada, Y.; Yamada, T. Molecular marker dissection of ryegrass plant development and its response to growth environments and foliage cuts. *Crop Sci.* **2011**, *51*, 600–611. [CrossRef]
42. Pearson, A.; Cogan, N.O.I.; Baillie, R.C.; Hand, M.L.; Bandaranayake, C.K.; Erb, S.; Wang, J.; Kearney, G.A.; Gendall, A.R.; Smith, K.F.; *et al.* Identification of QTLs for morphological traits influencing waterlogging tolerance in perennial ryegrass (*Lolium perenne* L.). *Theor. Appl. Genet.* **2011**, *122*, 609–622. [CrossRef] [PubMed]
43. Turner, L.; Cairns, A.; Armstead, I.; Thomas, H.; Humphreys, M.W.; Humphreys, M.O. Does fructan have a functional role in physiological traits? Investigation by quantitative trait locus mapping. *New Phytol.* **2008**, *179*, 765–777. [CrossRef] [PubMed]
44. Andersen, J.; Jensen, L.; Asp, T.; Lübberstedt, T. Vernalization response in perennial ryegrass (*Lolium perenne* L.) involves orthologues of diploid wheat (*Triticum monococcum*) Vrn1 and rice (*Oryza sativa*) HD1. *Plant Mol. Biol.* **2006**, *60*, 481–494. [CrossRef] [PubMed]
45. Jokela, V.; Virkajarvi, P.; Tanskanen, J.; Seppanen, M.M. Vernalization, gibberellic acid and photoperiod are important signals of yield formation in timothy (*Phleum pratense*). *Physiol. Plant.* **2014**, *152*, 152–163. [CrossRef] [PubMed]
46. Fjellheim, S.; Boden, S.; Trevaskis, B. The role of seasonal flowering responses in adaptation of grasses to temperate climates. *Front. Plant Sci.* **2014**, *5*. [CrossRef] [PubMed]
47. Ghesquière, M.; Hazard, L.; Betin, M. Breeding for management adaptation in perennial ryegrass (*Lolium perenne* L.). II. Genetic variability and heritability of leaf morphogenesis components. *Agronomie* **1994**, *14*, 267–272. [CrossRef]
48. Edwards, D.; Cooper, J.P. The genetic control of leaf development in *Lolium*. II. Response to selection. *Heredity* **1963**, *18*, 307–317. [CrossRef]
49. Reeder, L.; Sleper, D.; Nelson, C. Response to selection for leaf area expansion rate of tall fescue. *Crop Sci.* **1984**, *24*, 97–100. [CrossRef]
50. Shinozuka, H.; Cogan, N.O.I.; Spangenberg, G.C.; Forster, J.W. Quantitative trait locus (QTL) meta-analysis and comparative genomics for candidate gene prediction in perennial ryegrass (*Lolium perenne* L.). *BMC Genet.* **2012**, *13*, 101–113. [CrossRef] [PubMed]

51. Barre, P.; Cameleyre, I.; Huyghe, C.; Gastal, F.; Lübberstedt, T.; INRA, Lusignan, France. Unpublished work. 2007.
52. Xu, S.Z. Theoretical basis of the beavis effect. *Genetics* **2003**, *165*, 2259–2268. [PubMed]
53. Bernardo, R. Molecular markers and selection for complex traits in plants: Learning from the last 20 years. *Crop Sci.* **2008**, *48*, 1649–1664. [CrossRef]
54. Yamada, T.; Jones, E.S.; Cogan, N.O.I.; Vecchies, A.C.; Nomura, T.; Hisano, H.; Shimamoto, Y.; Smith, K.F.; Hayward, M.D.; Forster, J.W. QTL analysis of morphological, developmental, and winter hardiness-associated traits in perennial ryegrass. *Crop Sci.* **2004**, *44*, 925–935. [CrossRef]
55. Pauly, L.; Flajoulot, S.; Garon, J.; Julier, B.; Beguier, V.; Barre, P. Detection of favorable alleles for plant height and crown rust tolerance in three connected populations of perennial ryegrass (*Lolium perenne* L.). *Theor. Appl. Genet.* **2012**, *124*, 1139–1153. [CrossRef] [PubMed]
56. Armstead, I.; Turner, L.; Marshall, A.; Humphreys, M.; King, I.; Thorogood, D. Identifying genetic components controlling fertility in the outcrossing grass species perennial ryegrass ( *Lolium perenne* L.) by quantitative trait loci analysis and comparative genetics. *New Phytol.* **2008**, *178*, 559–571. [CrossRef] [PubMed]
57. Studer, B.; Jensen, L.B.; Hentrup, S.; Brazauskas, G.; Kölliker, R.; Lübberstedt, T. Genetic characterisation of seed yield and fertility traits in perennial ryegrass (*Lolium perenne* L.). *Theor. Appl. Genet.* **2008**, *117*, 781–791. [CrossRef] [PubMed]
58. Barillot, R.; Frak, E.; Combes, D.; Durand, J.-L.; Escobar-Gutierrez, A.J. What determines the complex kinetics of stomatal conductance under blueless par in *Festuca arundinacea*? Subsequent effects on leaf transpiration. *J. Exp. Bot.* **2010**, *61*, 2795–2806. [CrossRef] [PubMed]
59. Forde, B.J. Effect of various environments on the anatomy and growth of perennial ryegrass and cocksfoot. *N. Z. J. Bot.* **1966**, *4*, 13.
60. Duru, M.; Ducrocq, H. Growth and senescence of the successive grass leaves on a tiller. Ontogenic development and effect of temperature. *Ann. Bot.* **2000**, *85*, 635–643. [CrossRef]
61. Verdenal, A.; Combes, D.; Escobar-Gutierrez, A.J. A study of ryegrass architecture as a self-regulated system, using functional-structural plant modelling. *Funct. Plant Biol.* **2008**, *35*, 911–924. [CrossRef]
62. Gastal, F.; Barre, P.; Carré, S. What are the cellular components underlying the genetic diversity of leaf growth in *Lolium perenne* L.? In Sustainable Use of Genetic Diversity in Forage and Turf Breeding, Proceedings of the EUCARPIA 2009, La Rochelle, France, 11–14 May 2009; Huyghe, C., Ed.; Springer: Berlin, Germany, 2010; pp. 95–100.
63. Sugiyama, S. Developmental basis of interspecific differences in leaf size and specific leaf area among C3 grass species. *Funct. Ecol.* **2005**, *19*, 916–924. [CrossRef]
64. Casey, I.; Brereton, A.; Laidlaw, A.; McGilloway, D. Effects of sheath tube length on leaf development in perennial ryegrass (*Lolium perenne* L.). *Ann. Appl. Biol.* **1999**, *134*, 251–257. [CrossRef]
65. Wilson, R.E.; Laidlaw, A.S. The role of the sheath tube in the development of expanding leaves in perennial ryegrass. *Ann. Appl. Biol.* **1985**, *106*, 385–391. [CrossRef]
66. Andrieu, B.; Hillier, J.; Birch, C. Onset of sheath extension and duration of lamina extension are major determinants of the response of maize lamina length to plant density. *Ann. Bot.* **2006**, *98*, 1005–1016. [CrossRef] [PubMed]
67. Verdenal, A.; Combes, D.; Escobar-Gutiérrez, A.J. Programmable and self-organised processes in plant morphogenesis: The architectural development of ryegrass. In *Studies on Complexity: Morphogenetic Engineering*; Doursa, R.S., Michel, O., Eds.; Springer: Berlin, Germany, 2012; pp. 501–517.
68. Verdenal, A. De la Simulation de la Morphogénèse de L'appareil Aérien du Ray-Grass Anglais (*Lolium Perenne* L.). Exploration d'un Schéma Cybernétique Inspiré du Concept D'auto-Organisation et Applications. Ph.D. Thesis, The Université de Poitiers, Poitiers, France, December 2009; p. 190.
69. Migault, V.; Combes, D.; Barre, P.; Gueye, B.; Louarn, G.; Escobar-Gutiérrez, A.J. Improved modelling of ryegrass foliar growth. In Proceedings of the Plant Growth Modeling, Simulation, Visualization and Applications—PMA12, Shanghai, China, 31 October–3 November 2012; IEEE: Beijing, China, 2012; pp. 282–288.

# 10

# Beneficial Effects of Temperate Forage Legumes that Contain Condensed Tannins

**Jennifer W. MacAdam [1,*] and Juan J. Villalba [2]**

1 Department of Plants, Soils and Climate, Utah State University, 4820 Old Main Hill, Logan, UT 84322, USA
2 Department of Wildland Resources, Utah State University, 5230 Old Main Hill, Logan, UT 84322, USA; juan.villalba@usu.edu
* Author to whom correspondence should be addressed; jennifer.macadam@usu.edu

Academic Editor: Cory Matthew

**Abstract:** The two temperate forage legumes containing condensed tannins (CT) that promote ruminant production are birdsfoot trefoil (*Lotus corniculatus* L.; BFT) and sainfoin (*Onobrychis viciifolia* Scop.; SF). Both are well-adapted to the cool-temperate climate and alkaline soils of the Mountain West USA. Condensed tannins comprise a diverse family of bioactive chemicals with multiple beneficial functions for ruminants, including suppression of internal parasites and enteric methane. Birdsfoot trefoil contains 10 to 40 $g{\cdot}CT{\cdot}kg^{-1}$ dry matter (DM), while SF contains 30 to 80 $g{\cdot}CT{\cdot}kg^{-1}$ DM. Our studies have focused on these two plant species and have demonstrated consistently elevated rates of gain for beef calves grazing both BFT and SF. Novel results from our BFT research include carcass dressing percentages and consumer sensory evaluations equivalent to feedlot-finished steers and significantly greater than grass-finished steers, but with omega-3 fatty acid concentrations equal to grass-finished beef. We have further demonstrated that ruminants fed BFT or SF will consume more endophyte-infected tall fescue (*Schedonorus arundinaceus* (Schreb.) Dumort.) forage or seed than ruminants fed a non-CT forage legume. There is great potential value for sustainable livestock production in the use of highly digestible, nitrogen-fixing legumes containing tannins demonstrated to improve ruminant productivity.

**Keywords:** proanthocyanidin; birdsfoot trefoil (*Lotus corniculatus* L.); sainfoin (*Onobrychis viciifolia* Scop.)

---

## 1. Introduction: Structural Diversity and Function of Condensed Tannins:

Condensed tannins (CT) are a heterogeneous family of highly-reactive, carbon-based secondary compounds of high molecular weight [1] that bind to proteins with great affinity [2,3], precipitating them from solution [4]. These biomolecules, also known as polyphenolics or proanthocyanidins, cause responses in herbivores that vary with the diversity of their chemical structures [5,6]. Such diversity is in part due to the fact that CT are polymers of flavan-3-ol subunits, which offer the possibility of a wide array of molecular weights [7,8], ranging from dimers and trimers to oligomers with multiple subunits [9]. The structure of CT also depends on the degree of polymerization of these subunits, stereochemistry, the hydroxylation pattern on the "A" and "B" rings of the flavan-3-ol subunits and the presence or absence of the 3-hydroxyl group in the molecule [6,8]. In addition, CT are amphipathic molecules with aromatic (hydrophobic) rings and hydroxyl (hydrophilic) groups, allowing for bonds with different molecules, from proteins and minerals to polysaccharides and other plant secondary compounds [10].

The abundance of CT in the leaves of forest trees and early findings on the negative relationship between CT concentration and levels of insect herbivory led to the proposal that CT evolved as a quantitative chemical defense against insect herbivory [11]. This early view was reinforced by the fact that there are no compelling findings suggesting that CT have a fundamental role in plant physiological processes [12,13]. Nevertheless, several recent studies have failed to show a consistent protective role for CT against herbivory (*i.e.*, [14,15]), and thus, the role of CT as a defensive chemical has been challenged. For instance, it has been noted that the vision of CT as toxins is at odds with the fact that humans have enjoyed CT in drinks and foods for thousands of years [4]. Moreover, recent evidence suggests that at appropriate doses, CT may enhance the nutrition and health of consumers, as well as the quality of milk and meat products [16–18].

In ruminants, the binding and precipitation of dietary proteins by CT shifts the site of protein digestion from the rumen to the intestines and nitrogen excretion from urine to feces [19]. Such shifts can improve the nutrition of ruminants [20] in feeds that have a greater concentration of soluble protein than is required for carbohydrate utilization by rumen microbes, which is the case with most well-managed forage legumes. Moreover, improving the protein nutrition of ruminants enhances immune responses [21,22], which may increase resistance to gastrointestinal nematodes [23]; increasing the essential and branched-chain amino acids reaching the small intestine also improves reproductive efficiency in ruminants [24]. By consuming CT-containing forages, herbivores alleviate bloat [25], reduce methane emissions [26,27] and reduce internal parasites [21]. For instance, CT-containing legumes, like sainfoin (SF; *Onobrychis viciifolia* Scop.), have antiparasitic activity against helminth nematodes [28], and sheep appear to recognize this benefit of SF during grazing by increasing their selection of this forage when challenged by parasitic burdens [29]. Further, when herbivores consume feeds containing CT, their meat is lighter in color, with a greater concentration of antioxidants and omega-3 compared with omega-6 fatty acids, reduced "pastoral" flavor [30] and is generally more desirable for human consumption [16,17,31,32].

For this review, we will focus on the beneficial effects of CT and CT-containing temperate legumes in ruminant diets and the potential interactions that can enhance such effects. Both birdsfoot trefoil (BFT; *Lotus corniculatus* L.) and SF perform well under irrigation in the high pH soils and cool-temperate climate of the Mountain West USA, and both are beneficial for ruminant productivity [4,18].

## 2. Condensed Tannins in Temperate Forage Legumes

The CT synthesized in forage plant species differ in molecular weight and the type and proportion of subunit (see the comprehensive appendix of [4]). The most commonly-cultivated temperate forage legumes are alfalfa (*Medicago sativa* L.) and white clover (*Trifolium repens* L.). Alfalfa expresses CT in its seed coat, and white clover expresses CT in its flowers. The selection of white clover for elevated floral CT [33] resulted in white clover shoot CT concentrations as great as 12 $g{\cdot}kg^{-1}$ dry matter (DM). Alfalfa and white clover have recently been genetically engineered to express CT in their leaves [34], but commercial release of CT-containing cultivars is anticipated to require nine years or more [35,36]. Two forage plant species that grow well in the Mountain West USA and that naturally contain significant concentrations of CT in their leaves are BFT, which contains 10 to 40 $g{\cdot}CT{\cdot}kg^{-1}$ DM, and SF, which contains 30 to 80 $g{\cdot}CT{\cdot}kg^{-1}$ DM [9].

### 2.1. Grass and Legume Forages

Grazing-based cattle and sheep production is commonly carried out on grass or mixed grass-legume pastures. The rapid digestibility and greater accessibility of proteins in legumes, such as alfalfa and white clover, can result in bloat if these forages constitute more than 25% of a pasture mixture [37] or if they occur in patches that can be selected by ruminants. However, CT concentrations as little as 5 $g{\cdot}kg^{-1}$ DM can prevent bloat [38], so pastures composed of CT-containing legumes can be grazed without restriction. Ruminant intake is negatively correlated with forage fiber concentration, and intake is greater for ruminants fed legumes than grasses [39]; ruminant production on forages is a function of intake, which is typically limited by rumen fill and fiber digestion [40]. At similar stages of forage maturity, legumes have less fiber than grasses (e.g., [41]) and can be digested more readily. The protein content of forage legumes is often greater than the dietary protein requirement of ruminants, and when the available protein exceeds the available carbohydrate (*i.e.*, energy) in the rumen, amino acids will be used for energy [42]. This results in an undesirable generation of ammonia that is excreted in urine or milk [43].

### 2.2. *The Beneficial Role of Some Condensed Tannins in Ruminant Digestion*

Condensed tannins prevent bloat by binding to and precipitating proteins, which reduces protein concentration in the rumen and increases rumen bypass or "undegradable" protein [44]. The CT expressed by many forages can bind to salivary proteins, reducing palatability and intake; CT can also form indigestible complexes with rumen microbes and cell wall carbohydrates, reducing the rate of rumen digestion and intake [45]. As a result, forages, such as big trefoil (*Lotus pedunculatus* Cav.), can prevent bloat, but also reduce ruminant productivity [46]. The CT in BFT is present in a relatively low concentration and is not reported to inhibit intake or digestion [47,48]. Birdsfoot trefoil CT precipitates excess plant proteins in the rumen, preventing bloat, but it does not suppress post-ruminal digestion of proteins or absorption in the small intestine [4,49]. This has resulted in consistently better productivity of sheep fed BFT compared with alfalfa [50,51] and cattle fed BFT compared with non-CT legumes [52,53].

## 3. Ruminant Production on CT Legumes in the Western U.S.

### 3.1. *Conventional Beef Production*

The conventional North American beef production system consists of a low-input, grazing- and hay-based cow-calf phase lasting 11 months to 12 months and a high-input feedlot phase in which a ration comprised primarily of grain is fed for three to four months, for a total of 14 months to 16 months from birth to slaughter. The resulting grain-finished beef is juicy and tender; the intramuscular fat content of the longissimus muscle of beef graded choice ranges from 4% to 8% [54]. Alternatively, weaned calves that are never fed grain, antibiotics or hormones may be certified as grass-fed beef [55] and typically are slaughtered at 18 months to 24 months [56]. The longissimus intramuscular fat concentration of grass-finished beef is approximately half that of grain-finished beef [57].

### 3.2. *Beef Production on a CT-Containing Forage Legume*

Birdsfoot trefoil is productive and persistent in the northern Mountain West [58], typically continuing to increase in DM yield for two years after planting [59]. In production-scale swards, mature BFT stands averaged approximately 7000 $kg{\cdot}ha^{-1}$ per year in the second and third year after planting [60], about two-thirds the DM yield of grazing types of alfalfa [59]. In a study carried out at Utah State University (USU), yearling Angus (*Bos taurus*) steers that were finished for 111 days (3.7 months) gained 0.55 $kg{\cdot}day^{-1}$ on introduced grass pastures, 0.95 $kg{\cdot}day^{-1}$ on BFT pastures or 1.74 $kg{\cdot}day^{-1}$ on a concentrate diet [61]. Steers that averaged 451 kg when treatments were initiated weighed 512 kg, 557 kg and 644 kg at slaughter when finished on grass, BFT and concentrate, respectively. Stocking densities were low; post-grazing pasture DM did not fall below 1700 $kg{\cdot}ha^{-1}$. Expressed as the number of cattle required to produce one billion pounds of beef (Table 1; adapted from [62]), estimates from USU studies for cattle finished on grass (USU grass-finished (GFD); 3.4 million) were similar to those reported for current model predictions (GFD; 3.6 million) [62]. The estimate of the number of cattle required to produce one billion pounds of beef with BFT (USU BFT), however, was 2.9 million cattle, which compared favorably with Capper's [62] estimate for feedlot finishing in a U.S. conventional system that includes growth-enhancing feed additives (2.7 million) or in a feedlot system that does not include growth-enhancing feed additives (natural; 3.0 million). Age at slaughter ranged from 14.6 months for feedlot-finished beef to 22.6 months for grass-finished beef in Capper's [62] estimate, while the USU cattle were all slaughtered at 18 months. This comparison illustrates the importance of dressing percentage, which was 62% for BFT-finished beef and 64% for feedlot-finished beef. The dressing percentage for grass-finished beef in both reports was similar: 57.5% for GFD and 57% for USU GFD [62]. Time to finishing for all cattle in the USU study was 18 months, while time to finishing varied from 14.6 months for feedlot-finished cattle to 22.6 months for grass-finished cattle in Capper's [62] estimate.

**Table 1.** Capper [62] calculated the number of days to slaughter, final weights, dressing percentage and the number of cattle required to produce 1 billion kg beef for three U.S. beef production systems: CON (feedlot concentrate-finished), NAT (CON without growth-enhancing feed additives) and GFD (grass-finished). Results from Pitcher [61] have been compared in columns labeled USU (Utah State University). Treatments were USU NAT (CON without growth-enhancing feed additives), USU BFT (birdsfoot trefoil-finished) and USU GFD (grass-finished). Output is the number of cattle required to produce 1 billion kg red meat.

| | **Time in System (day)** | | | | | |
|---|---|---|---|---|---|---|
| | **CON** | **USU NAT** | **NAT** | **USU BFT** | **GFD** | **USU GFD** |
| Pre-weaned beef calf | 207 | 215 | 207 | 215 | 207 | 215 |
| Stocker | 123 | 216 | 159 | 216 | 159 | 216 |
| Yearling finishing | 110 | 111 | 110 | 111 | 313 | 111 |
| Total in days | 440 | 542 | 476 | 542 | 679 | 542 |
| Total in months | 14.6 | 18 | 15.8 | 18 | 22.6 | 18 |
| | **Weight (kg)** | | | | | |
| | **CON** | **USU NAT** | **NAT** | **USU BFT** | **GFD** | **USU GFD** |
| Pre-weaned beef calf | 245 | 289 | 245 | 289 | 226 | 289 |
| Stocker | 122 | 162 | 122 | 162 | 67 | 162 |
| Yearling finishing | 204 | 193 | 163 | 106 | 192 | 61 |
| Total weight | 571 | 644 | 530 | 557 | 486 | 512 |
| Dressing percentage (%) | 63.8 | 58.1 | 63.3 | 62.1 | 57.5 | 57.0 |
| Cattle required for $1 \times 10^9$ kg red meat | 2,745,005 | 2,672,625 | 2,980,715 | 2,891,034 | 3,585,836 | 3,426,535 |

The underlining signifies that the numbers in the next row are the sums of the first three rows in both the upper and lower halves of the table.

Cattle from the three USU treatments in Table 1 were slaughtered, and steaks from the longissimus muscle were subjected to consumer sensory evaluation. For the characteristics tenderness, juiciness, degree of fattiness and overall preference, steaks from BFT-finished steers were comparable to steaks from grain-finished steers and preferred to steaks from grass-finished steers [63]. The fatty acid composition of steaks from these treatments was also determined, and the meat from grass- and BFT-finished steers had an equivalent ratios of omega-6 (n-6) to omega-3 (n-3) fatty acid concentrations, and both were less than the n-6:n-3 of grain-finished beef, with greater n-3, as well as reduced n-6 in BFT-finished beef [64]. Likewise, beef from cattle fed CT-containing SF had greater marbling scores, quality grades (select *versus* standard) and backfat thicknesses than alfalfa-fed animals. Steaks from cattle finished on CT-containing SF were redder in color than steaks from cattle finished on alfalfa and contained more unsaturated fatty acids [65].

In a 2014 grazing study, the enteric methane emissions of beef cows grazing BFT were half those of cows grazing meadow bromegrass (*Bromus riparius* Rehm.) [61]; we believe these are the first data comparing enteric methane emissions of grazed perennial grasses and CT legumes. Along with the reduction of excess rumen protein due to CT, BFT has a relatively great concentration of non-fibrous carbohydrate [66] that could contribute to the elevated gain and dressing percentage observed for the BFT-finishing treatment.

### *3.3. Dairy Production on a CT-Containing Forage Legume*

In a study carried out on a commercial organic dairy in southeast Idaho USA, an 8-ha sprinkler-irrigated grass pasture was subdivided, and 4-ha were cultivated and seeded with 'Norcen' BFT in late summer of 2011. Data were collected on establishment in autumn of 2011 and spring of

2012 [67], and forage and milk production data were collected in 2012 and 2013. Grass pastures were comprised of *Lolium perenne* L., *Dactylis glomerata* L., *S. arundinaceus*, *Elymus repens* (L.) Gould and *T. repens* L. Nine Holstein dairy cows rotationally grazed either the BFT or the grass pasture. Cows were also fed 2.27 kg of barley each day, which included a vitamin and mineral supplement, and moved to fresh paddocks after each milking (every 12 h). Intake was estimated with a rising plate meter as the difference between pre-grazing and post-grazing DM. Milk production was measured at the beginning of the study and every two weeks by collecting milk from each cow at four successive (two morning and two evening) milkings.

Forage intake and milk production were significantly greater on BFT in 2012 and 2013 (Table 2). A medium cheddar cheese was made from the milk of cows grazing grass and BFT pastures and compared with cheese made at the same time from the milk of non-organic cows fed a total mixed ration (TMR). Cheese made from the milk of pasture-fed cows was significantly greater in omega-3 fatty acids and conjugated linoleic acid than cheese made from the milk of TMR-fed cows. The omega-3 fatty acid concentration of cheese made from the milk of BFT-fed cows was also significantly greater than that of cheese made from the milk of grass-fed cows (Figure 1).

Birdsfoot trefoil is a tap-rooted legume, like alfalfa, giving it an agronomic advantage over grasses [68] even under irrigation in an environment that has low humidity and elevated evapotranspiration, such as the semi-arid Mountain West. The BFT was inoculated with *Mesorhizobium loti*, so it was receiving nitrogen from a symbiotic relationship with soil bacteria. Legumes typically have more crude protein and less neutral detergent fiber than grasses [69], which increases the rate of forage digestion, resulting in greater intake [70]. Dairy intake and production on BFT pastures likely benefitted from the combined effects of plant adaptation, morphology, nutritive value and CT concentration.

**Table 2.** Mean intake and milk production on pastures in 2012 and 2013.

| Year | BFT Intake | SEM | Grass Intake | SEM | BFT Milk Production | SEM | Grass Milk Production | SEM |
|---|---|---|---|---|---|---|---|---|
| | | kg/ha | | | | kg/cow/day | | |
| **2012** | 1603 | 235 | 773 | 149 | 30 | 1.4 | 25 | 1.6 |
| **2013** | 2183 | 256 | 1301 | 181 | 35 | 0.7 | 30 | 2.1 |

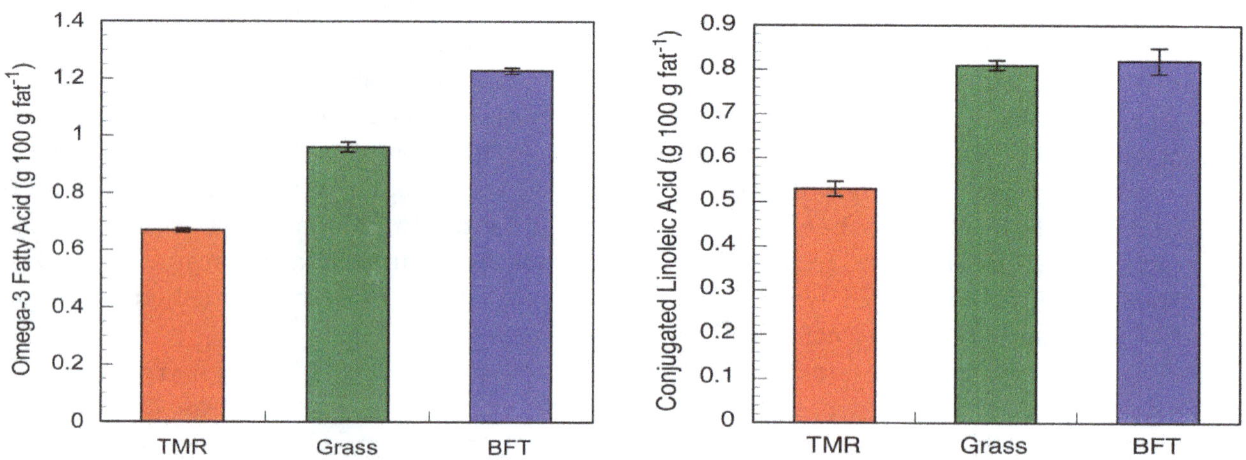


**Figure 1.** Omega-3 (left) and conjugated linoleic fatty acid concentrations (right) of cheese made from the milk of organic dairy cows grazing mixed grass pastures or birdsfoot trefoil (BFT) pastures, compared with cheese made from the milk of non-organic cows fed a total mixed ration (TMR).

## 4. Chemical Interactions of Condensed Tannins within a Dietary Context

The aforementioned analysis describes recent findings on the positive effects of CT and CT-containing legumes on ruminant nutrition and product quality. However, ruminants ingest a diversity of chemicals during grazing, from those present in a single forage to those ingested with a diverse diet. Thus, there is potential for multiple interactions among all of the different chemicals ingested within and between meals; this is particularly true for CT, since as stated above, these biomolecules are highly reactive.

### 4.1. Interaction with Proteins

The CT-protein interaction has been given considerable attention since the basis for the role of CT in chemical defense has been attributed to their ability to precipitate plant proteins and gastrointestinal enzymes [5,71], despite the fact that the CT-protein bond may be prevented or reversed [71].

The formation of CT complexes involves covalent and non-covalent bonding with other molecules [5]. During oxidative coupling, covalent bonds are formed by the conversion of phenols to quinones or semi-quinones, and these bonds are not reversible [2]; however, the CT-binding mechanism in a dietary context typically involves non-covalent forces (*i.e.*, hydrogen bonds and hydrophobic interactions), which are reversible [10,72]. Condensed tannins that interact predominantly via hydrogen bonds form stronger complexes with dietary proteins than those based on hydrophobic interactions. Therefore, CT that form hydrogen bonds with proteins are more likely to result in ruminal escape proteins, while CT that interact with proteins via hydrophobic bonds form weaker complexes [4]. Condensed tannin-protein complexes formed by hydrogen bonds are stable in the rumen at pH values close to neutrality, but may become unstable in the more acidic (e.g., pH 2.5 to 3.0) abomasum. The protein released from the disrupted complex provides amino acids for digestion and absorption in the small intestine at pH 8.0 to 9.0 [49,73].

Weaker CT-protein bonds may also mean an increased likelihood for interactions of CT with chemical structures present in the broad array of plant species ingested by herbivores, such as in pastures with diverse forage species composition [74]. Such interactions may lead to antagonistic effects that reduce the bioavailability of CT or to synergistic effects that increase their bioactivity [74]. As an example, the negative post-ingestive effects of different CT may be attenuated due to complexation when they are consumed in the same meal. It has been observed that DM intake by sheep is enhanced as the number of CT-containing shrubs in the diet increases relative to single shrub diets [75]. Likewise, phenolic compounds have been shown to have antagonistic interactions, making some of them (e.g., resveratrol) much less bioavailable when other phenols are present in the diet [76]. In contrast, other nutrient or plant chemical interactions may increase the activity or bioavailability of CT. Synergistic effects have been observed between increments in dietary calcium concentration and the bioavailability of resveratrol, a phenolic compound found in grapes and berries [77].

### 4.2. Interactions with Carbohydrates

It has been reported that greater CT concentrations depress rumen digestion of both readily fermentable and structural carbohydrates [44,46,78]. These effects are likely to be caused by an inactivation of extracellular microbial and mammalian enzymes through the formation of CT-enzyme complexes [79], rather than by a direct CT-carbohydrate interaction. In addition, CT can potentially interfere with the adhesion process of microbial bacteria onto forage cell walls, which can also lead to a depression in structural carbohydrate digestion [79].

### 4.3. Interactions with Saponins

Steroidal and triterpenoid saponins are a family of chemical compounds consisting of an isoprenoidal-derived aglycone (sapogenin), covalently linked to one or more sugar moieties [80]. Saponins are generally regarded as plant defenses, although several valuable pharmacological properties have recently been identified, such as anti-cancer, immunomodulatory and cholesterol-lowering activities [81]. In addition, saponins have been shown to reduce methane emissions [82] and to control gastrointestinal nematodes in ruminants [83]. The anthelmintic effects of CT and saponins can be explained through their toxic effects on parasites [84].

Consuming a diversity of plant secondary compounds like saponins and CT may reduce the overall toxic effect of the mix, as the formation of gastrointestinal complexes could reduce the absorption and activity of single toxic compounds [85]. In support of this, it has been found that intestinal bonding of CT and saponins results in moderated toxic effects, representing a mechanism that allows herbivores to consume more nutrients when offered diverse foods [86]. Condensed tannins and saponins cross-react and bind in the gastrointestinal tract, nullifying the effects of both compounds [86]. For example, goats increase intake when shrubs contain a combination of CT and saponins relative to when animals are offered single shrubs [75], and sheep offered a choice between saponin- and CT-containing rations ate more feed than animals only offered CT or saponins in single rations [87]. However, the *in vivo* antiparasitic effect of the CT-saponin combination was found to be less than that of the single rations [87]. Thus, a reduction in the negative impacts of CT or saponins on the herbivore (*i.e.*, through inactivation) can be carried over to the endoparasite, resulting in beneficial effects for both the herbivore and the parasite. In other words, the complexation (and potential inactivation) of CT and saponins can be beneficial to the ruminant relative to nutrient intake, but negative in regard to its effects on parasite loads. This suggests that the value of the interaction between saponins and CT may depend on whether concern is with nutrient intake or antiparasitic activity.

*4.4. Interactions with Alkaloids*

Condensed tannins are known to complex not only with proteins and saponins, but also with alkaloids [88]. Alkaloids are nitrogen-based secondary compounds, and the strong binding capacity of CT for nitrogen-containing compounds [89] may be responsible for this interaction. Stable complexes between alkaloids and CT made alkaloids less available in the gastrointestinal tract, thus reducing their toxic effects [90]. A CT-containing legume like BFT ingested prior to eating endophyte-infected (E+) tall fescue (*S. arundinaceus* or reed canarygrass (*Phalaris arundinacea* L.), both alkaloid-containing forage grasses, enabled lambs and calves to consume more tall fescue or reed canarygrass (and therefore more energy and protein) than lambs and calves fed the same grasses without the legumes [91–93]. Likewise, lambs offered CT-containing supplements ingested more alkaloid-containing rations than lambs offered just the alkaloid-containing rations [94,95]. Lambs offered a choice of alfalfa with an elevated saponin concentration, BFT containing CT and E+ tall fescue containing alkaloids preferred alfalfa [96]. However, lambs receiving intraruminal infusions of saponins increased their preference for BFT and E+ tall fescue, and lambs receiving intraruminal infusions of CT increased their preference for E+ tall fescue. Overall, these results support the notion that CT complex with other secondary compounds and attenuate negative post-ingestive effects of individual components of the complex.

Recent evidence further supported the interaction of CT and alkaloids: polyethylene glycol, a polymer that selectively binds to CT, reduced the benefits of SF CT on a basal diet containing E+ tall fescue in sheep [90]. These benefits included an increase in the total amount of nutrients ingested and improvements in some physiological parameters indicative of fescue toxicosis, such as reduced rectal temperatures, increased numbers of leukocytes and lymphocytes and increased plasmatic concentrations of globulin and prolactin compared with control lambs that consumed E+ tall fescue without CT-containing SF. Radial diffusion assays [97] in which tannins with or without ergot alkaloids were precipitated as they diffused through agar containing bovine serum albumin demonstrated that ergotamine, an alkaloid from E+ tall fescue, reduced the protein binding capacity of CT from SF and BFT (B. Goff, personal communication). This *in vitro* evidence supports *in vivo* observations of increased intake of E+ tall fescue seed by lambs consuming SF, but reduced intake by lambs consuming the non-CT forage legume cicer milkvetch (*Astragalus cicer* L.) [98].

*4.5. Interactions with Terpenes*

Terpenes are a large and diverse class of carbon-based secondary compounds, biosynthetically derived from isoprene subunits [99] and produced by a variety of plants, particularly woody species. Evidence suggests that CT bind to terpenes and reduce their bioavailability. When offered feeds with higher levels of terpenes and CT, sheep consumed more feed containing terpenes if they first consumed feed with CT [100]. Likewise, sheep with a preference for terpene-containing sagebrush (*Artemisia tridentata* Nutt.) consumed considerably more CT-containing bitterbrush (*Purshia tridentata* (Pursh) DC) than sheep with less preference for sagebrush [101]. These findings are consistent with the hypothesis that terpenes and CT interact in the digestive system of sheep. Finally, lambs offered CT- and terpene-containing feeds consumed more than when they were offered only one feed [102], which again suggests the formation of CT-terpene complexes with reduced toxicity.

## 5. Conclusions

Condensed tannins are a family of plant secondary compounds with diverse structures and multiple impacts on herbivores that can range from detrimental to beneficial. Condensed tannins can form complexes with proteins, as well as with other plant secondary compounds. The temperate forage legumes BFT and SF have CT concentrations ranging from 10 $g \cdot kg^{-1}$ to 80 $g \cdot kg^{-1}$ and protein concentrations ranging from 12 $g \cdot kg^{-1}$ to 25 $g \cdot kg^{-1}$. Their CT are generally beneficial to ruminants by eliminating bloat, suppressing internal parasites, reducing enteric methane emissions and increasing the quantity of protein that is absorbed from the intestines. Our studies have demonstrated greater productivity of beef cattle fattened on BFT compared with other forages and greater milk production of commercial dairy cows grazing BFT compared with cows grazing grass pastures in mid-summer. We have also demonstrated the attenuation of negative tall fescue endophyte effects when CT were ingested along with E+ tall fescue. One of the next challenges in the study of CT-containing forages is to relate CT biochemical structures to CT activities within the ruminant gut, particularly to balance beneficial effects on protein digestion with suppressive effects on parasites and toxic secondary compounds.

**Acknowledgments:** This research was supported by the Utah Agricultural Experiment Station, Utah State University, and approved as journal paper Number 8795.

**Author Contributions:** The authors contributed equally to the content of this article.

## References

1. Mueller-Harvey, I.; Caygill, J.C. Tannins: Their nature and biological significance. In *Secondary Plant Products: Antinutritional and Beneficial Actions in Animal Feeding*; Nottingham University Press: Nottingham, UK, 1999.
2. Jones, E.T.; Mangan, J.L. Complexes of the condensed tannins of sainfoin (*Onobrychis viciifolia* Scop.) with fraction 1 leaf protein and with submixillary mucoprotein, and their reversal by polyethylene glycol and pH. *J. Sci. Food Agric.* **1977**, *28*, 126–136. [CrossRef]
3. Hagerman, A.E.; Robbins, C.T.; Weerasuriya, Y.; Wilson, T.C.; McArthur, C. Tannin chemistry in relation to digestion. *J. Range Manag.* **1992**, *45*, 57–62. [CrossRef]
4. Mueller-Harvey, I. Unravelling the conundrum of tannins in animal nutrition and health. *J. Sci. Food Agric.* **2006**, *86*, 2010–2037. [CrossRef]
5. Zucker, W.V. Tannins: Does structure determine function? An ecological perspective. *Am. Nat.* **1983**, *121*, 335–365. [CrossRef]
6. Dixon, R.A.; Xie, D.; Sharma, S.B. Tansley review. Proanthocyanidins—A final frontier in flavonoid research? *New Phytol.* **2005**, *165*, 9–28. [CrossRef] [PubMed]
7. Kennedy, J.A.; Saucier, C.; Glories, Y. Grape and wine phenolics: History and perspective. *Am. J. Enol. Vitic.* **2006**, *57*, 239–248.
8. Xie, D.; Dixon, R.A. Proanthocyanidin biosynthesis—Still more questions than answers? *Phytochemistry* **2005**, *66*, 2127–2144. [CrossRef] [PubMed]
9. Adams, D.O. Phenolics and ripening in grape berries. *Am. J. Enol. Vitic.* **2006**, *57*, 249–256.
10. Hanlin, R.L.; Hrmova, M.; Harbertson, J.F.; Downey, M.O. Review: Condensed tannin and grape cell wall interactions and their impact on tannin extractability into wine. *Aust. J. Grape Wine Res.* **2010**, *16*, 173–188. [CrossRef]

11. Feeny, P. Effect of oak leaf tannins on larval growth of the winter moth *Operopthera brumata*. *J. Insect Physiol.* **1968**, *14*, 805–817. [CrossRef]
12. McKey, D. The distribution of secondary compounds within plants. In *Herbivores, Their Interaction with Secondary Plant Metabolites*; Rosenthal, G.W., Janzen, D.H., Eds.; Academic Press: New York, NY, USA, 1979; pp. 55–133.
13. Waghorn, G.C. Beneficial and detrimental effects of dietary condensed tannins for sustainable sheep and goat production—Progress and challenges. *Anim. Feed Sci. Technol.* **2008**, *147*, 116–139. [CrossRef]
14. Ayres, M.P.; Clausen, T.P.; MacLean, S.F., Jr.; Redman, A.M.; Reichardt, P.B. Diversity of structure and antiherbivore activity in condensed tannins. *Ecology* **1997**, *78*, 1696–1712. [CrossRef]
15. Schweitzer, J.A.; Madritch, M.D.; Bailey, J.K.; LeRoy, C.J.; Fischer, D.G.; Rehill, B.J.; Lindroth, E.L.; Hagerman, A.E.; Wooley, S.C.; Hart, S.C.; *et al.* From genes to ecosystems: The genetic basis of condensed tannins and their role in nutrient regulation in a *Populus* model system. *Ecosystems* **2008**, *11*, 1005–1020. [CrossRef]
16. Priolo, A.; Bella, M.; Lanza, M.; Galofaro, V.; Biondi, L.; Barbagallo, D.; Ben Salem, H.; Pennisi, P. Carcass and meat quality of lambs fed fresh sulla (*Hedysarum coronarium* L.) with or without polyethylene glycol or concentrate. *Small Rum. Res.* **2005**, *59*, 281–288. [CrossRef]
17. Priolo, A.; Vasta, V.; Fasone, V.; Lanza, C.M.; Scerra, M.; Biondi, L.; Bella, M.; Whittington, F.M. Meat odour and flavour and indoles concentration in ruminal fluid and adipose tissue of lambs fed green herbage or concentrates with or without tannins. *Animal* **2009**, *3*, 454–460. [CrossRef]
18. Waghorn, G.C.; McNabb, W.C. Consequences of plant phenolic compounds for productivity and health of ruminants. *Proc. Nutr. Soc.* **2003**, *62*, 383–392. [CrossRef] [PubMed]
19. Woodward, S.L.; Waghorn, G.C.; Watkins, K.A.; Bryant, M.A. Feeding birdsfoot trefoil (*Lotus corniculatus*) reduces the environmental impacts of dairy farming. *Proc. N. Z. Soc. Anim. Prod.* **2009**, *69*, 179–183.
20. Barry, T.N.; McNeill, D.M.; McNabb, W.C. Plant secondary compounds: Their impact on nutritive value and upon animal production. In Proceedings of the XIX International Grassland Conference, Sao Paulo, Brazil, 11–21 February 2001; Wageningen Academic Publishers: Wageningen, The Netherlands, 2001; pp. 445–452.
21. Min, B.R.; Hart, S.P. Tannins for suppression of internal parasites. *J. Anim. Sci.* **2003**, *81*, E102–E109.
22. Niezen, J.H.; Charleston, W.A.G.; Robertson, H.A.; Shelton, D.; Waghorn, G.C.; Green, R. The effect of feeding sulla (*Hedysarum coronarium*) or lucerne (*Medicago sativa*) on lamb parasite burdens and development of immunity to gastrointestinal nematodes. *Vet. Parasit.* **2002**, *105*, 229–245. [CrossRef]
23. Min, B.R.; Pomroy, W.E.; Hart, S.P.; Sahlu, T. The effect of short-term consumption of a forage containing condensed tannins on gastro-intestinal nematode parasite infections in grazing wether goats. *Small Rum. Res.* **2004**, *51*, 279–283. [CrossRef]
24. Min, B.R.; Fernandez, J.M.; Barry, T.N.; McNabb, W.C.; Kemp, P.D. The effect of condensed tannins in *Lotus corniculatus* upon reproductive efficiency and wool production in ewes during autumn. *Anim. Feed Sci. Technol.* **2001**, *92*, 185–202. [CrossRef]
25. Waghorn, G.C. Beneficial effects of low concentrations of condensed tannins in forages fed to ruminants. In *Microbial and Plant Opportunities to Improve Lignocellulose Utilization by Ruminants*; Akin, D.E., Ljungdahl, L.G., Wilson, J.R., Harris, P.J., Eds.; Elsevier Science Publisher: New York, NY, USA, 1990.
26. Chung, Y.-H.; McGeough, E.J.; Acharya, S.; McAllister, T.A.; McGinn, S.M.; Harstad, O.M.; Beauchemin, K.A. Enteric methane emission, diet digestibility, and nitrogen excretion from beef heifers fed sainfoin or alfalfa. *J. Anim. Sci.* **2013**, *91*, 4861–4874. [CrossRef]
27. Woodward, S.L.; Waghorn, G.C.; Laboyrie, P.G. Condensed tannins in birdsfoot trefoil (*Lotus corniculatus*) reduce methane emissions from dairy cows. *Proc. N. Z. Soc. Anim. Prod.* **2004**, *64*, 160–164.
28. Paolini, V.; Dorchies, P.; Hoste, H. Effects of sainfoin hay on gastrointestinal nematode infections in goats. *Vet. Rec.* **2003**, *152*, 600–601. [CrossRef] [PubMed]
29. Villalba, J.J.; Miller, J.; Hall, J.O.; Clemensen, A.K.; Stott, R.; Snyder, D.; Provenza, F.D. Preference for tanniferous (*Onobrychis viciifolia*) and non-tanniferous (*Astragalus cicer*) forage plants by sheep in response to challenge infection with *Haemonchus contortus*. *Small Rum. Res.* **2013**, *112*, 199–207. [CrossRef]
30. Schreurs, N.M.; Lane, G.A.; Tavendale, M.H.; Barry, T.N.; McNabb, W.C. Pastoral flavor in meat product from ruminant fed fresh forages and it amelioration by forage condensed tannins. *Anim. Feed Sci. Technol.* **2008**, *146*, 193–221. [CrossRef]

31. Vasta, V.; Nudda, A.; Cannas, A.; Lanza, M.; Priolo, A. Alternative feed resources and their effects on the quality of meat and milk from small ruminants. *Anim. Feed Sci. Technol.* **2008**, *147*, 223–246. [CrossRef]
32. Luciano, G.; Monahan, F.J.; Vasta, V.; Biondi, L.; Lanza, M.; Priolo, A. Dietary tannins improve lamb meat colour stability. *Meat Sci.* **2009**, *81*, 120–125. [CrossRef] [PubMed]
33. Burggraaf, V.T.; Kemp, P.D.; Thom, E.R.; Waghorn, G.C.; Woodfield, D.R.; Woodward, S.L. Agronomic evaluation of white clover selected for increased floral condensed tannin. *Proc. N. Z. Grassl. Assoc.* **2003**, *65*, 139–145.
34. Hancock, K.R.; Collette, V.; Fraser, K.; Greig, M.; Xue, H.; Richardson, K.; Jones, C.; Rasmussen, S. Expression of the R2R3-MYB transcription factor TaMYB14 from *Trifolium arvense* activates proanthocyanidin biosynthesis in the legumes *Trifolium repens* and *Medicago sativa*. *Plant Physiol.* **2012**, *159*, 1204–1220. [CrossRef] [PubMed]
35. McCaslin, M. The commercial potential for genetic engineering in alfalfa. In Proceedings of the 38th North American Alfalfa Improvement Conference, Sacramento, CA, USA, 27–31 July 2002; pp. 16–18. Available online: http://www.naaic.org/Meetings/National/2002meeting/2002Abstracts/McCaslinSymposium.pdf (accessed on 12 May 2015).
36. University of Nebraska-Lincoln Ag BioSafety Education Center. Available online: http://agbiosafety.unl.edu/education/timeline.htm (accessed on 15 May 2015).
37. Majak, W.; McAllister, T.A.; McCartney, D.; Stanford, K.; Cheng, K.J. *Bloat in Cattle*; Alberta Agriculture and Rural Development: Edmonton, AB, Canada, 2003.
38. Barry, T.N.; McNabb, W.C. Review article: The implications of condensed tannins on the nutritive value of temperate forages fed to ruminants. *Br. J. Nutr.* **1999**, *81*, 263–272. [CrossRef] [PubMed]
39. Van Soest, P.J. Symposium on factors influencing the voluntary intake of herbage by ruminants: Voluntary intake in relation to chemical composition and digestibility. *J. Anim. Sci.* **1965**, *24*, 834–843.
40. Jung, H.G.; Allen, M.S. Characteristics of plant cell walls affecting intake and digestibility of forages by ruminants. *J. Anim. Sci.* **1995**, *73*, 2774–2790. [PubMed]
41. Wen, L.; Kallenbach, R.L.; Williams, J.E.; Roberts, C.A.; Beuselinck, P.R.; McGraw, R.L.; Benedict, H.R. Performance of steers grazing rhizomatous and nonrhizomatous birdsfoot trefoil in pure stands and in tall fescue mixtures. *J. Anim. Sci.* **2002**, *80*, 1970–1976. [PubMed]
42. Hoekstra, N.J.; Schulte, R.P.O.; Struik, P.C.; Lantinga, E.A. Pathways to improving the N efficiency of grazing bovines. *Eur. J. Agron.* **2007**, *26*, 363–374. [CrossRef]
43. Satter, L.D.; Roffler, R.E. Nitrogen requirement and utilization in dairy cattle. *J. Dairy Sci.* **1975**, *58*, 1219–1237. [CrossRef]
44. Barry, T.N.; Manley, T.R. Interrelationships between the concentrations of total condensed tannin, free condensed tannin and lignin in *Lotus* sp. and their possible consequences in ruminant nutrition. *J. Sci. Food Agric.* **1986**, *37*, 248–254. [CrossRef]
45. Reed, J.D. Nutritional toxicology of tannins and related polyphenols in forage legumes. *J. Anim. Sci.* **1995**, *73*, 1516–1528. [PubMed]
46. Barry, T.N.; Duncan, S.J. The role of condensed tannins in the nutritional value of *Lotus pedunculatus* for sheep. 1. Voluntary intake. *Br. J. Nutr.* **1984**, *51*, 493–504. [CrossRef]
47. Ramírez-Restrepo, C.A.; Barry, T.N. Alternative temperate forages containing secondary compounds for improving sustainable productivity in grazing ruminants. *Anim. Feed Sci. Technol.* **2005**, *120*, 179–201. [CrossRef]
48. Ramírez-Restrepo, C.A.; Barry, T.N.; López-Villalobos, N. Organic matter digestibility of condensed tannin-containing *Lotus corniculatus* and its prediction *in vitro* using cellulase/hemicellulase enzymes. *Anim. Feed Sci. Technol.* **2006**, *125*, 61–71. [CrossRef]
49. Waghorn, G.C.; Ulyatt, M.J.; John, A.; Fisher, M.T. The effect of condensed tannins on the site of digestion of amino acids and other nutrients in sheep fed on *Lotus corniculatus* L. *Br. J. Nutr.* **1987**, *57*, 115–126. [CrossRef]
50. Douglas, G.B.; Wang, Y.; Waghorn, G.C.; Barry, T.N.; Purchas, R.W.; Foote, A.G.; Wilson, G.F. Liveweight gain and wool production of sheep grazing *Lotus corniculatus* and lucerne (*Medicago sativa*). *N. Z. J. Agric. Res.* **1995**, *38*, 95–104. [CrossRef]
51. Marten, G.C.; Jordan, R.M. Substitution value of birdsfoot trefoil for alfalfa-grass in pasture systems. *Agron. J.* **1979**, *71*, 55–59. [CrossRef]

52. MacAdam, J.W.; Ward, R.E.; Griggs, T.C.; Min, B.R.; Aiken, G.E. Average daily gain and blood fatty acid composition of cattle grazing the non-bloating legumes birdsfoot trefoil and cicer milkvetch in the Mountain West. *Prof. Anim. Sci.* **2011**, *27*, 574–583.
53. Marten, G.C.; Ehle, F.R.; Ristau, E.A. Performance and photosensitization of cattle related to forage quality of four legumes. *Crop Sci.* **1987**, *27*, 138–145. [CrossRef]
54. Smith, S.B.; Johnson, B.J. *Marbling: Management of Cattle to Maximize the Deposition of intramuscular Adipose Tissue*; Cattlemen's Beef Board and National Cattlemen's Beef Association: Centennial, CO, USA, 2014.
55. American Grassfed Association. *Grassfed and Grass Pastured Ruminant Standards*; American Grassfed Association: Denver, CO, USA, 2014.
56. Shattuck, K. Where corn is king, a new regard for grass-fed beef. Available online: http://healthimpactnews.com/2013/why-grass-fed-beef-is-good-for-your-health/ (accessed on 13 May 2015).
57. Van Elswyk, M.E.; McNeill, S.H. Impact of grass/forage feeding *versus* grain finishing on beef nutrients and sensory quality: The U.S. experience. *Meat Sci.* **2014**, *96*, 535–540.
58. Grabber, J.H.; Riday, H.; Cassida, K.A.; Griggs, T.C.; Min, D.H.; MacAdam, J.W. Yield, morphological characteristics, and chemical composition of European- and Mediterranean-derived birdsfoot trefoil cultivars grown in the colder continental United States. *Crop Sci.* **2014**, *54*, 1893–1901. [CrossRef]
59. MacAdam, J.W.; Griggs, T.C. *Irrigated Birdsfoot Trefoil Variety Trial: Forage Yields*; Utah State University Cooperative Extension Service: Logan, UT, USA, 2013.
60. Hunt, S.R.; MacAdam, J.W.; Griggs, T.C. Seeding rate, oat companion crop and planting season effects on organic establishment of irrigated birdsfoot trefoil in the Mountain West USA. *Crop Sci.* **2015**, in press.
61. Pitcher, L.R. Beef Average Daily Gain and Enteric Methane Emissions on Birdsfoot Trefoil, Cicer Milkvetch and Meadow Brome Pastures. Master's Thesis, Utah State University, Logan, UT, USA, December 2015.
62. Capper, J.L. Is the grass always greener? Comparing the environmental impact of conventional, natural and grass-fed beef production systems. *Animals* **2012**, *2*, 127–143. [CrossRef]
63. Chail, A.; Legako, J.F.; Martini, S.; MacAdam, J.W. Consumer sensory evaluation of forage and conventional feedlot finished beef ribeye steaks. *Meat Sci.* **2014**, *101*, 120–121. [CrossRef]
64. Chail, A.; Legako, J.F.; Martini, S.; Ward, R.; MacAdam, J.W. Comparison of proximate composition, pH and fatty acids of beef ribeye steaks from forage and conventional feedlot finished cattle. In Proceedings of the Reciprocal Meat Conference, Lincoln, NB, USA, 12–17 June 2015; Available online: https://guidebook.com/guide/24176/poi/3363090/?pcat=132880 (acccessed on 15 May 2015).
65. Maughan, B.; Provenza, F.D.; Tansawat, R.; Maughan, C.; Martini, S.; Ward, R.; Clemensen, A.; Song, X.; Cornforth, D.; Villalba, J.J. Importance of grass-legume choices on cattle grazing behavior, performance, and meat characteristics. *J. Anim. Sci.* **2014**, *92*, 2309–2324. [CrossRef] [PubMed]
66. MacAdam, J.W.; Griggs, T.C. *Irrigated Birdsfoot Trefoil Variety Trial: Forage Nutritive Value*; Utah State University Cooperative Extension Service: Logan, UT, USA, 2013.
67. Hunt, S.R.; MacAdam, J.W.; Reeve, J.R. Establishment of birdsfoot trefoil (*Lotus corniculatus*) pastures on organic dairy farms in the Mountain West USA. *Org. Agric.* **2015**, *5*, 63–77. [CrossRef]
68. Berdahl, J.D.; Karn, J.F.; Hendrickson, J.R. Dry matter yields of cool-season grass monocultures and grass-alfalfa binary mixtures. *Agron. J.* **2001**, *93*, 463–467. [CrossRef]
69. Smith, L.W.; Goering, H.K.; Gordon, C.H. Relationships of forage compositions with rates of cell wall digestion and indigestibility of cell walls. *J. Dairy Sci.* **1972**, *55*, 1140–1147. [CrossRef]
70. Crampton, E.W.; Donefer, E.; Lloyd, L.E. A nutritive value index for forages. *J. Anim. Sci.* **1960**, *19*, 538–544.
71. Robbins, C.T.; Hanley, T.A.; Hagerman, A.E.; Hjeljord, O.; Baker, D.L.; Schwartz, C.C.; Mautz, W.W. Role of tannins in defending plants against ruminants: Reduction in protein availability. *Ecology* **1987**, *68*, 98–107. [CrossRef]
72. Renard, C.M.; Baron, A.; Guyot, S.; Drilleau, J.F. Interactions between apple cell walls and native apple polyphenols: Quantification and some consequences. *Int. J. Biol. Macromol.* **2001**, *29*, 115–125. [CrossRef]
73. Mangan, J.L. Nutritional effects of tannins in animal feeds. *Nutr. Res. Rev.* **1988**, *1*, 209–231. [CrossRef]
74. Villalba, J.J.; Provenza, F.D.; Gibson, N.; López-Ortíz, S. Veterinary medicine: The value of plant secondary compounds and diversity in balancing consumer and ecological health. In *Sustainable Food Production Includes Human and Environmental Health*; Springer Netherlands: Dordrecht, The Netherlands, 2014; Volume 3, pp. 165–190.

75. Rogosic, J.; Estell, R.E.; Skobic, D.; Martinovic, A.; Maric, S. Role of species diversity and secondary compound complementarity on diet selection of Mediterranean shrubs by goats. *J. Chem. Ecol.* **2006**, *32*, 1279–1287. [CrossRef] [PubMed]
76. Pinelo, M.; Manzocco, L.; Nuñez, M.J.; Nicoli, M.C. Interaction among phenols in food fortification: Negative synergism on antioxidant capacity. *J. Agric. Food Chem.* **2004**, *52*, 1177–1180. [CrossRef] [PubMed]
77. Liang, L.; Tajmir-Riahi, H.A.; Subirade, M. Interaction of β-lactoglobulin with resveratrol and its biological implications. *Biomacromolecules* **2008**, *9*, 50–56. [CrossRef] [PubMed]
78. Barry, T.N.; Manley, T.R.; Duncan, S.J. The role of condensed tannins in the nutritional value of *Lotus pedunculatus* for sheep. *Br. J. Nutr.* **1986**, *55*, 123–137. [CrossRef] [PubMed]
79. Bae, H.D.; McAllister, T.A.; Yanke, J.; Cheng, K.J.; Muir, A.D. Effects of condensed tannins on endoglucanase activity and filter paper digestion by *Fibrobacter succinogenes* S85. *Appl. Environ. Microbiol.* **1993**, *59*, 2132–2138. [PubMed]
80. Augustin, J.M.; Kuzina, V.; Andersen, S.B.; Bak, S. Molecular activities, biosynthesis and evolution of triterpenoid saponins. *Phytochemistry* **2011**, *72*, 435–457. [CrossRef] [PubMed]
81. Guçlu-Ustundag, O.; Mazza, G. Saponins: Properties, applications and processing. *Crit. Rev. Food Sci. Nutr.* **2007**, *47*, 231–258. [CrossRef]
82. Hu, W.L.; Liu, J.X.; Ye, J.A.; Wu, Y.M.; Guo, Y.Q. Effect of tea saponin on rumen fermentation *in vitro*. *Anim. Feed Sci. Technol.* **2005**, *120*, 333–339. [CrossRef]
83. Sinha Babu, S.P. Saponins and its possible role in the control of helminth parasites. In *Recent Progress in Medicinal Plants*; Sharma, S.K., Govil, J.N., Singh, V.K., Eds.; Studium Press: New Delhi, India, 2005; Volume 10, pp. 405–418.
84. Lozano, G.A. Parasitic stress and self-medication in wild animals. *Adv. Study Behav.* **1998**, *27*, 291–317.
85. Freeland, W.J.; Janzen, D.H. Strategies in herbivory by mammals: The role of plant secondary compounds. *Am. Nat.* **1974**, *108*, 269–286. [CrossRef]
86. Freeland, W.J.; Calcott, P.H.; Anderson, L.R. Tannins and saponin: Interaction in herbivore diets. *Biochem. Syst. Ecol.* **1985**, *13*, 189–193. [CrossRef]
87. Copani, G.; Hall, J.O.; Miller, J.; Priolo, A.; Villalba, J.J. Plant secondary compounds as complementary resources: Are they always complementary? *Oecologia* **2013**, *172*, 1041–1049. [CrossRef] [PubMed]
88. Okuda, T.; Mori, K.; Shiota, M. Effects of interaction of tannins and coexisting substances. III Formation and solubilization of precipitates with alkaloids. *J. Pharm. Soc. Jpn.* **1982**, *102*, 854–858.
89. Charlton, A.J.; Davis, A.L.; Jones, D.P.; Lewis, J.R.; Davies, A.P.; Haslam, E.; Williamson, M.P. The self-association of the black tea polyphenol theaflavin and its complexation with caffeine. *J. Chem. Soc.* **2000**, *2*, 317–322. [CrossRef]
90. Catanese, F.; Distel, R.A.; Villalba, J.J. Effects of supplementing endophyte-infected tall fescue with sainfoin and polyethylene glycol on the physiology and ingestive behavior of sheep. *J. Anim. Sci.* **2014**, *92*, 744–757. [CrossRef] [PubMed]
91. Owens, J.; Provenza, F.D.; Wiedmeier, R.D.; Villalba, J.J. Influence of saponins and tannins on intake and nutrient digestion of alkaloid-containing foods. *J. Sci. Food Agric.* **2012**, *92*, 2373–2378. [CrossRef] [PubMed]
92. Lyman, T.D.; Provenza, F.D.; Villalba, J.J.; Wiedmeier, R.D. Cattle preferences differ when endophyte-infected tall fescue, birdsfoot trefoil, and alfalfa are grazed in different sequences. *J. Anim. Sci.* **2011**, *89*, 1131–1137. [CrossRef] [PubMed]
93. Lyman, T.D.; Provenza, F.D.; Villalba, J.J.; Wiedmeier, R.D. Phytochemical complementarities among endophyte-infected tall fescue, reed canarygrass, birdsfoot trefoil and alfalfa affect cattle foraging. *Animal* **2012**, *6*, 676–682. [CrossRef] [PubMed]
94. Lyman, T.D.; Provenza, F.D.; Villalba, J.J. Sheep foraging behavior in response to interactions among alkaloids, tannins and saponins. *J. Sci. Food Agric.* **2008**, *88*, 824–831. [CrossRef]
95. Owens, J.; Provenza, F.D.; Wiedmeier, R.D.; Villalba, J.J. Supplementing endophyte-infected tall fescue or reed canarygrass with alfalfa or birdsfoot trefoil increases forage intake and digestibility by sheep. *J. Sci. Food Agric.* **2012**, *92*, 987–992. [CrossRef]
96. Villalba, J.J.; Provenza, F.D.; Clemensen, A.K.; Larsen, R.; Juhnke, J. Preference for diverse pastures by sheep in response to intraruminal administrations of tannins, saponins and alkaloids. *Grass Forage Sci.* **2011**, *66*, 224–236. [CrossRef]

97. Hagerman, A.E. Radial diffusion method for determining tannin in plant extracts. *J. Chem. Ecol.* **1987**, *13*, 437–449. [CrossRef] [PubMed]
98. Villalba, J.J.; Spackman, C.; Goff, B.; Klotz, J.L.; MacAdam, J.W. Interaction between a tannin-containing legume and endophyte-infected tall fescue seed on lambs' feeding behavior and physiology. In Proceedings of the American Society of Animal Science Annual Meeting, Orlando, FL, USA, 12–16 July 2015. Available online: http://m.jtmtg.org/abs/t/62691 (accessed on 15 February 2015).
99. Tholl, D. Terpene synthases and the regulation, diversity and biological roles of terpene metabolism. *Curr. Opin. Plant Biol.* **2006**, *9*, 297–304. [CrossRef] [PubMed]
100. Mote, T.; Villalba, J.J.; Provenza, F.D. Foraging sequence influences the ability of lambs to consume foods containing tannins and terpenes. *Appl. Anim. Behav. Sci.* **2008**, *113*, 57–68. [CrossRef]
101. Seefeldt, S.S. Consequences of selecting Ramboulliet ewes for Mountain Big Sagebrush (*Artemisia tridentata* ssp. *vaseyana*) dietary preference. *Rangeland Ecol. Manag.* **2005**, *58*, 380–384. [CrossRef]
102. Villalba, J.J.; Provenza, F.D.; GouDong, H. Experience influences diet mixing by herbivores: Implications for plant biochemical diversity. *Oikos* **2004**, *107*, 100–109. [CrossRef]

# 11

# Temperature Impact on the Forage Quality of Two Wheat Cultivars with Contrasting Capacity to Accumulate Sugars

**Máximo Lorenzo [1], Silvia G. Assuero [2,*] and Jorge A. Tognetti [2,3]**

1 INTA, Estación Experimental Balcarce, C.C. 276, Balcarce 7620, Argentina; lorenzo.maximo@inta.gob.ar
2 Laboratorio de Fisiología Vegetal, Facultad de Ciencias Agrarias, Universidad Nacional de Mar del Plata, C.C. 276, Balcarce 7620, Argentina
3 Comisión de Investigaciones Científicas de la Provincia de Buenos Aires, La Plata 1900, Argentina; jtognetti2001@yahoo.com.ar
* Author to whom correspondence should be addressed; assuero.silvia@inta.gob.ar

Academic Editor: Cory Matthew

**Abstract:** Wheat is increasingly used as a dual-purpose crop (for forage and grain production) worldwide. Plants encounter low temperatures in winter, which commonly results in sugar accumulation. High sugar levels might have a positive impact on forage digestibility, but may also lead to an increased risk of bloat. We hypothesized that cultivars with a lower capacity to accumulate sugars when grown under cold conditions may have a lower bloat risk than higher sugar-accumulating genotypes, without showing significantly lower forage digestibility. This possibility was studied using two wheat cultivars with contrasting sugar accumulation at low temperature. A series of experiments with contrasting temperatures were performed in controlled-temperature field enclosures (three experiments) and growth chambers (two experiments). Plants were grown at either cool (8.1 °C–9.3 °C) or warm (15.7 °C–16.5 °C) conditions in field enclosures, and at either 5 °C or 25 °C in growth chambers. An additional treatment consisted of transferring plants from cool to warm conditions in the field enclosures and from 5 °C to 25 °C in the growth chambers. The plants in the field enclosure experiments were exposed to higher irradiances (*i.e.*, 30%–100%) than those in the growth chambers. Our results show that (i) low temperatures led to an increased hemicellulose content, in parallel with sugar accumulation; (ii) low temperatures produced negligible changes in *in vitro* dry matter digestibility while leading to a higher *in vitro* rumen gas production, especially in the higher sugar-accumulating cultivar; (iii) transferring plants from cool to warm conditions led to a sharp decrease in *in vitro* rumen gas production in both cultivars; and (iv) light intensity (in contrast to temperature) appeared to have a lower impact on forage quality.

**Keywords:** *Triticum aestivum* L.; dual purpose; cellulose; hemicellulose; lignin; crude protein; *in vitro* rumen gas production; *in vitro* dry matter digestibility

---

## 1. Introduction

Wheat is increasingly cultivated as a dual-purpose crop in several main wheat areas of the world, including the USA southern Great Plains [1,2], Australia [3,4], China [5] and the Argentinean Pampas region [6–8]. The reasons for this expansion are mainly the capacity of wheat to provide forage early in winter without excessively decreasing grain production. This practice increases the profitability at the whole-farm system level and additionally reduces the risk associated with both price and climate variability [9–15].

Wheat is often considered a high quality, cool season forage when consumed at earlier developmental stages due to the high digestibility of young leaf blades, which is in turn associated with a low lignin content [16]. In general, forage is considered high quality when the *in vitro* digestibility of dry matter (IVDMD) is higher than 600 $g \cdot kg^{-1}$ DM [17]. Accordingly, values higher than 800 $g \cdot kg^{-1}$ DM IVDMD have been reported for wheat at the pre-stem elongation stage [16,18]. Nevertheless, several reports have related the intake of wheat and other annual winter grasses to bloat risk due to high levels of rapidly fermentable components (*i.e.*, soluble protein and sugars [19–21]). Pasture bloat takes place when the grazing animal's capacity to expel gases produced by fermentation is exceeded [22], and gases become trapped in bio-film complexes [21,23].

*In vitro* rumen gas production has been positively correlated with plant protein fractions and IVDMD when incubated with mixed rumen microorganisms [24]. However, this correlation is not necessarily straightforward. The concentration of soluble protein and sugars in wheat leaves may vary, depending on genotypic and environmental conditions [25–28]. Exposure of grasses to low temperature induces a steady accumulation of both components, while reversion to non-chilling conditions determines a very rapid decline in their concentration [25,29]. Considerable variation in the capacity to accumulate sugars and proteins exists among wheat cultivars: cultivars which undergo deeper cold-acclimation (winter hardy cultivars) are able to accumulate substantially higher amounts of compatible solutes in their cells compared with less hardy cultivars [26–28]. Because of the transient nature of solute accumulation under cold conditions, the ratio between rapidly fermentable non-structural carbohydrates and proteins, and structural components of grass cells may vary with temperature, and thus wheat pastures might present a variable bloat risk while maintaining a constantly high IVDMD.

In addition to temperature, light intensity may also play a role in determining IVDMD and bloat risk. A reduction in light intensity has been associated with reduced forage quality in some evergreen species [30]. However, there are conflicting reports regarding the influence of light intensity on lignin content, even though most studies suggest that higher intensities favor an increase in lignin levels [31].

In the present work, we studied the forage composition of two wheat cultivars with contrasting capacity to accumulate solutes when grown under cold conditions, in parallel with IVDMD and *in vitro* gas rumen production, as affected by temperature and light intensity. A set of experiments with contrasting temperatures was conducted in both field enclosures (high irradiance, three experiments) and growth chambers (low irradiance, two experiments) to test the following hypotheses: (i) low temperature increases the concentration of soluble and structural components of wheat leaf blades; (ii) low temperature increases *in vitro* rumen gas production, without a significant effect on forage digestibility; (iii) the effect of low temperature on *in vitro* rumen gas production is stronger in a cultivar with a higher capacity to accumulate solutes; and (iv) low temperature effects on *in vitro* rumen gas production and IVDMD are enhanced under higher light intensity conditions.

## 2. Experimental Section

### 2.1. Plant Material

Two wheat (*Triticum aestivum* L.) cultivars were selected for their contrasting morpho-physiological responses to low temperature, which have been described elsewhere [29,32,33]. Briefly, ProINTA Pincén is a winter hardy wheat that reduces its growth more and accumulates higher sugar concentration than Buck Patacón under low temperatures. In all experiments, seeds were soaked in tap water for 24 h at ambient temperature prior to sowing.

### 2.2. Experimental Layout

#### 2.2.1. Field Enclosure Experiments

Three experiments were conducted in field enclosures during the winter seasons of 2005, 2006 and 2008 at the Facultad de Ciencias Agrarias campus (Universidad Nacional de Mar del Plata, Balcarce,

Argentina, 37°45′47.94″ S, 58°17′38.82″ W, 130 m a.s.l.) under a natural photoperiod. The enclosures were constructed of pipe structures covered with polyethylene film (100 μm thick) (Figure S1). Plants were grown up to the fourth fully expanded leaf stage in polyethylene containers with a 0.1-m diameter and a 0.6-m height filled with a uniform mixture of soil (topsoil of a Typic Argiudol) and vermiculite (1:1 v/v) located in an excavation within the enclosures in order to maintain the top of the containers at soil level. Twenty-four containers were placed in each enclosure (12 for each cultivar, from which three were monitored during development and harvested, three were used for water status determination and the rest were used as borders). The substrate was saturated at sowing with $\frac{1}{2}$-strength, and irrigated daily thereafter with $\frac{1}{4}$-strength Hoagland's solution [34]. Seeds (12 per container) were germinated and seedlings were thinned to 6 plants per container after emergence. Two electrical fan heaters with a thermostatic control (set to turn on under 16 °C) were located at opposite corners of one of the enclosures (warm treatment) at sowing. Accordingly, two electrical fans were located at opposite corners of the other enclosure (cool treatment) where the roof permanently covered the plants while the sidewalls were opened during the diurnal period and closed during the night. Air temperature was measured using thermistors and recorded using a data logger (Meteo, Cavadevices, Buenos Aires, Argentina) every 30 min. Thermistors were protected by shields to prevent absorption of solar radiation. The fourth channel of the data logger was used to record the photosynthetically active radiation. In the three experiments, the mean temperatures measured in the cool environment were very similar (ranging between 8.1 °C–9.3 °C), as were those of the warm enclosures (15.7 °C–16.5 °C) (Table 1). Daily mean air temperature dynamics during the 2006 field enclosure experiment, as well as the air temperatures recorded on one typical day of the same experiment, are shown in Figure S2 to illustrate the temperature conditions in the field enclosures. The average photosynthetic daily light integral (DLI) values diverged between cool and warm because of the different duration of the growing periods (Table 1). In 2008, a third treatment consisting of transferring plants from cool to warm conditions at the third leaf stage was applied. Plants were harvested early in the morning following a developmental criterion (*i.e.*, when 100% of plants attained the third fully expanded leaf stage for the cool and warm treatments, or the fourth fully expanded leaf stage for the cool-warm treatment); therefore, the harvest dates differed between the treatments.

**Table 1.** Average (± SD) daily mean air temperature and photosynthetic daily light integral (DLI) in the cool and warm field enclosures for the 2005 (sown on 20 June), 2006 (sown on 19 June) and 2008 (sown on 12 June) experiments. The cool-warm (C-W) data correspond to the post-transferred period only.

| Field Enclosure | Mean air temperature (°C) | | | DLI (mol photons $m^{-2} \cdot day^{-1}$) | | |
|---|---|---|---|---|---|---|
| | Cool | C-W | Warm | Cool | C-W | Warm |
| 2005 | 8.5 ± 2.8 | – | 16.5 ± 1.7 | 12.7 ± 5.7 | – | 17.8 ± 7.8 |
| 2006 | 9.3 ± 3.2 | – | 15.7 ± 2.2 | 12.7 ± 3.4 | – | 11.1 ± 4.1 |
| 2008 | 8.7 ± 3.3 | 16.6 ± 2.3 | 16.3 ± 1.9 | 14.0 ± 5.3 | 12.2 ± 3.9 | 11.9 ± 3.6 |

2.2.2. Growth Chamber Experiments

Two complete independent experiments were carried out in growth chambers at either 5 °C ± 0.5 °C or 25 °C ± 1 °C, under otherwise similar environmental conditions: 200 μmol photon $m^{-2} \cdot s^{-1}$ (photosynthetically active radiation, PAR) at the canopy level provided by fluorescent lamps (Osram Lumilux 21–840), 50% ± 10% relative humidity and a 12-h photoperiod. The average photosynthetic DLI (for both experiments) was 8.5 ± 0.1 and 8.6 ± 0.1 (mol photons $m^{-2} \cdot day^{-1}$) at 5 °C and 25 °C, respectively. Twenty-four plastic containers (0.1-m diameter, 0.3-m depth) filled with vermiculite and saturated with $\frac{1}{2}$-strength Hoagland's solution [34] were placed in the chamber (12 for each cultivar, from which three were monitored during development and harvested, three were used for water status determination and the rest were used as borders). Seeds (twelve per container) were germinated and seedlings were thinned to 6 plants per container after emergence. Plants were harvested as described above for the field experiment.

The time in days and the thermal time from sowing to harvest for both field enclosures and growth chamber experiments are shown in Table S1.

### 2.3. Determinations

#### 2.3.1. Plant Development

The number of fully expanded leaves was recorded at least twice a week.

#### 2.3.2. Relative Water Content

The relative water content (RWC) at harvest was determined on the youngest fully expanded leaf of the mainstem as described by Equiza *et al.* [32]. Sampling was performed early in the morning in parallel with harvesting for other determinations. In all experiments (the field enclosures and the growth chambers), irrespective of cultivars and temperature treatments, the RWC values at harvest were higher than 96%. Therefore, differences in the concentrations of the cell components, when expressed per unit of fresh mass, are not attributable to variation in water status among cultivars or treatments.

#### 2.3.3. Dry matter content and sugar concentration

The dry matter content was expressed as $g{\cdot}DM{\cdot}kg^{-1}$ FM. The total sugar concentration (TSC) in the leaf blades (mainly fructan, sucrose and monosaccharides) was quantified spectrophotometrically according to the phenol-sulfuric acid procedure [35]. Briefly, oven-dried leaf blades were ground, weighed and extracted in boiling distilled water ($10\ mg{\cdot}DM{\cdot}mL^{-1}$) for 10 min. The mixtures were centrifuged at 1000 g, and supernatants were used for analysis. The reaction mixture contained 0.57 mL of a 5% phenol solution and 2.85 mL of $H_2SO_4$ in a total volume of 4 mL. The mixture was stirred and incubated for 20 min in a bath at 25 °C, agitated and after 15 min at ambient temperature, the absorbance at 490 nm was read using a UV-1700 PharmaSpec spectrophotometer (Shimadzu Corp., Kyoto, Japan). A glucose solution was used as a standard. All samples were run in duplicate, and the values are expressed on a fresh mass (FM) basis.

#### 2.3.4. Cell Wall Components

The neutral detergent fiber (NDF) and acid detergent fiber (ADF) contents were determined using F57 filter bags (ANKOM A200, ANKOM Technology Corp., Fairport, NY, USA) according to Komareck *et al.* [36] and Komareck *et al.* [37], respectively. The lignin (ADL) content was determined using the acid detergent fiber permanganate lignin method [38]. The cellulose (ADF-ADL) and hemicellulose (NDF-ADF) contents were estimated by the difference. All values are expressed on a FM basis.

#### 2.3.5. Crude Protein

The crude protein (CP) concentration was determined from the nitrogen levels ($CP = 6.25 \times N$) using a LECO FP-528 (LECO Corporation, St. Joseph, MI, USA) nitrogen auto-analyzer [39]. The values are expressed on a FM basis.

#### 2.3.6. True *in Vitro* Dry Matter Digestibility (IVDMD)

This procedure followed the ANKOM-DAISY procedure [40]. Samples (0.5 g DW) were weighed directly into F57 filter bags that were sealed with a heater and placed in a Daisy$^{II}$ Incubator (ANKOM Technology Corp., Fairport, NY, USA) digestion jar. Buffered rumen fluid was prepared according to Goering and Van Soest [38] and transferred into the jars containing the bags. The jars were then placed in the Daisy$^{II}$ Incubator at 39 °C, with continuous rotation. After 48 h of incubation in buffered rumen fluid, the bags were gently rinsed under cold tap water and placed in an ANKOM$^{200}$ Fiber Analyzer to remove microbial debris and any remaining soluble fractions using neutral detergent solution so that true digestibility could be determined. Incubations were performed in duplicate.

#### 2.3.7. *In Vitro* Rumen Gas Production

Fresh wheat leaf blade samples were cut into 5-mm long pieces prior to all *in vitro* experiments. *In vitro* rumen gas production was determined following the general procedure described previously by Fay *et al.* [41] and Min [21,24], with modifications. The method consisted of measuring a syringe plunger displacement (ml) in 0–6-h incubation periods over a period of 28 h. Total *in vitro* rumen gas production was corrected to blank incubations (*i.e.*, no ruminal fluid). The rumen fluid was collected from a cannulated steer continuously receiving an alfalfa diet, mixed and strained through four layers of cheesecloth and flushed with $CO_2$ gas for *in vitro* rumen incubation. The *in vitro* rumen incubation procedure consisted of placing 2.5 g of minced fresh forage in 100-mL volumetric flasks containing 50 mL of rumen fluid diluted with artificial saliva [42], buffered to pH 6.8, saturated with $CO_2$ gas and maintained at 39 °C. Luer-type syringes (30 mL) with a 50/18 hypodermic needle, previously lubricated with distilled water to ensure consistent plunger resistance and movement to avoid gas losses, were inserted into the flask rubber stoppers. All gases were collected from the *in vitro* rumen incubation for gas production analyses. *In vitro* incubation was undertaken in duplicate.

### *2.4. Experimental Design and Statistics*

A completely randomized design with three replicates (containers) per combination of two cultivars and two or three growth temperatures (depending on the experiment) was used. The temperature effect on plant carbon status (dry matter content and total sugar concentration), cell wall components (cellulose, hemicellulose and lignin), crude protein and *in vitro* rumen gas production at 28 h was analyzed using two-way ANOVA (Statistica 7, StatSoft Inc., Tulsa, OK, USA). Means were separated using Tukey's test at a significance level of 5%. No attempt was made to compare the effect of light intensity because of differences in temperature conditions between the field enclosure and the growth chamber experiments.

## 3. Results

### *3.1. Forage Composition*

#### 3.1.1. Forage Dry Matter Content

Significantly higher leaf blade dry matter content (DMC) values were found in the cool than in the warm environments for both cultivars (Table 2); the increase induced by lower temperatures was more pronounced in winter hardy Pincén than in Patacón (between 23%–29% and 15%–19%, respectively, for field enclosures, and averaging 47% and 16%, respectively, for growth chambers). Accordingly, Pincén had a higher DMC in cool environments. Conversely, no significant differences were found between the two cultivars under warm growing conditions. Transferring the plants from cool to warm conditions resulted in a significant decrease in DMC in Pincén but not in Patacón for both the field enclosures and growth chamber experiments.

Similar DMC values between the field enclosures and the growth chamber experiments were found for Pincén, while in Patacón, the growth chamber values were approximately 12% lower than their counterparts in the field enclosures.

It is well known that during cold acclimation of grasses, cellular dry matter content increases due to a transient deposition of many solutes, including non-structural carbohydrates, proteins, amino acids, *etc.*, while the cell water content may not be affected [43]. Since the RWC in our experiments was close to saturation (*i.e.*, higher than 96%) and was unaffected by temperature treatments or cultivars, the changes in DMC reflected the variation in the C concentration, not the plant water status. Because not all components are accumulated in the same proportion, the concentration of a component that accumulates less than the average could be seen as diminishing when expressed on a dry matter basis. For this reason, the concentrations of the different forage components listed below are expressed on a fresh mass basis, as in similar experiments reported elsewhere [44].

### 3.1.2. Total Sugar Concentration (TSC)

In general, the TSC results were similar to those of the DMC, with higher leaf blade TSC in cool than in warm treatments for both cultivars in all experiments (Figure 1), but the cold-induced increases were larger than for the DMC (between 155%–167% and 83%–100%, for Pincén and Patacón, respectively). Within each experiment and under cool conditions, Pincén showed the highest values.

For both cultivars, similar TSC values between the field enclosures and the growth chamber experiments were attained under the cooler environments. On the other hand, plants grown in growth chambers under warm conditions had TSC values that were approximately 40% lower than their counterparts in the field enclosures.

**Table 2.** Dry matter content (DMC, g $kg^{-1}$ FM) of leaf blades of wheat cv. Pincén and cv. Patacón grown in cool (8.1 °C–9.3 °C) or warm (15.7 °C–16.5 °C) field enclosures in the 2005, 2006 and 2008 experiments, and in growth chambers at either 5 °C or 25 °C. Plants were harvested at the 3rd fully expanded leaf stage, or at the 4th fully expanded leaf stage for plants that were transferred from cool to warm and from 5 °C to 25 °C in the 2008 field enclosure experiment and in the second growth chamber experiment, respectively. Values are the means (± SE) of three replicates. Within each experiment, different letters indicate significant differences ($p < 0.05$).

| Field enclosure | Pincén | | | Patacón | | |
|---|---|---|---|---|---|---|
| | Cool | Cool-Warm | Warm | Cool | Cool-Warm | Warm |
| 2005 | 179 ± 6.9 a | N.D. | 146 ± 7.3 c | 161 ± 8.3 b | N.D. | 140 ± 6.7 c |
| 2006 | 176 ± 7.1 a | N.D. | 141 ± 3.6 c | 159 ± 8.5 b | N.D. | 140 ± 6.0 c |
| 2008 | 175 ± 2.3 a | 162 ± 6.7 b | 136 ± 2.9 c | 159 ± 6.1 b | 155 ± 4.6 b | 134 ± 8.9 c |
| Growth Chamber | 5 °C | 5 °C–25 °C | 25 °C | 5 °C | 5 °C–25 °C | 25 °C |
| Experiment 1 | 189 ± 10.3 a | N.D. | 128 ± 7.3 c | 141 ± 8.7 b | N.D. | 123 ± 8.6 c |
| Experiment 2 | 183 ± 9.0 a | 156 ± 5.6 b | 125 ± 6.1 d | 142 ± 2.3 c | 133 ± 8.9 cd | 121 ± 8.9 d |

N.D.: Not determined.

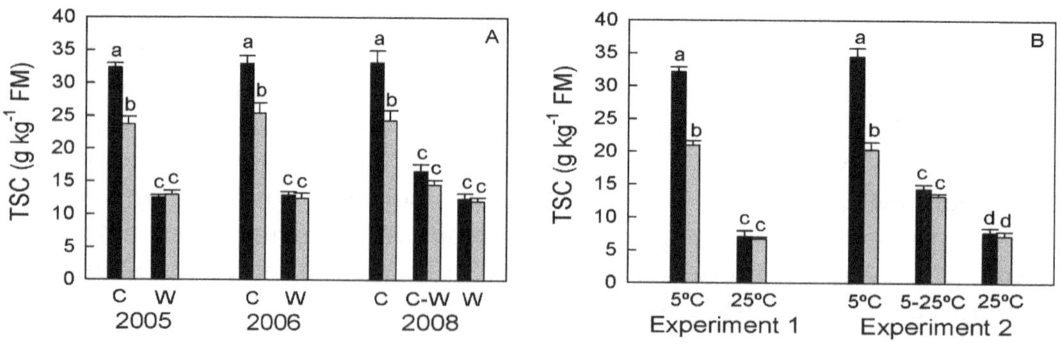


**Figure 1.** Total sugar concentration (TSC, g·$kg^{-1}$ FM) of leaf blades of wheat cv. Pincén (black bars) and cv. Patacón (grey bars). (**A**) plants grown in cool (C, 8.1 °C–9.3 °C) or warm (W, 15.7 °C–16.5 °C) field enclosures in the 2005, 2006 and 2008 experiments. (**B**) plants grown in growth chambers at 5 °C or 25 °C. Plants were harvested at the 3rd fully expanded leaf stage, or at the 4th fully expanded leaf stage for plants that were transferred from cool to warm (C-W) and from 5 °C to 25 °C in the 2008 field enclosure experiment and in the second growth chamber experiment, respectively. Vertical bars indicate SE ($n = 3$). Within each experiment, different letters indicate significant differences ($p < 0.05$).

### 3.1.3. Structural Carbohydrates and Lignin

Cellulose and hemicellulose were the main cell wall components, ranging between 18 and 37 g·kg$^{-1}$ FM, and 12 and 49 g·kg$^{-1}$ FM, respectively (Figure 2). The lignin content was generally low, ranging between 1.2 and 3.5 g·kg$^{-1}$ FM.

For both cultivars, the cellulose content of the leaf blades increased slightly under cooler conditions (between 13% and 25%, and 8% and 45% for cool *vs.* warm conditions for Pincén and Patacón, respectively, Figure 2A,B) except in the 2005 field enclosure experiment when no significant differences were found between temperatures. In the field enclosures, transferring plants from cool to warm conditions resulted in a slight (4%–7%) but significant decrease in the cellulose content of the leaf blades. The values obtained in the growth chamber experiments tended to be similar or higher than their counterparts in the field enclosures.

Hemicellulose varied most among the temperature treatments and cultivars (Figure 2C,D). Similar to cellulose, the hemicellulose values were generally higher under cool conditions (between 51% and 177%, and between 24% and 95% higher than the warm condition values for Pincén and Patacón, respectively) with the sole exception of Patacón in the 2006 experiment (−11%). The hemicellulose values of the transferred plants decreased and approached those of the warm-grown plants. The hemicellulose values in the growth chamber experiments tended to be lower than those in the field enclosures, particularly under warm conditions.

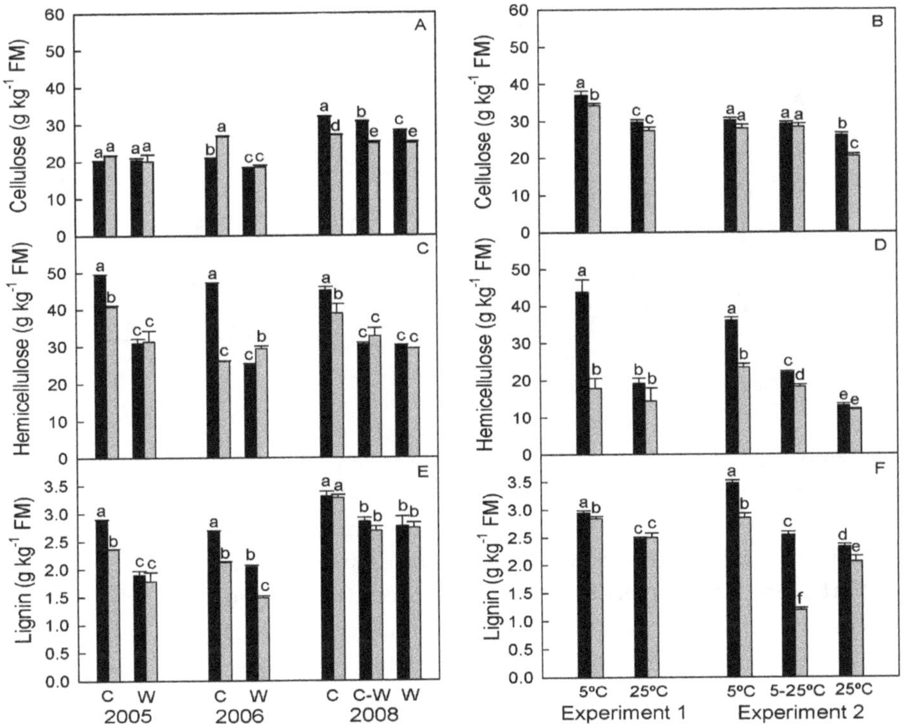


**Figure 2.** Cellulose (g·kg$^{-1}$ FM, **A,B**), hemicellulose (g·kg$^{-1}$ FM, **C,D**) and lignin (g·kg$^{-1}$ FM, **E,F**) contents of leaf blades of wheat cv. Pincén (black bars) and cv. Patacón (grey bars). (**A,C,E**): plants grown in cool (C, 8.1 °C–9.3 °C) or warm (W, 15.7 °C–16.5 °C) field enclosures in the 2005, 2006 and 2008 experiments. (**B,D,F**): plants grown in growth chambers at 5 °C or 25 °C. Plants were harvested at the 3rd fully expanded leaf stage, or at the 4th fully expanded leaf stage for plants that were transferred from cool to warm (C-W) and from 5 °C to 25 °C in the 2008 field enclosure experiment and the second growth chamber experiment, respectively. Vertical bars indicate SE ($n = 3$). Within each experiment, different letters indicate significant differences ($p < 0.05$).

The lignin content of the leaf blades was higher under cool conditions (between 19% and 53%, and 12% and 44% higher than the warm condition values for Pincén and Patacón, respectively, Figure 2E,F). Transferring plants from cool to warm conditions resulted in a 14%–58% reduction in lignin concentrations, which approached the values of warm-grown plants or were even lower in one case (Figure 2F). The lignin values in the growth chamber were similar to those for the field enclosures, with the exception of the transferred plants of Patacón in Experiment 2, which, for unknown reasons, presented a rather low value.

#### 3.1.4. Crude Protein Concentration

The crude protein (CP) concentration in the leaf blades ranged between 37 and 54 g·kg$^{-1}$ FM, and 32 and 48 g·kg$^{-1}$ FM for Pincén and Patacón, respectively. The values for winter hardy Pincén were significantly higher under cooler conditions in both the field enclosures and the growth chamber experiments, except for the 2006 experiment when the difference was not significant (Figure 3). In contrast, no cold-induced increase in CP concentration was observed in Patacón except for the 2006 experiment. Transferring Pincén plants from cool to warm conditions did not significantly modify the CP concentration in either the field enclosures or the growth chamber experiments. No straightforward trend was observed for the CP concentration in Patacón. In general, similar values were found in the field enclosure and the growth chamber experiments, except that for Experiment 2, somewhat higher values were observed, especially for Pincén.

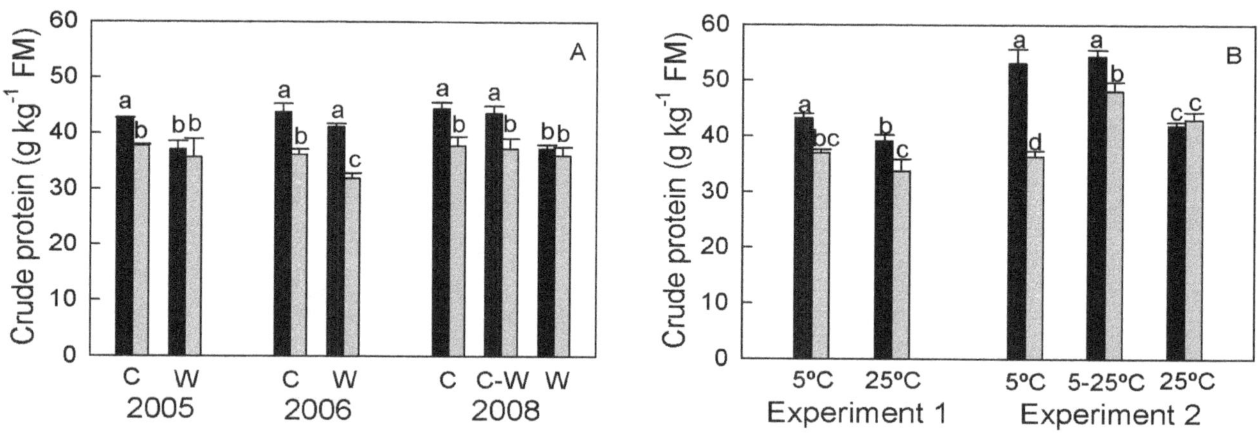


**Figure 3.** Crude protein content (g·kg$^{-1}$ FM) of the leaf blades of wheat cv. Pincén (black bars) and cv. Patacón (grey bars). (**A**) plants grown in cool (C, 8.1 °C–9.3 °C) or warm (W, 15.7 °C–16.5 °C) field enclosures in the 2005, 2006 and 2008 experiments. (**B**) plants grown in growth chambers at 5 °C or 25°C. Plants were harvested at the 3rd fully expanded leaf stage, or at the 4th fully expanded leaf stage for plants that were transferred from cool to warm (C-W) and from 5 °C to 25 °C in the 2008 field enclosure experiment and in the second growth chamber experiment, respectively. Vertical bars indicate SE ($n$ = 3). Within each experiment, different letters indicate significant differences ($p < 0.05$).

### 3.2. *Forage Quality*

#### 3.2.1. True *in Vitro* Dry Matter Digestibility (IVDMD)

The IVDMD values were consistently high (above 75%, Table 3) irrespective of temperature, light environment, and cultivar. Although in some experiments significant differences were found among treatments, a straightforward pattern was not observed. In addition, the actual differences were small, even between the most contrasting treatments (approximately 70 and 40 g·kg$^{-1}$ DM for the field enclosures and growth chambers, respectively).

**Table 3.** True *in vitro* dry matter digestibility (IVDMD, g·kg$^{-1}$ DM) of leaf blades of wheat cv. Pincén and cv. Patacón grown in cool (8.1 °C–9.3 °C) or warm (15.7 °C–16.5 °C) field enclosures in the 2005, 2006 and 2008 experiments, and in two growth chamber experiments at 5 °C or 25 °C. Plants were harvested at the 3rd fully expanded leaf stage, or at the 4th fully expanded leaf stage for plants that were transferred from cool to warm and from 5 °C to 25 °C in the 2008 field enclosure experiment and in the second growth chamber experiment, respectively. Values are the means (± SE) of three replicates. Within each experiment, different letters indicate significant differences ($p < 0.05$).

| Field Enclosure | Pincén | | | Patacón | | |
|---|---|---|---|---|---|---|
| | Cool | Cool-Warm | Warm | Cool | Cool-Warm | Warm |
| 2005 | 913 ± 1.6 a | N.D. | 917 ± 13.4 a | 894 ± 4.9 a | N.D. | 912 ± 14.0 a |
| 2006 | 951 ± 33.2 a | N.D. | 868 ± 13.4 a | 942 ± 28.4 a | N.D. | 905 ± 23.4 a |
| 2008 | 754 ± 38.4 b | 783 ± 4.3 ab | 820 ± 3.9 ab | 842 ± 7.5 a | 830±2.6 ab | 838 ± 2.9 ab |
| Growth chamber | 5 °C | 5 °C–25 °C | 25 °C | 5 °C | 5 °C–25 °C | 25 °C |
| Experiment 1 | 941 ± 1.6 c | N.D. | 972 ± 2.1 a | 953 ± 1.8 bc | N.D. | 962 ± 3.4 ab |
| Experiment 2 | 975 ± 4.6 a | 963±1.4 ab | 936 ± 2.3 d | 954 ± 2.1 bc | 910±3.5 e | 946 ± 3.1 cd |

N.D.: Not determined.

### 3.2.2. In Vitro Rumen Gas Production

*In vitro* rumen gas production analysis of leaf blades was performed for the 2008 field enclosure experiment and in the growth chamber Experiment 2. Curvilinear relationships were obtained between cumulative gas production and time up to 28 h of incubation irrespective of temperature, light environment and cultivar for plants grown at constant temperature (Figure 4).

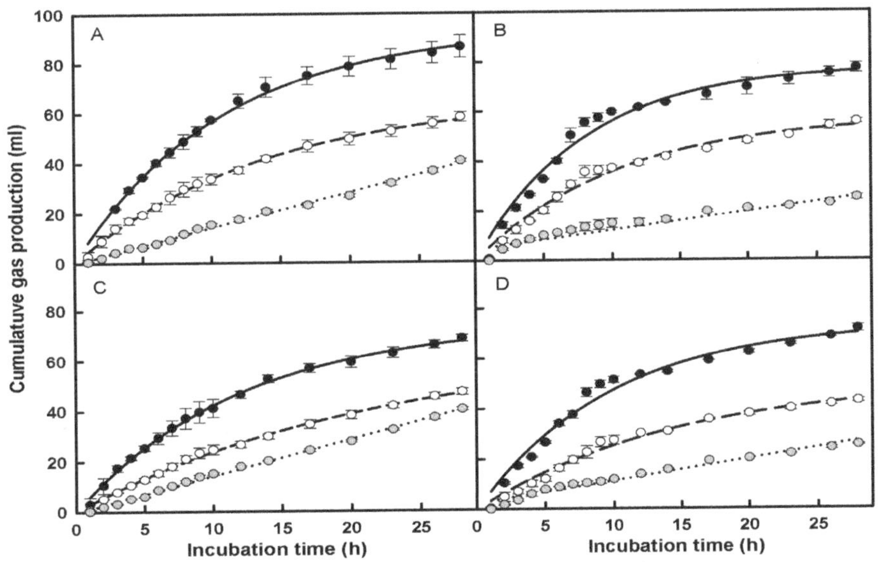


**Figure 4.** Cumulative gas production (ml) of leaf blades of wheat cv. Pincén (**A**,**B**) and cv. Patacón (**C**,**D**). (**A**,**C**): plants were grown in cool (8.1 °C–9.3 °C, black symbols, solid line) or warm (15.7 °C–16.5 °C, white symbols, dashed line) field enclosures in the 2008 experiment. (**B**,**D**): plants were grown in growth chambers at 5 °C (black symbols, solid line) or 25 °C (white symbols, dashed line) in Experiment 2. Plants were harvested at the 3rd fully expanded leaf stage except those that were transferred (grey symbols, dotted line) from cool to warm conditions in the 2008 field enclosure experiment (**A**,**C**) or from 5 °C to 25 °C (**B**,**D**) in the second growth chamber experiment, in which the plants were harvested at the 4th fully expanded leaf stage. Values correspond to the incubation of 2.5 g of minced fresh forage (mean ± SE of three replicates). The fitted models are $y = a \times (1 - \exp(-b \times t))$ [45], where $y$ is the volume of gas produced at time $t$, $a$ is the final asymptotic gas volume, and $b$ is the fractional rate of gas production and, for transferred plants, $y = a + b \times t$, where $a$ is the y-axis value at $t = 0$ and $b$ is the rate of gas production.

Different gas production profiles were observed between temperatures and cultivars under constant temperature. After the first 4–5 h, the cumulative gas production from plants grown under cool conditions was approximately 65%–70% (field enclosures) or 55%–78% (growth chambers) higher than the values observed in warm environments for both cultivars. Moreover, Pincén cumulative gas production after this period was always higher than that observed for Patacón (approximately 35% in the field enclosures and 17% in the growth chambers).

Linear relationships best fit the data of plants transferred from cool to warm conditions. Cumulative gas production was considerably reduced in transferred plants, *i.e.*, to values markedly lower than those of plants grown under warm conditions (−30% and −56% for Pincén, and −15% and −44% for Patacón for field enclosures and growth chamber experiments, respectively, at 28 h of incubation).

On the other hand, slight differences were found between the light environments for plants of either cultivar grown at constant temperatures (*i.e.*, at 28 h, the field enclosure values were between −3% and +11% of the equivalent treatments in the growth chambers).

## 4. Discussion

Plant components that affect forage quality, including readily digestible ones (soluble carbohydrates and crude protein, which are associated with bloat risk [19–21]) and non-digestible ones (lignin) generally increased under cooler conditions. This led to an increase in the dry matter content of the leaf blades, which was more pronounced in Pincén than in Patacón (Table 2). In the present work, soluble carbohydrates showed the most important changes while cellulose showed the least pronounced ones (Figure 1). In addition, changes were generally more marked in winter hardy cv. Pincén than in cv. Patacón. An increase in soluble carbohydrate concentration in cold temperatures is a very well known phenomenon in temperate grasses, which results from photosynthesis being less affected than growth by cold [46]. This response, which is associated with the capacity of plants to develop freezing tolerance [47–49], has been reported previously for the wheat cultivars studied here [25,32]. Transferring plants from cool to warm conditions led to a sharp decrease in TSC (Figure 1). This response may be the consequence of either a decreased TSC concentration in pre-existing leaves, or of a low concentration in the leaves developed under warm conditions, or more likely a combination of both. In any case, preliminary observations indicate that a peak in respiration within the first 24 h after transfer occurs in Pincén but not in Patacón [50].

Much less is known about the changes in the concentration of structural components due to temperature. An increase in cell wall thickness in grasses acclimated to cold has been reported [32,51]. Our finding of increased hemicellulose and lignin and, to a lesser extent, cellulose concentrations under cooler conditions (Figure 2) might be associated with this anatomical response. In plants transferred from cool to warm environments, the concentration of the different compounds tended to decrease toward values close to those of warm-grown plants. Since the proportion of the different components of leaf blades remained unmodified by ontogeny (at least up to the fifth leaf stage; data not shown) under continuous environmental conditions, the observed changes in forage composition in transferred plants could be solely attributable to the effect of temperature.

On the other hand, the light environment appeared to have only a marginal effect on the concentration of forage constituents. It has been argued that because of energy balance factors, low temperature and high light environments modify grass morpho-physiological characters in a similar manner (*i.e.*, promoting compactness of growth habit [52], and a comparable redox state of photosystem II [53]). Nevertheless, it has been noted that freezing tolerance (measured as LT50) depends on low temperature exposure that is independent of irradiance levels [54]; therefore, the observed marked increase in TSC due to cold rather than to light is in agreement with the corresponding winter hardiness of Pincén and Patacón. While cold promotes a strong C accumulation due to an altered balance between growth and C utilization [46], it has been shown that light intensity not only favors assimilate availability but also plant growth (*i.e.*, higher shoot fresh mass at harvest in the field enclosures compared with the growth chambers [28]), thus preventing a substantial carbohydrate accumulation. Consequently, within each temperature treatment (*i.e.*, cool and 5 °C or warm and 25 °C), plants grown under higher irradiances (field enclosures) were larger than those grown under low irradiances (growth chambers), as shown previously [28].

Both parameters of forage quality, *i.e.*, IVDMD and gas production, had markedly different responses to growing conditions and cultivar (Table 3 and Figure 4). A low correlation between both parameters was reported previously [55]. A first possible reason for the divergence between the two parameters could be the different incubation times. IVDMD was assessed at the end of a 48 h-incubation period while cumulative gas production assessed the dynamics of forage degradation up to 28 h, which is sensitive to changes in the proportion of forage constituents that differ in digestion rates. A 48 h-incubation period exceeds the ruminal retention time of high quality feeds in animals of high potential production (e.g., <24 h) [56]. However, a second major factor is that fresh tissue was used for analyzing gas production, in contrast with IVDMD, which uses dry matter. Thus, variation in the dry matter content of the tissue is likely to modify only *in vitro* rumen gas production. There is strong evidence indicating that voluntary consumption of fresh forage by ruminants is closely related to forage volume (which in turn is related to fresh mass), and not to dry mass [57,58]. Therefore, *in vitro* rumen gas production from fresh forage incubation may more closely reflect what is actually happening in the animal rumen.

Despite large variations in the concentration of the forage components, IVDMD was almost unaffected by changes in temperature, light environment or cultivar (Table 3). There are reports indicating that cool temperatures may increase IVDMD. For example, this response has been found in tall fescue [59], timothy [60,61], and six other temperate grasses [61]. However, in other cases, no consistent effect of temperature on IVDMD was found [62]. A possible explanation for the lack of IVDMD increase at low temperature here is that even under warm environments, the values were high (ranging from 820 to 917 $g \cdot kg^{-1}$ DM in the field enclosures and 936 to 972 $g \cdot kg^{-1}$ DM in the growth chambers). The high values of IVDMD were expected since we evaluated the forage quality of the leaf blades of very young wheat plants in this study. It is well known that IVDMD decreases with leaf age [63,64].

Contrasting results have been reported with respect to the effect of light on IVDMD [65]. It has been suggested that only shade-intolerant species have their quality reduced by shade, mainly because of a decrease in total soluble carbohydrate concentration [66]. In our experiments, plants from the growth chambers (*i.e.*, lower light intensity) tended to show higher IVDMD values than those from the field enclosures (Table 3).

On the other hand, *in vitro* rumen gas production was largely modified by temperature and, to a lesser extent, by cultivars (Figure 4). In general, variation in gas production was in agreement with the changes in TSC (Figure 1). This is expected since gas is produced as the result of fermentation by ruminal fluid. The only exception was observed in transferred plants, which exhibited the lowest gas production despite showing TSC values that were higher than those of warm grown plants. In this sense, it has been shown that a high proportion of TSC in cold-acclimated plants consists of fructans that are inserted into the lipid headgroup region of the plasma membrane and help to stabilize it under freezing stress [67–69]. Given the fact that fructans may represent as much as 80% of TSC in cold-acclimated wheat [70], a hypothesis for further study is that part of the fructans inserted in the plasma membranes during growth at low temperature could remain there for a certain time after plants have been subjected to warmer conditions. This category of fructans might be less easily fermented by ruminal enzymes, but still captured in the chemical quantification of TSC. Because gas production is associated with bloat risk, information from further research into these points could be useful for grazing management of dual-purpose wheat crops grown in environments with changing temperatures in the autumn-winter period, conditions commonly found in the Argentinean Pampas.

It is well known bloat risk is tightly related to high sugar levels [19–21], but in turn, the latter are required for high freezing tolerance [47–49]. Consequently, it appears that simultaneously improving wheat for both low bloat risk and freeze hardening may be difficult. However, if the hypothesis that fructan inserted into the plasma membranes is less readily fermented in the rumen is supported, then studying genotype-associated variation in fructan partitioning between soluble and membrane-bound fractions could provide useful information for breeding purposes.

## 5. Conclusions

In parallel with the expected increase in sugar accumulation, low temperatures led to an increase in hemicellulose and crude protein concentration. This response was more marked in the hardy cultivar Pincén. While negligible changes in response to temperature were observed in *in vitro* dry matter digestibility, *in vitro* rumen gas production was much higher at cooler temperatures, especially in the higher sugar accumulating cultivar. This effect was rapidly reversed in plants transferred from cool to warm conditions. In contrast to temperature, light intensity appeared to have a lower impact on forage quality. Future experiments should focus on the remobilization of cool-induced membrane-bound fructans and their association with rumen gas production in animals.

**Acknowledgments:** This work, which is part of Máximo Lorenzo's PhD thesis at Universidad Nacional de Mar del Plata (UNMdP), Argentina, is supported by a grant from UNMdP (AGR-360). The authors thank Patricio Fay for his assistance with the *in vitro* rumen gas production analyses.

**Author Contributions:** All co-authors contributed equally to this work.

## References

1. Hossaina, I.; Epplin, F.M.; Krenzer, E.G. Planting date influence on dual-purpose winter wheat forage yield, grain yield, and test weight. *Agron. J.* **2003**, *95*, 1179–1188. [CrossRef]
2. Butchee, J.D.; Edwards, J.T. Dual-purpose wheat grain yield as affected by growth habit and simulated grazing intensity. *Crop Sci.* **2013**, *53*, 1686–1692. [CrossRef]
3. Dove, H.; McMullen, G. Diet selection, herbage intake and liveweight gain in young sheep grazing dual-purpose wheats and sheep responses to mineral supplements. *Anim. Prod. Sci.* **2009**, *49*, 749–758. [CrossRef]
4. Kelman, W.M.; Dove, H. Growth and phenology of winter wheat and oats in a dual-purpose management system. *Crop Pasture Sci.* **2009**, *60*, 921–932. [CrossRef]
5. Tian, L.H.; Bell, L.W.; Shen, Y.Y.; Whish, J.P.M.; Nan, Z.B. Dual-purpose use of winter wheat in western China: Cutting time effects on forage production and grain yield. *Crop Pasture Sci.* **2012**, *63*, 520–528. [CrossRef]
6. Arzadun, M.J.; Arroquy, J.I.; Laborde, H.E.; Brevedan, R.E. Grazing pressure on beef and grain production of dual-purpose wheat in Argentina. *Agron. J.* **2003**, *95*, 1157–1162. [CrossRef]
7. Arzadun, M.J.; Arroquy, J.I.; Laborde, H.E.; Brevedan, R.E. Effect of planting date, clipping height, and cultivar on forage and grain yield of winter wheat in Argentinean Pampas. *Agron. J.* **2006**, *98*, 1274–1279. [CrossRef]
8. Peralta, N.; Abbate, P.E.; Marino, A. Effect of the defoliation regime on grain production in dual purpose wheat. *Agriscientia* **2011**, *28*, 1–11.
9. Morant, A.E.; Merchán, H.D.; Lutz, E.E. Comparación de la producción forrajera de cultivares de trigo para doble propósito. *Rev. Argent. Prod. Anim.* **1998**, *18*, 213–214.
10. Lutz, E.E.; Merchán, H.D.; Morant, A.E. Carne y grano de un trigo doble propósito en condiciones semiá-ridas. *Phyton (Buenos Aires)* **2000**, *67*, 195–200.
11. Lutz, E.E.; Merchán, H.D.; Morant, A.E. Estado de desarrollo de la planta de trigo (var. ProINTA Pincén) al momento de la última defoliación y su rendimiento en grano. *Phyton (Buenos Aires)* **2000**, *68*, 83–87.
12. Bainotti, C.T.; Gomes, D.; Masiero, B.; Salines, J.; Fraschina, J.; Bertram, N.; Navarro, C. Evaluación de Cultivares de trigo como Doble Propósito. Available online: http://agrolluvia.com/wp-content/uploads/2010/05/INTA-Marcos-Ju%C3%A1rez-Evaluaci%C3%B3n-de-cultivares-de-trigo-como-doble-prop%C3%B3sito1.pdf (accessed on 18 May 2015).

13. Morant, A.E.; Merchán, H.D.; Lutz, E.E. Evaluación de genotipos de trigos para doble propósito. Fecha de siembra y producción de grano. *Rev. Argic. Prod. Anim.* **2003**, *23*, 222–223.
14. Lutz, E.; Merchán, H.; Morant, A. Mezcla de variedades de trigo para doble propósito. *Phyton (Buenos Aires)* **2008**, *77*, 217–223.
15. Bell, L.W.; Moore, A.D. Mixed Crop-livestock Businesses Reduce Price- and Climate-induced Variability in Farm Returns: A Model-derived Case Study. Available online: http://aciar.gov.au/files/node/13992/mixed_crop_livestock_businesses_reduce_price_and__20972.pdf (accessed on 3 August 2015).
16. Walker, D.W.; West, C.P.; Bacon, R.K.; Longer, D.E.; Turner, K.E. Changes in forage yield and composition of wheat and wheat-ryegrass mixtures with maturity. *J. Dairy Sci.* **1990**, *73*, 1296–1303. [CrossRef]
17. Paterson, J.A.; Bowman, J.P.; Belyea, R.L.; Kerley, M.S.; Williams, J.E. The impact of forage quality and supplementation regimen on ruminant animal intake and performance. In *Forage Quality, Evaluation, and Utilization*; Fahey, G.C., Ed.; American Society of Agronomy, Crop Science Society of America, Soil Science Society of America: Madison, WI, USA, 1994; pp. 59–114.
18. Kelman, W.M.; Dove, H.; Flint, P. The Potential of Winter Wheat Cultivars and Breeding Lines for Use in Dual-purpose (Grain and Graze) Systems. Available online: http://www.regional.org.au/au/asa/2006/poster/systems/4613_kelmanw.htm (accessed on 18 May 2015).
19. Howarth, R.E.; Horn, G.W. Wheat pasture bloat of stocker cattle: A comparison with legume pasture bloat. In Proceedings of the National Wheat Pasture Symposium; Division of Agriculture, Oklahoma State University: Stillwater, OK, USA, 1984; pp. 24–25.
20. Horn, G.W. Growing cattle on winter wheat pasture: Management and herd health considerations. *Vet. Clin. North Am. Food A* **2006**, *22*, 335–356. [CrossRef] [PubMed]
21. Min, B.R.; Pinchak, W.E.; Mathews, D.; Fulford, J.D. *In vitro* rumen fermentation and *in vivo* bloat dynamics of steers grazing winter wheat to corn oil supplementation. *Anim. Feed Sci. Technol.* **2007**, *133*, 192–205. [CrossRef]
22. Mayland, H.F.; Cheeke, P.R.; Majak, W.; Goff, J.P. Forage-induced animal disorders. In *Forages*, 6th ed.; Nelson, C.J., Moore, K.M., Collins, M., Eds.; Blackwell Publication: Ames, IA, USA, 2007; Volume 2, pp. 687–707.
23. Malinowski, D.P.; Pitta, D.W.; Pinchak, W.E.; Min, B.R.; Emendack, Y.Y. Effect of nitrogen fertilisation on diurnal phenolic concentration and foam strength in forage of hard red wheat (*Triticum aestivum* L.) cv. Cutter. *Crop Pasture Sci.* **2011**, *62*, 656–665. [CrossRef]
24. Min, B.R.; Pinchak, W.E.; Fulford, J.D.; Puchala, R. Wheat pasture bloat dynamics, *in vitro* ruminal gas production, and potential bloat mitigation with condensed tannins. *J. Anim. Sci.* **2005**, *83*, 1322–1331. [PubMed]
25. Tognetti, J.A.; Calderón, P.L.; Pontis, H.G. Fructan metabolism: Reversal of cold acclimation. *J. Plant Physiol.* **1989**, *134*, 232–236. [CrossRef]
26. Tognetti, J.A.; Salerno, C.L.; Crespi, M.D.; Pontis, H.G. Sucrose and fructan metabolism of different wheat cultivars at chilling temperatures. *Physiol. Plant.* **1990**, *78*, 554–559. [CrossRef]
27. Equiza, M.A.; Miravé, J.P.; Tognetti, J.A. Differential root *versus* shoot growth inhibition and its relationship with carbohydrate accumulation at low temperature in different wheat cultivars. *Ann. Bot.* **1997**, *80*, 657–663. [CrossRef]
28. Lorenzo, M.; Assuero, S.G.; Tognetti, J.A. Low temperature differentially affects tillering in spring and winter wheat in association with changes in plant carbon status. *Ann. App. Biol.* **2015**, *166*, 236–248. [CrossRef]
29. Equiza, M.A.; Tognetti, J.A. Morphological plasticity of spring and winter wheats under changing temperatures. *Funct. Plant Biol.* **2002**, *29*, 1427–1436. [CrossRef]
30. Blair, R.M.; Alcaniz, R.; Harrell, A. Shade intensity influences the nutrient quality and digestibility of southern deer browse leaves. *J. Range Manag.* **1983**, *36*, 257–264. [CrossRef]
31. Moura, J.C.; Bonine, C.A.; de Oliveira Fernandes Viana, J.; Dornelas, M.C.; Mazzafera, P. Abiotic and biotic stresses and changes in the lignin content and composition in plants. *J. Integr. Plant Biol.* **2010**, *52*, 360–376. [CrossRef] [PubMed]
32. Equiza, M.A.; Miravé, J.P.; Tognetti, J.A. Morphological, anatomical and physiological responses related to differential shoot *vs.* root growth inhibition at low temperature in spring and winter wheat. *Ann. Bot.* **2001**, *87*, 67–76. [CrossRef]

33. Assuero, S.G.; Lorenzo, M.; Pérez, N.M.; Velázquez, L.; Tognetti, J.A. Tillering promotion by paclobutrazol in wheat and its relationship with plant carbohydrate status. *N. Z. J. Agric. Res.* **2012**, *55*, 347–358. [CrossRef]
34. Hoagland, D.R.; Arnon, D.I. The water-culture method for growing plants without soil. *Calif. Agric. Exp. Stn. Circ.* **1950**, *347*, 1–32.
35. Dubois, M.; Gilles, K.A.; Hamilton, J.K.; Rebers, P.A.; Smith, F. Colorimetric method for determination of sugars and related substances. *Anal. Chem.* **1956**, *28*, 350–356. [CrossRef]
36. Komareck, A.R.; Robertson, J.B.; van Soest, P.J. Comparison of the filter bag technique to conventional filtration in the Van Soest NDF analysis of 21 feeds. In Proceedings of the National Conference on Forage Quality, Evaluation and Utilization, Lincoln, NE, USA, 13–15 April 1994; Fahey, G.C., Ed.; Nebraska University: Lincoln, NE, USA, 1994.
37. Komareck, A.R.; Robertson, J.B.; Van Soest, P.J. A comparison of methods for determining ADF using the filter bag technique *versus* conventional filtration. *J Dairy Sci.* **1993**, *77*, 24–26.
38. Goering, H.K.; van Soest, P.J. Forage fiber analyses (Apparatus, Reagents, Procedures and Some Applications). In *USDA-ARS Agricultural Handbook 379*; US Government Printing Office: Washington, DC, USA, 1970; p. 20.
39. Horneck, D.A.; Miller, R.O. Determination of total nitrogen in plant tissue. In *Handbook of Reference Methods for Plant Analysis*; Kalra, Y.P., Ed.; CRC Press: London, UK, 1998; pp. 75–83.
40. ANKOM Tecnology. Analytical Methods *in vitro* True Digestibility Method (IVTD-Daisy). Available online: https://ankom.com/sites/default/files/document-files/Method_3_Invitro_0805_D200%2CD200I.pdf (accessed on 3 August 2015).
41. Fay, J.P.; Cheng, K.-J.; Hanna, M.R.; Howarth, R.E.; Costerton, J.W. *In vitro* digestion of boat-safe and boat-causing legumes by rumen microorganisms: Gas and foam production. *J Dairy Sci.* **1980**, *63*, 1273–1281. [CrossRef]
42. McDougall, E.I. Studies on ruminant saliva. 1. The composition and output of sheep's saliva. *Biochem. J.* **1948**, *43*, 99–109. [CrossRef] [PubMed]
43. Tanino, K.; Weiser, C.J.; Fuchigami, L.H.; Chen, T.H. Water content during abscisic acid induced freezing tolerance in bromegrass cells. *Plant Physiol.* **1990**, *93*, 460–464. [CrossRef] [PubMed]
44. Wanner, L.A.; Junttila, O. Cold-induced freezing tolerance in Arabidopsis. *Plant Physiol.* **1999**, *120*, 391–400. [CrossRef] [PubMed]
45. Ørskov, E.R.; McDonald, I. The estimation of protein degradability in the rumen from incubation measurements weighted according to rate of passage. *J. Agric. Sci.* **1979**, *92*, 499–503. [CrossRef]
46. Pollock, C.J. The response of plants to temperature change. *J. Agric. Sci.* **1990**, *115*, 1–5. [CrossRef]
47. Levitt, J. Responses of plants to environmental stress. In *Chilling, Freezing, and High Temperature Stresses*, 2nd ed.; Academic Press: New York, NY, USA, 1980; p. 447.
48. Tarkowski, Ł.P.; van den Ende, W. Cold tolerance triggered by soluble sugars: A multifaceted countermeasure. *Front. Plant Sci.* **2015**, *6*, 203.
49. Van den Ende, W. Multifunctional fructans and raffinose family oligosaccharides. *Front. Plant Sci.* **2013**, *4*, 247. [PubMed]
50. Panelo, J.S.; Redi, W.I.; Lorenzo, M.; Tognetti, J. Efecto del Incremento de la Temperatura Sobre la Fotosíntesis y la Respiración en Plantas de Trigo Aclimatadas a Bajas Temperaturas. Available online: http://fisiologiavegetal.org/fv2014/abstract-index/abstracts/#905 (accessed on 21 September 2014).
51. Huner, N.P.A.; Palta, J.P.; Li, P.H.; Carter, J.V. Anatomical changes in leaves of Puma rye in response to growth at cold-hardening temperatures. *Bot. Gaz.* **1981**, *142*, 55–62. [CrossRef]
52. Huner, N.P.A.; Oquist, G.; Sarhan, F. Energy balance and acclimation to light and cold. *Trends Plant Sci.* **1998**, *3*, 224–230. [CrossRef]
53. Ndong, C.; Danyluk, J.; Huner, N.P.; Sarhan, F. Survey of gene expression in winter rye during changes in growth temperature, irradiance or excitation pressure. *Plant Mol. Biol.* **2001**, *45*, 691–703. [CrossRef] [PubMed]
54. Gray, G.R.; Chauvin, L.P.; Sarhan, F.; Huner, N.P. Cold acclimation and freezing tolerance (A complex interaction of light and temperature). *Plant Physiol.* **1997**, *114*, 467–474. [PubMed]
55. Getachew, G.; Robinson, P.H.; DePeters, E.J.; Taylor, S.J. Relationships between chemical composition, dry matter degradation and *in vitro* gas production of several ruminant feeds. *Anim. Feed Sci. Technol.* **2004**, *111*, 57–71. [CrossRef]

56. Lopez-Guisa, J.M.; Satter, L.D. Effect of forage source on retention of digesta markers applied to corn gluten meal and brewers grains for heifers. *J. Dairy Sci.* **1991**, *74*, 4297–4304. [CrossRef]
57. John, A.; Ulyatt, M.J. Importance of dry matter content to voluntary intake of fresh grass forages. *Proc N. Z. Soc. Anim. Prod.* **1987**, *47*, 13–16.
58. Cabrera Estrada, J.I.; Delagarde, R.; Faverdin, P.; Peyraud, J.L. Dry matter intake and eating rate of grass by dairy cows is restricted by internal, but not external water. *Anim. Feed Sci. Technol.* **2004**, *114*, 59–74. [CrossRef]
59. Allinson, D.W. Influence of photoperiod and thermoperiod on the IVDMD and cell wall components of tall fescue. *Crop Sci.* **1971**, *11*, 456–458. [CrossRef]
60. Bertrand, A.; Tremblay, G.F.; Pelletier, S.; Castonguay, Y.; Bélanger, G. Yield and nutritive value of timothy as affected by temperature, photoperiod and time of harvest. *Grass Forage Sci.* **2008**, *63*, 421–432. [CrossRef]
61. Thorvaldsson, G.; Tremblay, G.F.; Tapani Kunelius, H. The effects of growth temperature on digestibility and fibre concentration of seven temperate grass species. *Acta Agric. Scand. Sect. B* **2007**, *57*, 322–328. [CrossRef]
62. Crasta, O.R.; Cox, W.J.; Cherney, J.H. Factors affecting maize forage quality development in the northeastern USA. *Agron. J.* **1997**, *89*, 251–256. [CrossRef]
63. Agnusdei, M.G.; di Marco, O.N.; Nenning, F.R.; Aello, M.S. Leaf blade nutritional quality of rhodes grass (*Chloris gayana*) as affected by leaf age and length. *Crop Pasture Sci.* **2012**, *62*, 1098–1105. [CrossRef]
64. Di Marco, O.N.; Harkes, H.; Agnusdei, M.G. Calidad de agropiro alargado (*Thinopyrum ponticum*) en estado vegetativo en relación con la edad y longitud de las hojas. *RIA* **2013**, *39*, 105–110.
65. Reynolds, S.G. *Pasture-Cattle-Coconut Systems*; FAO RAPA Publication: Bangkok, Thailand, 1995; p. 668.
66. Samarakoon, S.P.; Wilson, J.R.; Shelton, H.M. Growth, morphology and nutritive quality of shaded *Stenotaphrum secundatum*, *Axonopus compressus* and *Pennisetum clandestinum*. *J Agric. Sci.* **1990**, *114*, 161–169. [CrossRef]
67. Livingston, D.P., III; Hincha, D.K.; Heyer, A.G.; Norio, S.; Noureddine, B.; Shuichi, O. The relationship of fructan to abiotic stress tolerance in plants. In *Recent Advances in Fructooligosaccharides Research*; Norio, S., Noureddine, B., Shuichi, O., Eds.; Research Signpost: Kerala, India, 2007; pp. 181–199.
68. Valluru, R.; van den Ende, W. Plant fructans in stress environments: Emerging concepts and future prospects. *J. Exp. Bot.* **2008**, *59*, 2905–2916. [CrossRef] [PubMed]
69. Livingston, D.P., III; Hincha, D.K.; Heyer, A.G. Fructan and its relationship to abiotic stress tolerance in plants. *Cell Mol. Life Sci.* **2009**, *66*, 2007–2023. [CrossRef] [PubMed]
70. Vágújfalvi, A.; Kerepesi, I.; Galiba, G.; Tischner, T.; Sutka, J. Frost hardiness depending on carbohydrate changes during cold acclimation in wheat. *Plant Sci.* **1999**, *144*, 85–92. [CrossRef]

# 12

# Growth Strategy of Rhizomatous and Non-Rhizomatous Tall Fescue Populations in Response to Defoliation

**Racheal H. Bryant [1,*], Cory Matthew [2] and John Hodgson [2]**

[1] Faculty of Agriculture and Life Sciences, Lincoln University, P.O. Box 85084, Lincoln 7647, New Zealand
[2] Institute of Agriculture and Environment, Massey University, Private Bag 11-222, Palmerston North 4442, New Zealand; C.Matthew@massey.ac.nz (C.M.); johnruth@inspire.net.nz (J.H.)
* Author to whom correspondence should be addressed; Racheal.Bryant@lincoln.ac.nz

Academic Editor: Les Copeland

**Abstract:** The aim of this study was to determine the morphology of rhizome production, in two contrasting rhizomatous (R) and non-rhizomatous (NR) tall fescue (*Schedonorus arundinaceus* (Schreb.) Dumort) populations, and to assess whether rhizome production is associated with changed biomass allocation or plant growth pattern. Growth of R and NR populations was compared, under hard defoliation (H, 50 mm stubble), lax defoliation (L, 100 mm stubble), or without defoliation (U, uncut). Populations were cloned and grown in a glasshouse and defoliated every three weeks, with destructive harvests performed at 6, 12 and 18 weeks. R plants allocated more biomass to root and less to pseudostem than NR plants. Plant tiller numbers were greatly reduced by defoliation, and R and NR populations differed in leaf formation strategy. R plants had narrower leaves than NR, but their leaves were longer, because of greater leaf elongation duration. R plants were more plastic than NR plants in response to defoliation. Ultimately, biomass allocation to rhizomes did not differ between populations but R plants exhibited a subtle shift in distribution of internode length with a few longer internode segments typically located on secondary and tertiary tillers.

**Keywords:** leaf appearance rate; site filling; plant morphology; *Festuca arundinacea*; relative tiller appearance

---

## 1. Introduction

Tall fescue (*Schedonorus arundinaceus* (Schreb.) Dumort) is a genetically diverse species which can exhibit a wide range of variation in tiller size, leaf appearance rate and leaf elongation rate [1]. Such variation has been the basis for a number of studies of ecophysiology of the species. For example, lines of plants selected for contrasting leaf elongation rate (LER) showed consistent morphological characteristics. Plants with high LER had high yield per tiller, large tiller size and few tillers per plant while low LER plants had small tillers with higher tiller number per plant [2,3].

Such behavioural differences arise ultimately from differences in expression of the same basic morphology. Tillers are formed from axillary buds. Therefore, each new leaf which appears has the potential to produce a tiller from an axillary bud. In young plants the appearance of new leaves and tillers is synchronous at corresponding sites across successive tiller generations. However, as plants mature, leaf and tiller appearance becomes non-synchronous between generations, and many sites they do not form tillers. In tall fescue plants, buds at these sites often remain viable and typically form a spherical shape [4]. Site filling is one measure of the frequency of tiller production from undeveloped buds and is calculated from the ratio between tiller appearance and leaf appearance rates [5]. The example mentioned previously illustrates that different plant populations may prioritize the use of assimilates differently, either in favour of leaf extension or tiller appearance. Defoliation, especially more severe defoliation, results in rapid depletion of non-structural carbohydrate reserves in the plant [6], and can modify such responses.

Another feature of tall fescue morphology is the formation, by internode elongation, of stolons and rhizomes. Although stolons and rhizomes are anatomically similar [7], rhizome development is the elongation of internodes to form horizontal shoots below the soil surface. Rhizomes have reduced scale-like leaves at each node and their tips turn upwards, forming new tillers some distance from the parent plant with laminate leaves above ground. Stolons are formed from basal internode elongation as new tillers form adjacent to the parent plant, without a prior period of horizontal growth. Because of their anatomical similarity, rhizomes and stolons have been referred to collectively as true stems [8], to emphasise their anatomical distinctness from pseudostems formed from rolled leaf sheaths and holding leaf laminae to the light.

It has been reported [4] that rhizomes were more likely to occur at down-facing axillary buds. These buds tended to be those which had been dormant for a period and had taken on a spherical shape. Stolon and rhizome formation on a per plant basis has also been quantified in Grasslands Roa tall fescue [9], however, factors determining rhizome initiation and details of site of initiation and numbers and lengths of internodes comprising each rhizome are largely unknown.

The objectives of this study were to quantify the morphology of differences in rhizome production between rhizomatous and non-rhizomatous tall fescue populations, in order to better understand their visually dramatic contrast in degree of rhizome formation, and to determine whether effects on tiller and leaf production attributable to the energy cost of rhizome production could be observed.

## 2. Experimental Section

### 2.1. Plant Preparation and Management

Two populations of tall fescue (*Schedonorus arundinaceus* (Schreb.) Dumort, previously *Festuca arundinacea* (Schreb.)) were compared; a standard non-rhizomatous (NR) population and a strongly rhizomatous (R) population. The standard NR population was derived from a cross between USA turf cultivars Tribute and Rebel, and exhibited no visually obvious rhizomatous behavior. The second population, a third generation cross from germplasm of a local ecotype collected in Spain and Portugal, had been selected specifically for rhizomatous behavior visually evident when the original germplasm was collected.

In April 1997 sods from established two year old swards of both populations were washed free of soil, divided into clones of single rooted tillers and planted into 500 mL plastic pots. Use of planting material from established swards was to ensure that immediate expression of rhizomes would occur in the rhizomatous population [10]. Two weeks later developing plants were transplanted into 5 L pots of a 50:50 sterile sand and soil mixture and supplemented with 2.5 g/L slow release fertiliser (NPK 15:4.8:10.8 Osmocote® Pro Everris, Geldermalsen, The Netherlands). Plants were maintained in a temperature controlled glasshouse (16 to 24 °C), under natural daylight, at the Massey University Plant Growth Unit in Palmerston North between April and September 1997 (40°20′ S).

The experimental design was a $2 \times 3 \times 3$ factorial, comprising the R and NR populations, three defoliation treatments and three destructive harvests. There were four replicates, giving 72 pots in total. The three defoliation treatments were: hard defoliation (H, 50 mm stubble); lax defoliation (L, 100 mm stubble); and undefoliated (U). Defoliation treatments began when all pots contained a minimum of five tillers and subsequently every three weeks over 15 weeks. Plants were grown on for a further three weeks after the last defoliation, giving six three-week regrowth intervals in total.

### 2.2. Measurements

Non-destructive measurements of leaf length (one random tiller per pot) and total tiller number per plant were performed weekly on marked tillers and leaves of a randomly chosen adult tiller.

From the weekly leaf length measurements it was possible to calculate leaf appearance rate (LAR), leaf elongation (LER) and senescence rates (SR) in units of mm per tiller per day. Tiller numbers per plant were counted once a week and site filling was determined from the ratio between LAR and relative tiller appearance rate (RTA) which was then used to calculate site usage [11].

Three destructive harvests, were carried out at six week intervals, coinciding with the conclusion of each alternate three week regrowth period. At destructive harvests, designated plants were removed from pots and washed free from soil. Primary tillers were removed sequentially from the original clone, recording the nodal position of each successive tiller removed, and associated daughter tillers categorized as either secondary or tertiary tillers. The original intention was to record numbers of tillers forming stolons and rhizomes, as in the study of Hume and Brock [9], but it was found that no clear distinction could be made between stolons and rhizomes. This was due to a continuous gradation in morphology from tillers with a small amount of internode elongation at one or two phytomers to tillers with fully formed rhizomes 100 mm or more in length. This phenomena has been depicted and described previously [4]. Accordingly, true stem length for each internode of all tillers was measured and stolons and rhizomes are not differentiated in the data presented here. After measurement of true stem length, tillers were dissected into leaf, pseudostem, true stem (stolon and rhizome), and root for primary, secondary, and tertiary tillers, and dry weights obtained for each herbage component.

### 2.3. *Statistical Analysis*

Population and defoliation effects on tiller component dry weight, tiller number, LAR, LER, SR were compared by a factorial two way analysis of variance (ANOVA) model using the general linear model (GLM) procedure in Statistical Analysis System (SAS Institute Inc, Cary, NC, USA) to identify population (R, NR) and defoliation (H, L, U) effects. Data from different harvests were analysed separately. RTAR values and their standard errors were obtained by regression of log (plant tiller number) on time. Due diligence was exercised with respect to statistical assumptions. Repeat measures analyses were not required for multiple harvest dates since a different potted plant was destructively harvested on each date and data for each harvest date was analysed separately; comparison of R and NR populations is not subject to variance heterogeneity concerns because only two treatments are being compared. In one case where plant size differences from grazing effects were felt serious, $\log_e$ transformation was employed. There were no obvious reasons for non-normal distribution of residuals such as numbers of zeroes in the data set needing to be addressed, and normal probability plots in Minitab v. 10.5 (Minitab Pty Ltd., Sydney, Australia) for a selection of reported variables were near linear with $P$-values > 0.05 for the Anderson-Darling $A^2$ statistic.

## 3. Results

### 3.1. *Plant Yield and Tiller Production*

In both populations plant dry weight was reduced nearly tenfold by the H defoliation treatment ($p < 0.001$), with decrease in tiller dry weight (Table 1) and tiller number (Table 2) being approximately 80% and 40%, respectively. Total herbage weight was very similar for the two populations when no defoliation occurred, but defoliation reduced dry matter (DM) more dramatically for R than for NR population (L 68% and 42%, H 90% and 85%, $p < 0.05$; Table 1). However, despite a large reduction in plant leaf mass across defoliation treatments and a smaller reduction in the R population compared to NR population, the % allocation of plant biomass to leaf was remarkably conserved across grazing management treatments and between populations (Table 1). Interestingly, the R and NR populations did not differ significantly in true stem mass per plant although there was trend to higher % allocation of biomass to true stem in the R population. Rather, the differential response between populations to defoliation, as reflected by a significant defoliation × population interaction, was observed in the pseudostem and root fractions. The R population had less pseudostem and more root than the NR population but a marked tendency to reduce % biomass allocation to roots under more intensive defoliation pressure (Table 1).

**Table 1.** Contribution of tiller components to plant weight (g DM) of non-rhizomatous (NR) and rhizomatous (R) tall fescue populations undefoliated (U), or under lax (L) or hard (H) defoliation treatments.

| | | Leaf | Pseudostem | True Stem | Root | Root:Shoot | Total | Tiller Weight |
|---|---|---|---|---|---|---|---|---|
| NR | U | 17.9 (43) | 11.4 (27) | 3.6 (8) | 9.2 (22) | 0.28 | 41.9 | 0.22 |
| | L | 8.8 (41) | 5.9 (28) | 2.4 (11) | 4.7 (20) | 0.26 | 21.7 | 0.16 |
| | H | 2.7 (43) | 1.6 (27) | 0.7 (11) | 1.3 (21) | 0.27 | 6.4 | 0.06 |
| Mean | | 9.77 | 6.31 | 2.22 | 5.07 | 0.27 | 23.4 | 0.15 |
| R | U | 15.2 (38) | 8.1 (20) | 4.3 (11) | 12.7 (32) | 0.47 | 40.4 | 0.21 |
| | L | 6.2 (41) | 3.3 (22) | 1.9 (13) | 3.8 (25) | 0.33 | 15.3 | 0.09 |
| | H | 1.7 (42) | 1.0 (23) | 0.6 (15) | 0.8 (20) | 0.25 | 4.1 | 0.04 |
| Mean | | 7.73 | 4.13 | 2.29 | 5.77 | 0.35 | 19.9 | 0.12 |
| SE | | 0.80 | 0.47 | 0.32 | 0.77 | 0.032 | 2.01 | 0.0092 |
| Signif | Popln | ** | *** | NS | NS | ** | * | *** |
| | Cut | *** | *** | *** | *** | ** | *** | *** |
| | P × C | NS | * | NS | * | * | *,LT | ** |

NS, not significant; * $p < 0.05$; ** $p < 0.01$; *** $p < 0.001$. Percentage of total mass shown in parentheses. $^{LT}$ denotes analysis performed on log-transformed data to remove heterogeneity of error variance across treatments with a tenfold difference in plant dry weight.

**Table 2.** Tiller numbers per plant, categorized by position on the tiller axis, on non-rhizomatous (NR) and rhizomatous (R) populations undefoliated (U), or under lax (L) or hard (H) defoliation after a 12 or 18 week growing period.

| | | 12 Weeks | | | | 18 Weeks | | | |
|---|---|---|---|---|---|---|---|---|---|
| | | Primary | Secondary | Tertiary | Total | Primary | Secondary | Tertiary | Total |
| NR | U | 16 | 45 | 44 | 104 | 19 | 71 | 102 | 192 |
| | L | 15 | 42 | 20 | 77 | 16 | 57 | 57 | 130 |
| | H | 11 | 26 | 13 | 50 | 17 | 51 | 49 | 117 |
| Mean | | 14 | 38 | 25 | 77 | 17 | 60 | 69 | 146 |
| R | U | 13 | 35 | 31 | 79 | 17 | 72 | 100 | 189 |
| | L | 9 | 30 | 26 | 64 | 15 | 57 | 97 | 169 |
| | H | 8 | 23 | 19 | 49 | 11 | 38 | 56 | 105 |
| Mean | | 10 | 29 | 25 | 64 | 14 | 56 | 84 | 154 |
| SEM | | 7.8 | 1.3 | 4.0 | 4.1 | 1.6 | 5.4 | 9.8 | 12.8 |
| Sig. | Popln | *** | * | NS | NS | * | NS | NS | NS |
| | Cut | ** | ** | *** | *** | * | *** | *** | *** |
| | P × C | NS | NS | NS | NS | NS | NS | NS | NS |

NS, not significant; * $p < 0.05$; ** $p < 0.01$; *** $p < 0.001$. Percentage of total mass shown in parentheses.

The total number of tillers per plant did not differ between populations. However, some differences in growth patterns between the two populations became evident when tillers were categorised into primary, secondary or tertiary tillers. Decrease in tiller number following defoliation, regardless of category, was consistent for both populations. After 12 weeks the NR plants had, averaged across defoliation treatments, 40% more primary tillers ($p < 0.001$) and 31% more secondary tillers ($p < 0.05$), but similar numbers of tertiary tillers compared with R plants (Table 2). By week 18 there were more primary tillers on the NR, compared with R, population ($p < 0.05$). However, the effect of population on secondary and tertiary tillers was not significant, even though, numerically, the proportion of whole plant tillers accounted for by tertiary tillers was greater for the R population. Differences between populations in leaf appearance and site filling (see below, Table 3) led to a higher relative tiller appearance rate in the R than the NR population. As a result, the average number of tillers per plant at the end of the experiment was similar for both populations (Table 2).

Relative tiller appearance (RTA) expressed as tillers/tiller/day was calculated for each population, both as the average for the entire experimental period and as the average for the final six weeks of measurement (Table 3). Throughout the observation period rhizomatous tall fescue had the highest

RTA and this was attributed to a higher site filling ratio ($F_S$) or equivalently higher site usage (Table 3). Over the final six weeks of measurement, RTA began to decline (Table 3), particularly for U and L defoliated plants, which by this time, had comparatively high numbers of tillers (Table 2).

**Table 3.** Effect of defoliation on relative tiller appearance (for initial and final stages of development, ± standard error), and calculated rates of site filling and site usage. Data are for non-rhizomatous (NR) or rhizomatous (R) populations undefoliated (U) or under lax (L) or hard (H) defoliation.

| | | Relative Tiller Appearance (No. obs) (Tillers/Tiller/Day) | | | Site Filling [1] | Site Usage [2] |
|---|---|---|---|---|---|---|
| | | **Overall** | **Initial** | **Final** | | |
| **NR** | **U** | 0.022 ± 0.0011 (17) | 0.030 ± 0.0011 (11) | 0.016 ± 0.0008 (6) | 0.16 | 0.17 |
| | **L** | 0.021 ± 0.0005 (17) | N/A | 0.014 ± 0.0009 (6) | 0.14 | 0.15 |
| | **H** | 0.017 ± 0.0006 (17) | N/A | 0.020 ± 0.0011 (6) | 0.18 | 0.20 |
| **R** | **U** | 0.025 ± 0.0007 (17) | 0.028 ± 0.006 (11) | 0.018 ± 0.0012 (6) | 0.30 | 0.35 |
| | **L** | 0.024 ± 0.0004 (17) | N/A | 0.024 ± 0.0008 (6) | 0.30 | 0.35 |
| | **H** | 0.021 ± 0.0007 (17) | N/A | 0.027 ± 0.0013 (6) | 0.30 | 0.35 |

[1] Ratio between tiller appearance and leaf appearance rates; [2] percentage of tiller sites which produce tillers; N/A is not applicable.

Mean tiller weight, including both above and below ground mass, was 0.208 g DM across treatments and did not differ between populations. However, an interaction ($p < 0.001$) between population, defoliation treatment, and tiller position revealed that while tiller weight was less for tertiary tillers, and defoliation further reduced tiller weight, the R population was more negatively affected by defoliation than the NR treatment. In particular, the most pronounced reduction in tiller weight occurred on primary tillers on the R population (Figure 1).

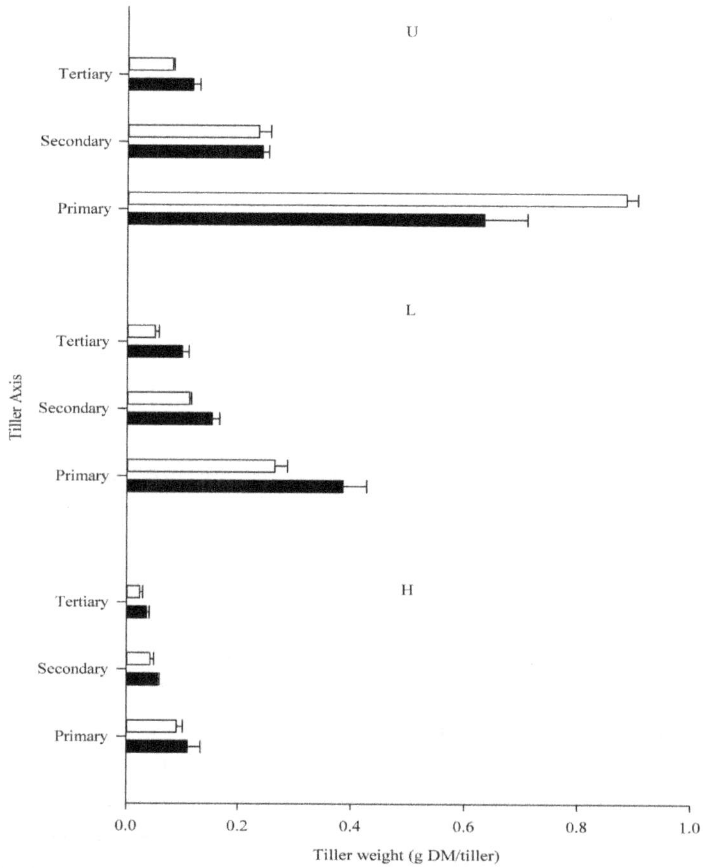


**Figure 1.** Mean ± SE for tiller weights at the primary, secondary and tertiary tiller axes of non-rhizomatous (■) and rhizomatous (□) tall fescue populations when plants were undefoliated (U), or received lax (L) defoliation to 100 mm stubble, or hard (H), defoliation to 50 mm stubble every three weeks for 18 weeks.

### 3.2. *Leaf Production Strategies*

Leaf area per plant, on average, was slightly greater ($p = 0.09$) for the NR compared with the R population (1044 *vs.* 776 ± 107 $cm^2$ respectively). Defoliation reduced leaf area from 1711 $cm^2$ in U plants to 271 $cm^2$ in H defoliation treatments ($p < 0.001$; Table 4). The R plants had fewer green leaves per tiller (2–3 *cf.* 3–4), and a lower leaf appearance rate (LAR) than NR plants (Table 4). However, the faster leaf elongation rate of the R population was largely offset by production of narrower leaves than NR plants (5.9 *vs.* 8.0 mm respectively; $p < 0.001$).

The R population had leaves 91% longer than the NR population in the U treatment and just under 60% longer than NR in L and H defoliation treatments (population × defoliation interaction $p < 0.05$; Table 4). One trait contributing to longer leaf length of the R population in undefoliated plants was greater LER (20.1 *cf.* 16.2 mm/tiller/day in undefoliated plants, $p < 0.01$). As might be expected from defoliation effects on tiller dry weight noted above, both populations responded to increased defoliation severity by reduced leaf elongation rate. However in contrast to the nearly 80% reduction in tiller dry weight in H defoliated plants compared to U plants (Table 1), the corresponding leaf elongation rate reduction was less than 60% (Table 4).

**Table 4.** Effect of defoliation regime and population on leaf area, leaf elongation rate (LER) per tiller per day, leaf senescence per tiller per day (SR), leaf appearance rate (LAR), number of live leaves per tiller (NL), tiller total leaf length (TiLL), and the longest live leaf per tiller (LoLL) for the final six weeks of observations. Data are for non-rhizomatous (NR) or rhizomatous (R) populations undefoliated (U) or under lax (L) or hard (H) defoliation.

| | | Leaf area ($cm^2$/Plant) | LER (mm/tlr/d) | SR (mm/tlr/d) | LAR (Leaves/d) | NL (Leaves/Tiller) | TiLL (mm) | LoLL (mm) |
|---|---|---|---|---|---|---|---|---|
| NR | U | 1805 | 16.2 | 16.6 | 0.10 | 5.8 | 846 | 191 |
| | L | 954 | 12.6 | 9.1 | 0.10 | 6.5 | 463 | 147 |
| | H | 375 | 7.1 | 4.2 | 0.11 | 5.5 | 203 | 71 |
| R | U | 1617 | 20.1 | 13.4 | 0.06 | 4.5 | 1247 | 364 |
| | L | 543 | 18.0 | 10.7 | 0.08 | 4.8 | 461 | 232 |
| | H | 166 | 7.9 | 4.2 | 0.09 | 4.0 | 198 | 113 |
| | SEM | 184.6 | 1.22 | 1.18 | 0.008 | 0.38 | 83.9 | 21.6 |
| Sig. | Popln | NS | ** | NS | *** | *** | NS | *** |
| | Defol. | *** | *** | *** | NS | NS | *** | *** |
| | P × D | NS | NS | NS | NS | NS | * | * |

SEM, Standard error of the mean for the interaction; *** $p < 0.001$; ** $p < 0.01$; * $p < 0.05$; NS, not significant, $p > 0.05$.

Dividing longest leaf length by LER indicates an estimated leaf elongation duration of 12 and 18 days for NR and R populations, respectively, when undefoliated. The duration of elongation reduced by 10% for NR populations and 25% for R when defoliated (Table 4). Furthermore, L defoliation resulted in a 63% reduction in leaf length in R populations but only a 45% reduction in leaf length in NR populations (Table 4). Hence, R populations exhibited a greater degree of plasticity for leaf length, forming longer leaves under laxer defoliation compared to NR populations, and reflected in a significant population × defoliation regime interaction for measures of leaf length.

### 3.3. *Rhizome Formation*

Undefoliated plants had a mean true stem length of 20 mm for R and 13 mm for NR populations ($p < 0.10$). As mentioned previously, difference in true stem formation between R and NR populations proved difficult to quantify because of the presence of numerous stems which were intermediate between stolons and rhizomes. Measures of plant morphology found to give insight into the respective differences were a distribution plot of true stem lengths per internode for each plant category (Y-axis) against serial rank (X-axis), compiled after sorting data by ascending order of internode length to determine serial rank for each internode (Figure 2), and total length of true stem per plant for primary, secondary, and tertiary tillers (Figure 3).

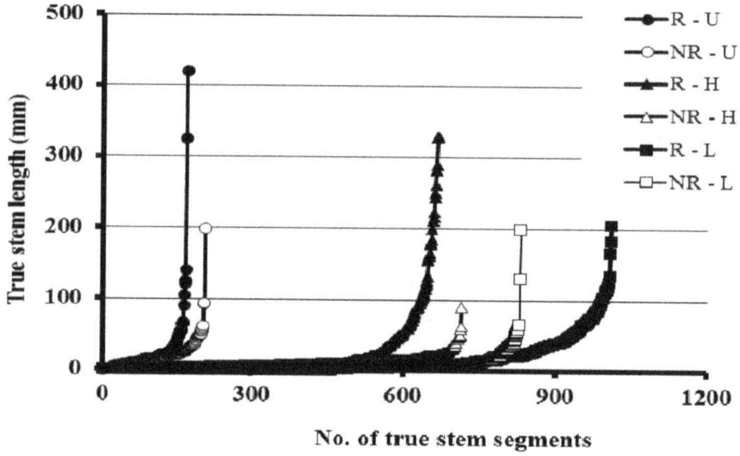


**Figure 2.** Cumulative distribution plot for internode true stem lengths in each of six plant population × defoliation categories. Each symbol represents a single internode with its length (mm) plotted on the Y-axis against its serial rank for length among all internodes of that plant category (X-axis) The total number and length of true stems on rhizomatous (R) and non-rhizomatous (NR) tall fescue plants subjected to hard defoliation (50 mm stubble, H), lax defoliation (100 mm stubble, L), or undefoliated (U).

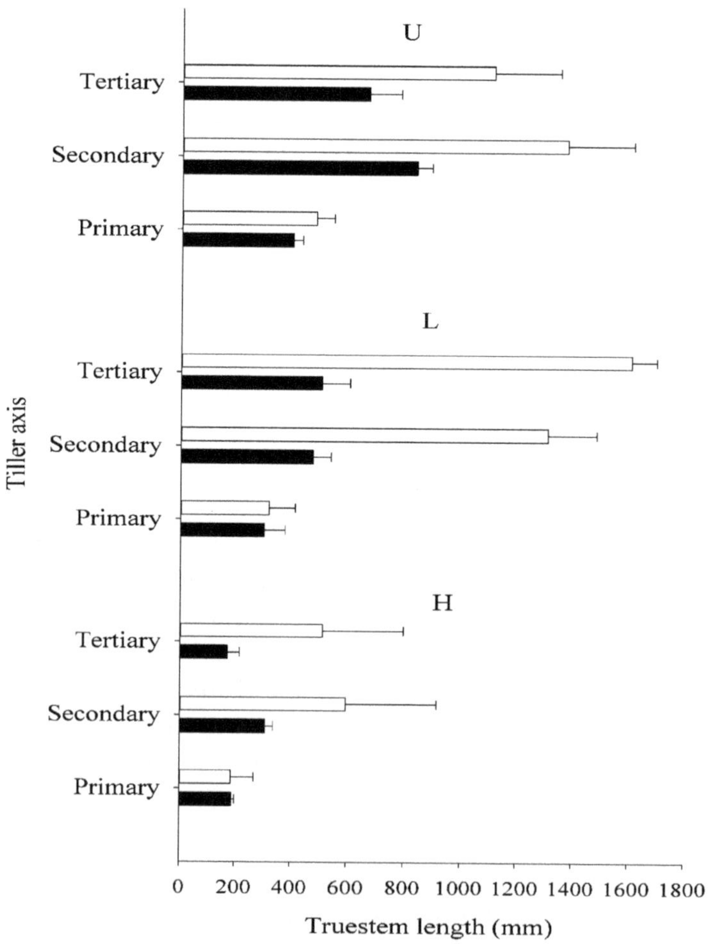


**Figure 3.** Mean ± SE for total true stem length of primary, secondary and tertiary tiller axes of non-rhizomatous (■) and rhizomatous (□) tall fescue populations when plants were undefoliated (U), or subject to lax (L) defoliation to 100 mm stubble, or hard (H) defoliation to 50 mm stubble every three weeks for 18 weeks.

It is seen in Figure 2 that the increase in true stem formation of the rhizomatous population was confined to approximately 20% of the internodes being somewhat longer than the comparably ranked internode of NR populations from the same defoliation regime, with a very small proportion of the internodes on R populations being dramatically longer than internodes for NR populations. Each rhizomatous plant possessed a small number of very long rhizomes (*i.e.*, truestem length greater than 150 mm) (Figure 2). Because of this distribution pattern it was hard to show statistically that R populations differed from NR populations at the whole plant level. However, when data are sorted according to the hierarchical position of the tiller (primary, secondary, or tertiary), it becomes clear that true stem development of the R population was much greater on secondary and tertiary tillers than on primary tillers (Figure 3). The longest of these rhizomes appeared on secondary tillers where true stem internode segments of up to 420, 330 and 210 mm length were recorded for U, H, and L defoliation, regimes respectively, for the R population (Figure 2).

In undefoliated R populations, rhizome or true stem formation was also greatest at secondary and tertiary node positions (Figure 3a), but the differences in total true stem length associated with tiller hierarchical position were much reduced at high defoliation intensity (Figure 3c).

Generally R and NR populations showed a similar response to defoliation treatments, with defoliation decreasing true stem formation in all categories of tillers (Figure 3). There was, however, a possible interaction ($p = 0.075$) between defoliation treatment and population for total stem length of tertiary tillers. Under the L defoliation treatment R population had a greater differential over the NR population for total stem length for undefoliated plants because these L-defoliated R plants maintained their tertiary tiller number near to that of undefoliated plants (Table 2) and those tertiary tillers had the highest total true stem lengths (Figure 3).

## 4. Discussion

There are two main ways in which a grass plant can express a comparatively greater size increase. The first is by increasing the size of currently formed tillers (e.g., through enhanced leaf elongation rate); the second is by producing more tillers (e.g., through increased site filling). The extent to which either of these occurs is a reflection of assimilate use. However, if physical boundaries are genetically predetermined then additional assimilates above tillering potential may be stored or used to increase underground biomass. Conversely, if photoassimilates are used to a greater degree for production of non-photosynthetic organs such as true stem or rhizome, it would be expected that the metabolic cost of this construction was reflected in a reduction in biomass of other plant organs. In this study, the data collection was structured so that population differences in allocation of biomass to tillering, leaf, true stem and root formation and the plant morphology basis for these differences could be measured under differing levels of assimilate removal by defoliation.

### *4.1. Morphology of NR and R Tall Fescue Populations When Undefoliated*

From the outset the two populations were visibly different in their appearance, with the NR population exhibiting a prostrate, leafy, clumpy growth habit while the R population was more upright and had a sparse, open growth habit. While tiller size and density can often be used to describe morphological differences, in the present study after 18 weeks growth from transplanted tiller ramets the undefoliated NR and R plants had similar total biomass (41.9 and 40.4 g DM/plant, respectively), similar mean tiller size (0.22 and 0.21 g DM/tiller), and similar tiller number (192 and 189 tillers/plant respectively). Instead of variation in plant or tiller size, the visible morphological differences between NR and R populations in this study arose from an enhancement of certain activities (especially rhizome formation) in secondary and tertiary tillers, compared to primary tillers, and from a variation in leaf production strategy with longer, narrower leaves in the R population.

Although the presence of rhizomes in tall fescue has been well documented [7,12–14], anatomical descriptions provide little insight towards structural development of rhizomatous plants. The complexity of morphological forms that caused difficulty distinguishing structures referred to in previous studies as stolons and rhizomes was not envisaged in the planning of this experiment. We observed that tiller initiation often began on nodes on the rhizomes themselves while the parent tiller still only possessed one or two scale leaves. These new tillers tended to grow vertically off lateral shoots, but because of their position beneath the soil, internode elongation was required for tillers to reach the surface. It soon became apparent that rhizome formation was not merely lateral expansion of older underground nodes, although this type of formation did tend to produce the longest rhizomes, but also a mechanism for allowing a succession of shoots to reach the soil surface at a distance from the original plant. In addition to this clarification, the results presented here appear to be unique among studies of stolon formation in tall fescue in that they provide a measure of biomass allocation to different organs within the plant and to tillers of primary, secondary, and tertiary hierarchical status. The biomass distribution data show an unexpected and rather complex picture. There is not a simple trade-off between forming stolons and other organs in the R population. Rather, increased true stem formation of R plants is scarcely statistically discernible at the whole plant level, but the R growth strategy is associated with increased biomass allocation to root formation and an energy saving through reduced investment in pseudostem, while leaf production is comparatively conserved, exhibiting only a modest reduction. Meanwhile, increased true stem formation in the R population is most evident at the periphery of a primary-secondary-tertiary hierarchical chain of tillers. There is a large body of literature on the dynamics of clonal plants. Contrasting rhizomatous and non-rhizomatous growth habits are typically referred to in ecology as guerrilla and phalanx growth strategies, e.g., [15], and some functional models have been explored identifying the potential foraging advantage of the guerilla strategy in environments with patchy root distribution [16]. However, the present information may indicate additional detail of a guerilla morphology. If true stem formation in the R population (guerrilla strategy) is an adaptation to reach and exploit nutrient patches in resource poor environments it is logical that this strategy would be associated with a greater investment in root mass. Further, a hypothesis for further investigation from the present data is that the R population had a greater assimilate transfer from parent to daughter tiller than the NR population; facilitating the spreading behavior, and building a compounding effect within a hierarchical chain of tillers. If this were true, it would explain very well the data pattern of increased true stem length in secondary and tertiary tillers of the R population observed in Figure 3. Notably, such a strategy whereby rhizomes are formed preferentially at the periphery of a plant will have a much lower energy cost than if rhizomes were formed routinely at every tiller, including primary tillers. The energy economy of this strategy is emphasized by the observation that a mere 3% increase in biomass allocation to true stem at the whole plant level (Table 1), with only a subtle change in distribution of individual internode lengths (Figure 1), when expressed preferentially in secondary and tertiary tillers (Figure 3) can result in visually dramatic differences in plant morphology.

Rhizomes are sometimes considered to function as storage organs, or as a survival strategy [17]. In this study it seems less likely that rhizomes functioned as storage organs as, at a whole plant level, allocation of plant DM to true stem formation was increased by only 2%–4% (NS) in the R population, compared to the NR.

With respect to the differences in leaf production strategy, undefoliated R plants had a 24% faster LER and a 50% longer duration of leaf elongation than NR plants and so produced much longer leaves, and consequently taller pseudostem (though counter-intuitively the slenderness of R tillers resulted in a lower % biomass allocation to pseudostem in the R population). However, the NR plants had wider leaves, and a more rapid leaf appearance interval meaning that more tiller buds were produced, even though as a result of lower site filling and site usage (Tables 3 and 4), this did not result in more tillers per plant. For instance, only a tenth of leaves which appeared on NR plants had developed tillers from their axillary buds in the final six weeks of development, compared with a third of axillary buds on R plants. The compensations between various processes such as leaf formation and tillering and the component contributing traits such as leaf appearance rate, leaf elongation rate, and leaf elongation duration in grass swards have been much discussed, both theoretically [18–20], and from actual data [21,22] but for all that any predictive synthesis remains elusive. It is clear that increased LER is a trait less easily offset by other trade-offs and often associated with increased productivity in population comparisons [22], but in this case high LER of the R population was largely offset by a reduced leaf width and was not associated with greater plant dry weight. A possibility for further investigation is that differing leaf production strategies of these NR and R populations are not directly related to the rhizome formation habit but an independent response to some other feature of the environments where they have respectively evolved. In this case a likely possibility is that the longer leaf length of the R population is an adaptation to intermittent defoliation in an extensively farmed environment whereas the shorter leaf length of the NR population is an adaptation to more intensive pastoral farming. It is known anecdotally, for example that cocksfoot (*Dactylis glomerata* L) becomes dominant in a grass mixture when kept undefoliated (and is called "Orchard grass" in USA), as it has long leaves that can shade and eventually eliminate neighbouring shorter leaved species in an "umbrella" effect, but does not dominate companion species in the same way when regularly grazed.

### *4.2. Response to Defoliation*

In both NR and R populations, plant size was sharply reduced by defoliation, but the % reduction was greater for the R than for the NR population, as reflected in the significant population × defoliation interaction for total plant dry weight (Table 1). However the growth strategy to achieve size reduction differed between populations and patterns of biomass allocation under defoliation also differed between populations. First, the R plants displayed plasticity, not displayed by NR plants, in showing greater reduction of % biomass allocation to roots and of leaf elongation duration and leaf length, under increased defoliation pressure. This pattern is logically consistent with the hypothesis that the R population has become adapted to intermittent grazing in the environment where it evolved. Thus plasticity resulted in continuing expansion of growing leaves when undefoliated but earlier cessation of leaf elongation when defoliated and which in parallel with that rationed energy by limiting root production after defoliation but continuing root production when undefoliated. This response would provide the guerrilla nutrient exploitation growth strategy with an intuitively desirable "pause when defoliated/resume when assimilate is available" dimension. One of the major reasons for reduced mean tiller size of the R population compared to NR is the higher proportion of tertiary tillers on R plants compared with NR plants which respectively accounted for 55% *versus* 43% of total tillers on defoliated plants (Table 2). By reason of their younger age tertiary tillers are smaller in size than their secondary counterparts (Figure 1). In spite of this, allocation of plant dry weight to true stem formation increased from 11% to 15% for R plants under defoliation pressure. An explanation for why there is increased tertiary tiller activity in R plants with a guerilla growth strategy would be advantageous and we hypothesize this could arise from a greater parent-to-daughter tiller photoassimilate allocation as was advanced above.

## 5. Conclusions

The present study adds new insight on the morphology of rhizome production in two tall fescue populations. The distinction between stolons and rhizomes in previous studies is shown to be somewhat arbitrary and in contrasting R and NR populations in this study rhizome formation was found to be much more complex than simple switching on of true stem formation at a particular category of bud sites. In the R population, biomass allocation to rhizomes was not statistically increased and the histogram of true stem internode lengths for the plant as a whole was only subtly shifted in the R compared to the NR population. R plants minimized any cost of increased true stem formation by prioritizing secondary and tertiary tillers within tiller hierarchies as sites for true stem formation, and exhibited some compensation through decreased allocation of biomass to pseudostem, while largely maintaining % biomass allocation to leaves. A hypothesis for further investigation is that this behavior is linked to greater parent-to-daughter photo-assimilate transfer within tiller hierarchies. In this study, compared to the NR population, the R population showed increased biomass allocation to root formation. There was also greater plasticity with reduction in leaf elongation duration resulting in shorter leaves and reduction in root formation under increased defoliation pressure. Logically this is consistent with adaptation to nutrient patch exploitation in a clonal species with a guerilla growth strategy.

**Acknowledgments:** The plant material for this research was kindly supplied by Alan Stewart of PGG Wrightson seeds.

**Author Contributions:** All three co-authors were involved in the design and analysis of the experiment. Racheal Bryant conducted the experiment and, alongside Cory Matthew, wrote the paper.

## References

1. Belesky, D.P.; Fedders, J.M. Does endophyte influence the regrowth of tall fescue? *Ann. Bot.* **1996**, *78*, 499–505.
2. Volenec, J.J.; Nelson, C.J. Responses of tall fescue leaf meristems to N fertilization and harvest frequency. *Crop Sci.* **1983**, *23*, 720–724. [CrossRef]
3. Zarrough, K.M.; Nelson, C.J.; Sleper, D.A. Interrelationships between rates of leaf appearance and tillering in selected tall fescue populations. *Crop Sci.* **1984**, *24*, 565–569. [CrossRef]
4. Brock, J.L.; Albrecht, K.A.; Hume, D.E. Stolons and rhizomes in tall fescue under grazing. *Proc. N. Z. Grassland Assoc.* **1997**, *59*, 93–98.
5. Neuteboom, J.H.; Lantinga, E.A. Tillering potential and relationship between leaf and tiller production in perennial ryegrass. *Ann. Bot.* **1989**, *63*, 265–270.
6. Richards, J.H. Physiology of plants recovering from defoliation. In Proceedings of the XVII International Grassland Congress, Palmerston North, New Zealand, 8–21 February 1993; pp. 85–94.
7. Jernstedt, J.A.; Bouton, J.H. Anatomy, morphology, and growth of tall fescue rhizomes. *Crop Sci.* **1996**, *36*, 407–411.
8. Matthew, C.; Assuero, S.G.; Black, C.K.; Sackville-Hamilton, N.R. Tiller dynamics of grazed swards. In *Grassland Ecophysiology and Grazing Ecology*; Lemaire, G., Hodgson, J., Moraes, A., Carvalho, P.C.F., Nabinger, C., Eds.; CAB International: New York, NY, USA, 2000; pp. 127–150.
9. Hume, D.E.; Brock, J.L. Morphology of tall fescue (*Festuca arundinacea*) and perennial ryegrass (*Lolium perenne*) plants in pastures under sheep and cattle grazing. *J. Agric. Sci.* **1997**, *129*, 19–31. [CrossRef]
10. De Battista, J.P.; Bouton, J.H. Greenhouse evaluation of tall fescue genotypes for rhizome production. *Crop Sci.* **1990**, *30*, 536–540. [CrossRef]
11. Matthew, C.; Yang, J.Z.; Potter, J.F. Determination of tiller and root appearance in perennial ryegrass (*Lolium perenne*) swards by observation of the tiller axis, and potential application in mechanistic modelling. *N. Z. J. Agric. Res.* **1998**, *41*, 1–11. [CrossRef]
12. Porter, H.L., Jr. Rhizomes in tall fescue. *Agron. J.* **1958**, *50*, 493–494. [CrossRef]
13. Lopez, R.R.; Matches, A.G.; Baldridge, J.D. Vegetative development and organic reserves of tall fescue under conditions of accumulated growth. *Crop Sci.* **1967**, *7*, 409–412. [CrossRef]
14. Bouton, J.H.; Smith, S.R., Jr.; Battista, J.P. Field screening for rhizome number in tall fescue. *Crop Sci.* **1992**, *32*, 686–689. [CrossRef]
15. Ye, X.H.; Yu, F.H.; Dong, M. A trade-off between guerrilla and phalanx growth forms in *Leymus secalinus* under different nutrient supplies. *Ann. Bot.* **2006**, *98*, 187–191. [CrossRef] [PubMed]
16. Herben, T.; Hara, T. Competition and spatial dynamics in clonal plants. In *The Ecology and Evolution of Clonal Plants*; Kroon, H., de van Groenendael, H., Eds.; Backhuys Publishers: Leiden, The Netherland, 1997; pp. 3331–3357.
17. Thom, E.R.; Sheath, G.W.; Bryant, A.M. Seasonal-variations in total nonstructural carbohydrate and major element levels in perennial ryegrass and paspalum in a mixed pasture. *N. Z. J. Agric. Res.* **1989**, *32*, 157–165.
18. Robson, M.J.; Ryle, G.J.A.; Woledge, J.; Nelson, C.J. The grass plant—Its form and function. In *The Grass Crop. The Physiological Basis of Production*; Jones, M.B., Lazenby, A., Eds.; Chapman and Hall: London, UK, 1988; pp. 25–83.

19. Chapman, D.F.; Lemaire, G. Morphogenetic and structural determinants of plant regrowth following defoliation. In Proceedings of the XVII International Grassland Congress, Palmerston North, New Zealand, 8–21 February 1993; pp. 95–104.
20. Nelson, C.J. Shoot morphological plasticity of grasses: Leaf growth *versus* tillering. In *Grassland Ecophysiology and Grazing Ecology*; Lemaire, G., Hodgson, J., Moraes, A., Carvalho, P.C.F., Nabinger, C., Eds.; CAB International: New York, NY, USA, 2000; pp. 101–126.
21. Van Loo, E.N. Tillering, leaf expansion, and growth of plants of 2 cultivars of perennial ryegrass grown using hydroponics at 2 water potentials. *Ann. Bot.* **1992**, *70*, 511–518.
22. Bahmani, I.; Hazard, L.; Varlet-Grancher, C.; Betin, M.; Lemaire, G.; Matthew, C.; Thom, E.R. Differences in tillering of long- and short-leaved perennial ryegrass genetic lines under full light and shade treatments. *Crop Sci.* **2000**, *40*, 1095–1102. [CrossRef]

# 13

# Variation in Response to Moisture Stress of Young Plants of Interspecific Hybrids between White Clover (*T. repens* L.) and Caucasian Clover (*T. ambiguum* M. Bieb.)

**Athole H. Marshall *, Matthew Lowe and Rosemary P. Collins**

Institute of Biological, Environmental and Rural Sciences, Aberystwyth University, Gogerddan, Aberystwyth, Ceredigion SY233EE, UK; mjl@aber.ac.uk (M.L.); rpc@aber.ac.uk (R.P.C.)

* Author to whom correspondence should be addressed; thm@aber.ac.uk

Academic Editor: Cory Matthew

**Abstract:** Backcross hybrids between the important forage legume white clover (*Trifolium repens* L.), which is stoloniferous, and the related rhizomatous species Caucasian clover (*T. ambiguum* M. Bieb), have been produced using white clover as the recurrent parent. The effect of drought on the parental species and two generations of backcrosses were studied in a short-term glasshouse experiment under three intensities of drought. Plants of Caucasian clover maintained a higher leaf relative water content and leaf water potential than white clover at comparable levels of drought, with the response of the backcrosses generally intermediate between the parents. Severe drought significantly reduced stolon growth rate and leaf development rate of white clover compared to the control, well-watered treatment, whilst differences between these two treatments in the backcross hybrids were relatively small. The differences between parental species and the backcrosses in root morphology were studied in 1 m long vertical pipes. The parental species differed in root weight distribution, with root weight of Caucasian clover significantly greater than white clover in the 0.1 m to 0.5 m root zone. The backcrosses exhibited root characteristics intermediate between the parents. The extent to which these differences influence the capacity to tolerate drought is discussed.

**Keywords:** white clover; interspecific hybrids; drought; leaf development rate; root weight distribution

---

## 1. Introduction

Changing climatic conditions mean that the growing demand for meat and milk based products must be met against a backdrop of rising global temperatures and changing patterns of precipitation [1]. Extreme weather events, including periods of drought, will increasingly become a major factor limiting crop productivity in many parts of the world, including the UK [2]. Adaptation of agriculture to predicted climate change scenarios is essential, with the development of improved plant varieties better able to tolerate periods of drought [1] increasingly a key objective of many plant breeding programmes [3]. Selection criteria that will lead to new improved varieties of wheat [4,5] and grain legumes [6,7] better able to cope with drought are being developed. Grassland systems face similar challenges from climate change, therefore the development of new varieties of forage grasses and legumes better able to tolerate periods of drought is crucial.

The most important forage legume component of temperate pastures is white clover (*Trifolium repens* L.) [8], a nitrogen fixing species that produces forage of high quality. It is an outbreeding, highly heterozygous allotetraploid ($2n = 4x = 32$) species and the wide genetic variation within its gene pool has been used successfully in the production of new varieties with improvements in many traits.

Less variation has been identified for traits such as drought tolerance, which have proved difficult to improve significantly by conventional selection methods [9]. Although some authors [10] showed differences between ten white clover cultivars with respect to their response to drought, others [11] found little variation in response to a drought stress gradient between six lines (three cultivars and three germplasm accessions). Selection for deeper, more extensive root systems has been recommended for better tolerance to intermittent drought [12]. Selection for thicker roots as an indirect selection criterion has, however, been unsuccessful [13], although selection for increased root weight ratio (proportion of total plant DM allocated to roots) was found to improve the growth and survival of white clover in drought prone environments [14].

Introgression of genes from closely related species has been used successfully to introduce desirable traits into white clover [15–18] including improved drought tolerance [19]. Caucasian or Kura Clover (*Trifolium ambiguum* M. Bieb) is a strongly rhizomatous perennial legume species with good drought tolerance and persistence [20]. It is considered to have a wider range of adaptation than white clover [21], although slow seedling establishment tends to reduce its competitiveness with grasses in mixtures [22]. The extensive root and rhizome system is thought to act as a nutrient store that can be remobilised and used for growth, thus allowing this species to persist under stressful conditions [23]. Hybrids have been developed between white clover and Caucasian clover with the objective of introgressing the rhizomatous trait from Caucasian clover into white clover [16] as a strategy for improving drought tolerance whilst retaining the desirable agronomic traits associated with the latter species. Fertile backcrosss (BC) hybrids (derived from backcrossing to white clover) have been produced and these are essentially like white clover, but with rhizomes as well as stolons. A drought experiment comparing the BC1 and BC2 hybrids with the white clover and Caucasian clover parents in deep soil bins [16] showed that the backcross hybrids maintained lower values of leaf relative water content (RWC) and leaf water potential than Caucasian clover, but higher levels than white clover at comparable levels of drought. The mechanism by which Caucasian clover maintains a higher leaf RWC is not known, nor is the extent to which this mechanism operates within the hybrids. However, previous studies have shown that the hybrids allocate a higher proportion of their total DM yield to roots than white clover *i.e.*, they maintain a higher root to shoot ratio [16]. Previous studies on white clover have shown that stolon growth and leaf development rate (LDR) are reduced by drought [24,25], but little is known about the effect of drought on these growth parameters in the backcross hybrids.

This study had the following objectives: firstly, to quantify the response of the backcross hybrids to drought; and secondly, to identify the extent to which ability to withstand drought may be related to differences in root depth distribution.

## 2. Materials and Methods

### *2.1. Experiment 1*

#### 2.1.1. Plant Material and Experimental Treatments

The *T. ambiguum* (Caucasian clover) accession Ah1254, collected in Turkey in 1971, and the *T. repens* (white clover) medium-leaved variety Menna were used in the hybridization programme. Fertile F1 plants were used as the basis for two generations of backcrossing to white clover as the recurrent parent. Details of the development of these backcrosses including methods of embryo rescue used in the development of the original hybrids and their morphological characterisation have been described previously [8,16]. Four genotypes within each of the white clover, Caucasian clover, BC1 and BC2 populations, selected based on their use in the development of the backcross populations, were cloned to provide six-plants of each genotype so that there were two clonal plants of each genotype available for each of three drought regimes. The genotypes of the BC1 and BC2 were selected on the basis of the presence of rhizomes and had been used in previous studies on forage yield and quality [16]. Clonal plants were obtained by removing a growing point with three nodes and planting in multi-compartment trays containing John Innes No. 3 compost. When they had produced at least three trifoliate leaves they were transplanted into 25 cm diameter × 27 cm deep pots filled with John Innes No. 3 compost. No rhizobia were added to the soil however nodules were observed on plant roots.

2.1.2. Drought Tolerance

There were three treatments: control (C) plants maintained at field capacity; moderate drought (M) plants maintained at 80% field capacity; severe drought (S) plants maintained at 65% field capacity. Field capacity was defined as the volume of water required for the soil within the pot to be saturated and was determined daily on the control plants. The M and S plants received 80% and 65% respectively of the quantity of water required by the C plants to maintain them at field capacity. This was repeated daily throughout the course of the experiment.

The experiment began when the plants were 3 months old, when they were cut to a height of 3cm above ground level. At 21 and 35 days after the start of the experiment, pre-dawn leaf water potential was measured. Two leaflets were sampled per plant and leaf water potential measured using a pressure bomb (Portable plant moisture system SKPM 1400/40; Skye Instruments Ltd., Llandrindod Wells, UK) using the method described previously [25]. After 21 and 35 days, leaf relative water content (RWC) was determined on three leaves per plant as described [16] using the formula

$$RWC = ((FW - DW) / (RW - DW)) \times 100$$

where FW = fresh weight, RW = rehydrated weight and DW = dry weight.

2.1.3. Plant Growth

Non-destructive measurements of stolon length and leaf development rate (LDR) were carried out on one rando mLy selected stolon per plant. At the beginning of the experiment the selected stolon was marked with an acrylic paint dot behind the youngest fully expanded leaf. After 7, 14, 21 and 28 days, stolon length from the tip of the growing point to the paint mark was measured and leaf development recorded using the criteria established by Carlson [26]: all leaves produced after the paint mark were given a score using the Carlson visual scale for leaf development, where 1.0 indicates a fully expanded leaf and 0.1 indicates a leaf just visible as it emerges. The sum of these scores was calculated for the measured stolon. The absence of stolons in Caucasian clover and the difficulty of measuring LDR in this species meant that this part of the experiment only compared white clover with the BC1 and BC2 hybrids. Thirty five days after the start of the experiment all plants were cut to a height of 3 cm above soil level. The leaf area of three leaves per plant was measured using a Delta-T-Devices leaf area meter and the dry weight of above ground material determined by drying for 12 h at 80 °C in a forced draught oven.

*2.2. Experiment 2*

Root Depth Distribution

Four clonal plants of each of the four genotypes of the populations used in Experiment 1 were obtained as described previously and planted into multi-compartment trays containing John Innes No. 3 compost. When they had produced three trifioliate leaves, they were transplanted into 1 m deep × 15 cm diameter plastic pipes with several drainage holes drilled in the base, into which was inserted a polythene tube filled with vermiculite. The pipes were placed vertically on a gravel bed in a glasshouse maintained at ambient temperature. The plants received 100 mL water daily and once a week received an additional 50 mL of a standard full-nutrient solution [27]. After ten weeks the polythene tube was removed from the pipe and the above ground foliage cut to ground level with hand held shears. The root column was removed and separated into 10 cm deep horizontal sections.

The roots within each section were removed by washing under running water. The dry weight of the above ground biomass and root biomass within each section were determined after drying at 80 °C for 24 h in a forced draught oven.

### 2.3. Data Analysis

Experiment 1 was established as a split-plot design with two replicate blocks, comprising drought treatments as whole plots and genotypes as sub-plots. Growth parameters (leaf water potential, leaf relative water content, leaf development rate, stolon growth rate, dry matter yield and leaf size) were analysed by analysis of variance (ANOVA) using GenStat® (VSN International, Hemel Hempstead, UK) Release 13 [28] to determine significant effects of population, genotype within population and drought, and their interactions. Experiment 2 was established as a split-plot design with four replicate blocks, comprising populations as whole plots and genotypes as sub-plots. Root dry weight at each depth was analysed separately by ANOVA as above to determine significant effects of population and genotype within population.

## 3. Results

### 3.1. Experiment 1

#### 3.1.1. Overall analysis

For most of the growth parameters measured there were significant effects of population and drought, and significant population × drought interactions, but no significant differences between genotypes within populations (Table 1). Consequently for all growth parameters only the population × drought means are presented.

**Table 1.** Significance levels for effect of drought, population and their interaction on plant growth parameters.

| Treatment | LWP | | Leaf RWC | | SGR | LDR | DM Yield | Leaf Area |
|---|---|---|---|---|---|---|---|---|
| | 21 Day | 35 Day | 21 Day | 35 Day | | | | |
| Drought (D) | NS | *** | ** | * | * | * | ** | * |
| Population (P) | NS | *** | *** | *** | *** | *** | *** | *** |
| D × P | *** | *** | *** | *** | NS | NS | *** | *** |

NS, not significant; * $p < 0.05$; ** $p < 0.01$; *** $p < 0.001$; Key to abbreviations: LWP—leaf water potential, Leaf RWC—leaf relative water content, SGR—stolon growth rate, LDR—leaf development rate, DM yield—dry matter yield.

#### 3.1.2. Plant Water Status

Results for the effects of the drought treatments and population on leaf water potential (LWP) are presented in Table 2. Twenty one days after the start of the experiment overall values of LWP were not affected by drought treatment, nor was there a difference between populations. However, there was a significant drought treatment × population interaction, such that LWP in white clover decreased more under the S drought treatment compared with LWP in Caucasian clover and the backcross hybrids. Thirty five days after the start of the experiment, drought treatment had a significant effect on overall values of LWP, which were greatly reduced under treatment S, followed by treatment M, and both were less than under the well-watered control treatment C. There was also a significant difference between populations and a significant drought × population interaction. As a result, LWP in white clover was significantly lower than in the other populations, and the magnitude of this reduction was greatest under the most severe drought treatment. Leaf RWC was significantly influenced by drought, population and there was a significant drought × population interaction when measured 21 and 35 days after the start of the experiment (Table 3). After 21 days, leaf RWC was lower under S than under M and C and in Caucasian clover was greater than that of white clover and the BC1 and BC2 hybrids. Leaf RWC of Caucasian clover was unaffected by moisture stress however in white clover and the BC1 and BC2 hybrids the leaf RWC was significantly lower under S than M and C. A similar result was observed after 35 days with the leaf RWC of Caucasian clover unaffected by moisture stress but the leaf RWC of white clover and the BC1 and BC2 hybrids significantly reduced under S in comparison with M and C.

**Table 2.** Leaf water potential (MPa) of Caucasian clover, white clover, BC1 and BC2 hybrids after 21 and 35 days at three levels of drought. C—-control treatment, M—-moderate moisture stress, S—-severe moisture stress.

| Population | Days after Start of Drought | | | | | |
|---|---|---|---|---|---|---|
| | 21 | | | 35 | | |
| | C | M | S | C | M | S |
| Caucasian Clover | −0.47 | −0.58 | −0.76 | −0.50 | −0.73 | −0.76 |
| White Clover | −0.27 | −0.51 | −1.35 | −0.39 | −0.85 | −2.00 |
| BC1 | −0.32 | −0.57 | −0.80 | −0.30 | −0.69 | −1.69 |
| BC2 | −0.26 | −0.41 | −0.94 | −0.32 | −0.62 | −1.49 |
| S.e.d. | | | | | | |
| Drought (D) | | 0.270 NS | | | 0.037 *** | |
| Population (P) | | 0.072 NS | | | 0.059 *** | |
| D × P | | 0.291 *** (0.124 ***) | | | 0.092 *** (0.097 ***) | |

NS not significant; *** $p < 0.001$; S.e.d in brackets to be used when comparing means with same level of drought.

**Table 3.** Leaf relative water content (%) of Caucasian clover, white clover, BC1 and BC2 hybrids after 21 and 35 days at three levels of drought. C—-control treatment, M—-moderate moisture stress, S—-severe moisture stress.

| Population | Days after Start of Drought | | | | | |
|---|---|---|---|---|---|---|
| | 21 | | | 35 | | |
| | C | M | S | C | M | S |
| Caucasian Clover | 93.1 | 94.3 | 93.1 | 92.2 | 92.9 | 92.0 |
| White Clover | 91.2 | 90.9 | 69.3 | 91.6 | 92.9 | 71.3 |
| BC1 | 94.1 | 93.2 | 76.5 | 92.7 | 92.5 | 76.4 |
| BC2 | 93.1 | 93.1 | 69.9 | 92.3 | 93.1 | 68.4 |
| S.e.d. | | | | | | |
| Drought (D) | | 1.23 ** | | | 1.96 * | |
| Population (P) | | 1.61 *** | | | 1.63 *** | |
| D × P | | 2.72 *** (2.79 ***) | | | 3.13 *** (2.82 ***) | |

S.e.d. in brackets to be used when comparing means with same level of drought. * $p < 0.05$; ** $p < 0.01$; *** $p < 0.001$.

### 3.1.3. Plant Growth

Stolon growth rate (SGR) and leaf development rate (LDR) were influenced by drought and population but there was no significant interaction (Table 4). Drought reduced SGR, and generally the SGR of white clover was significantly higher than the BC2 and both were higher than in the BC1. The LDR of white clover was significantly greater than the backcross hybrids which were not significantly different from each other. Drought treatment reduced LDR but only under S; under M and C it did not differ significantly. Leaf area was significantly influenced by drought, differed between populations and there was a significant drought × population interaction. Generally leaf area was reduced by drought and the leaf area of white clover was greater than the BC1 and BC2 hybrids with the leaf area of Caucasian clover smallest. Leaf area of white clover and Caucasian clover was reduced by the M treatment and the leaf area of white clover further significantly reduced under the S treatment, unlike Caucasian clover which showed no further reduction in leaf area. The BC1 and BC2 hybrids exhibited a similar response to the S treatment as white clover.

**Table 4.** Stolon growth rate (mm/7 days), leaf development rate (quantified using Carlson Scale) and leaf area (mm²) of Caucasian clover, white clover, BC1 and BC2 hybrids after 35 days at three levels of drought. C—control treatment, M—moderate moisture stress, S—severe moisture stress.

| Population | Stolon Growth Rate | | | Leaf Development Rate | | | Leaf Area (mm²) | | |
|---|---|---|---|---|---|---|---|---|---|
| | C | M | S | C | M | S | C | M | S |
| Caucasian Clover | - | - | - | - | - | - | 358.2 | 303.0 | 184.9 |
| White Clover | 3.8 | 4.0 | 1.1 | 9.9 | 10.0 | 5.6 | 280.4 | 241.9 | 178.6 |
| BC1 | 2.1 | 1.6 | 0.4 | 6.1 | 7.0 | 4.8 | 309.8 | 265.4 | 257.0 |
| BC2 | 3.1 | 2.3 | 0.6 | 6.4 | 7.0 | 4.3 | 322.2 | 248.3 | 220.8 |
| Drought (D) | 0.39 * | | | 0.38 * | | | 10.90 * | | |
| Population (P) | 0.32 *** | | | 0.43 *** | | | 10.19 *** | | |
| D × P | 0.65 NS (0.55 NS) | | | 0.73 NS (0.75 NS) | | | 18.77 *** (17.64 ***) | | |

NS, not significant; * $p < 0.05$; *** $p < 0.001$; S.e.d. in brackets to be used when comparing means with same level of drought.

Overall DM yield per plant was greater under C than in M and both greater than under the S treatment (Table 5). DM yield of white clover was significantly greater than the BC1 and BC2 hybrids and all had DM yields significantly greater than Caucasian clover reflecting the slow establishment of this species. There was also a significant drought × population interaction as drought had no significant effect on the DM yield of Caucasian clover but the DM yield of white clover and the BC1 and BC2 hybrids was significantly reduced by drought stress but white clover was reduced by a greater amount than the hybrids.

### 3.2. *Experiment 2*

There was a significant difference between populations in root dry weight to depths of 0.5 m and significant differences between genotypes within populations (Table 6). However, at depths below 0.5 m differences between populations were small and insignificant and are not shown. Root dry weight of white clover and Caucasian clover in the 0 to 0.1 m root zone was comparable (Figure 1). However, in subsequent zones, up to a depth of 0.5 m, the root dry weight of Caucasian clover was significantly greater than that of white clover (Figure 1). Apart from the 0 to 0.1 m root zone where the BC2 had the greatest root dry weight, the root dry weight of the BC1 and BC2 hybrids were not significantly different and were generally intermediate between the two parental species. Differences in root dry weight between genotypes of white clover, BC1 and BC2 hybrids were observed at depths of 0.1–0.4 m but no significant differences between genotypes of Caucasian clover were observed.

**Table 5.** Dry matter yield (g/plant) of Caucasian clover, white clover, BC1 and BC2 hybrids after 35 days at three levels of drought. C—control treatment, M—moderate moisture stress, S-severe moisture stress.

| Population | Moisture Level | | |
|---|---|---|---|
| | C | M | S |
| Caucasian Clover | 3.3 | 2.1 | 1.5 |
| White Clover | 30.8 | 20.2 | 4.4 |
| BC1 | 22.7 | 17.7 | 4.9 |
| BC2 | 26.9 | 16.1 | 4.4 |
| Drought (D) | | 0.98 ** | |
| Population (P) | | 0.75 *** | |
| P × S | | 1.49 *** (1.30 ***) | |

** $p < 0.01$; *** $p < 0.001$; S.e.d. in brackets to be used when comparing means with same level of drought.

**Table 6.** Significance levels for effect of population and genotype within population on root dry weight at different depths.

| Significance | Root Depth (m) | | | | |
|---|---|---|---|---|---|
| | 0.1 | 0.2 | 0.3 | 0.4 | 0.5 |
| Population | * | ** | *** | *** | *** |
| Genotypes within Population | *** | ** | ** | ** | NS |

NS, not significant; * $p < 0.05$; ** $p < 0.01$; *** $p < 0.001$.

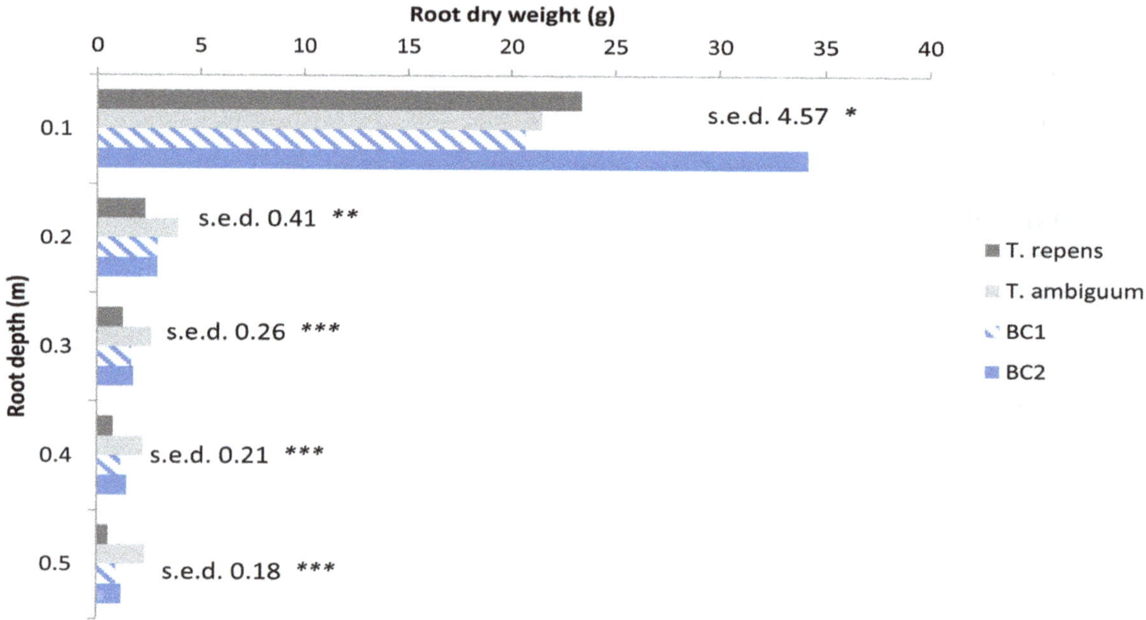


**Figure 1.** Root dry weight in 0.1 m sections of soil columns containing Caucasian clover, white clover, BC1 and BC2 hybrids. NS, not significant; * $p < 0.05$; ** $p < 0.01$; *** $p < 0.001$.

## 4. Discussion

### *4.1. Plant Water Status in Response to Drought*

It has been predicted that climate change will affect the distribution patterns of rainfall [29] and that this may have a negative impact on grassland productivity [30]. Development of forage varieties with a greater ability to tolerate drought is therefore an increasingly important target for forage plant breeding programmes [31]. Temperate pasture species such as white clover, whose agronomic yield consists of foliage, are considered to be more susceptible to drought stress than cereals or grain legumes [32]. One strategy to improve the drought tolerance of white clover is the development of interspecific hybrids between white clover and the drought tolerant, rhizomatous species Caucasian clover. Although the value of Caucasian clover as a drought tolerant species has been recognised in many Mediterranean areas, it has not been fully exploited due to its slow establishment [33].

Interspecific hybrids have been developed to introgress the rhizomatous trait into white clover without compromising the DM yield and forage quality of white clover [34,35]. Confirmation that Caucasian clover has a greater tolerance of drought than white clover has been clearly provided by the present experiment. Here, plants of Caucasian clover were found to maintain a higher leaf RWC and LWP than white clover under moderate and severe drought, confirming results from previous experiments carried out in deep soil bins [16]. Improved drought tolerance was also evident in the

BC1 and BC2 hybrids, which maintained higher values of LWP and RWC than white clover, although lower than Caucasian clover. Consequently, after 35 days of the experiment, the plants of white clover were visibly wilting whilst those of the BC1 and BC2 hybrids were still turgid. In this experiment, plants were grown individually in large pots with limited opportunity for development of adventitious roots and rhizomes and were relatively juvenile plants when measured. Therefore care is necessary in extrapolating results directly to a field situation. Nevertheless, the results are comparable with previous experiments where genotypes of these populations were grown in deep soil bins and growth was typical of that observed in a previous a field plot experiment, in which adventitious roots were able to develop [16].

### *4.2. Root Depth Distribution*

The objective of this experiment was to determine whether differences in response to drought in this germplasm might be associated with variation in root depth distribution. Plants tolerate drought through a range of mechanisms that include enhanced capture of soil moisture, limiting water loss and retention of cellular rehydration [36]. Rooting depth is one factor that influences soil moisture uptake. Drought resistant cultivars of species such as bean (*Phaseolus vulgaris* L.) [37] are those with thicker and deeper penetrating roots, and a higher proportion of total dry matter allocated to roots. In white clover, significant genetic variation has been observed for many root characters [38], some of which might be expected to confer improved drought tolerance. Some authors [39] for example, found significant genetic variation in seedling root system depth, and a high heritability for this character, within a collected population of white clover. These differences were related to dry matter yield under drought conditions, with genotypes from the "long" selection group producing higher yields than those with shorter roots. However, selection for thicker roots in white clover had varied success [13,14]. In the present experiment, Caucasian clover had a greater root weight than white clover at depths below 0.2 m. The BC1 and BC2 hybrids also had a greater root weight than white clover at depths below 0.2 m. This suggests that the introgression approach increased the allocation of resources to roots at lower depths thereby contributing to the improved drought tolerance of the hybrids compared to white clover.

When considering the effect of root system type on plant persistence in conditions of drought it is important to take into account both the timing of soil water deficits and the use of stored soil water in relation to crop phenology [40]. Thus, where soil water reserves exist at depth, the ability of plants to produce deep, extensive root systems is likely to be advantageous [41]. Conversely, where moisture reserves are confined to the upper layers of soil, then rooting depth becomes less important than the ability of plants to produce an efficient extraction system in these surface layers [41]. Although depth of rooting depends on soil type, cultivar and management, white clover is generally considered to be a shallow-rooted species, with most roots distributed in the top 0.1–0.2 m of soil [38]. This distribution pattern is likely to be detrimental to the ability of the species to persist under drought and suggests why the hybrids, with greater root weight than white clover below 0.2 m, are more drought tolerant. In addition, there is evidence that the growth of white clover root systems *per se* is adversely affected by drought conditions. For example, the production of new roots of white clover has been found to be greatly reduced under drought stress [42,43]. Studies [44] comparing root system development in seedlings of three legume species, *T. repens*, *Lotus corniculatus* and *Medicago lupulina*, under different levels of drought found that the depth of penetration of the root system of white clover was considerably more reduced by drought than that of the other species. Consequently, in deep, moist soils the shallow root system of white clover was found to be adequate to sustain vegetative growth and seed production, but in conditions of drought an inability to penetrate the soil profile effectively was found not only to reduce dry matter yield but also to lead to lower seed production per plant and, ultimately, to plant mortality [44]. The effect of drought on the root growth of the BC hybrids was not included in the present study but could be the focus of future work, particularly including an analysis of root growth in different soil types.

### 4.3. Plant Growth and Development

Persistence and dry matter yield of white clover within mixed swards is determined by stolon growth and development and the rate at which leaves are produced at stolon nodes [45]. In white clover, stolon growth and leaf development rate have been shown to be reduced by drought stress [24,25]. In one study [25], a moderate drought reduced leaf development rate by 40% and also reduced individual leaf area, effects that would have a considerable detrimental effect on DM yield if replicated in a mixed sward. A similar response to drought by white clover was observed in the present experiment, as severe drought significantly reduced stolon growth rate and production of new leaf area. Although the stolon growth, LDR and leaf area of the BC1 and BC2 hybrids were also reduced by severe drought the difference between growth in the well-watered control and the severe drought was, unlike white clover, relatively small. Both the moderate and severe drought treatment reduced the DM yield of white clover, BC1 and BC2 hybrids in comparison with the well-watered control. However, the greater drought tolerance in the hybrids was only reflected in a slight reduction in DM yield of the BC1 under moderate drought compared with white clover and the BC2. The reduction in DM yield of white clover, BC1 and BC2 hybrids was comparable under the severe drought treatment. Some similarity between white clover and the BC2 is not surprising since they are very closely related but it is possible that greater differences would have been observed had the experiment continued for longer than 35 days.

Although the BC1 and BC2 hybrids are essentially white clover-like in appearance, one of the surprising results was the lower stolon growth rate and of LDR both BC hybrids in the well-watered treatment compared with the white clover parent. No evidence of differences in stolon growth rate has been found during characterisation of this germplasm in spaced plant nurseries [8]. Further agronomic studies, carried out in field plots, have shown that the DM yield of the BC1 and BC2 hybrids when grown with *L. perenne* was slightly lower than in white clover in the first harvest year [34]. However, DM yield did improve in subsequent years and was comparable with white clover in the 2nd and 3rd harvest years, suggesting that any differences in stolon growth may not be significant for long term pasture performance. Caucasian clover is a persistent species, but slow to establish with consequently low initial DM yields and grows poorly in pastures compared with white clover [21], largely due to its rhizomatous growth. The low stolon growth rate and DR may be a consequence of the introgression of the rhizomatous trait from Caucasian clover and further selection for improved stolon growth rate may be required.

## 5. Conclusions

Analysis of plant water status (leaf relative water content, leaf water potential) and plant growth and development in response to different levels of drought revealed significant variation between Caucasian clover and white clover, with an intermediate response of backcross hybrids between these two species. The parental species differed in root depth distribution and in root weight distribution, with root weight of Caucasian clover significantly greater than white clover in the 0.1 m to 0.5 m root zone with the backcross hybrids exhibiting root characteristics intermediate between the parental species. It is suggested that differences in root distribution is a factor influencing the extent to which plants are able to tolerate drought, however, further studies are required to quantify the impact of the differences in root distribution and rhizomes in mature plants.

**Acknowledgments:** This research was funded by the Department for the Environment, Food and Rural Affairs through the Sustainable Livestock LINK programme.

**Author Contributions:** Athole Marshall contributed to the experimental analysis and writing of the manuscript, aided by Rosemary Collins. Matthew Lowe was responsible for management of the experiments and data collection.

## References

1. Foresight. *The Future of Food and Farming. Final Project Report*; The Government Office for Science: London, UK, 2011.

2. Hopkins, A.; Del Prado, A. Implications of climate change for grassland in Europe: Impacts, adaptations and mitigation options: A review. *Grass Forage Sci.* **2007**, *62*, 118–126. [CrossRef]
3. Abberton, M.T.; MacDuff, J.H.; Marshall, A.H.; Humphreys, M.W. The genetic improvement offorage grasses and legumes to enhance adaptation of grasslands to climate change. In Proceedings of the United National Climate Change Conference, Nusa Dua, Indonesia, 3–14 December 2007.
4. Foulkes, M.J.; Sylvester-Bradley, R.; Weightman, R.; Snape, J.W. Identifying physiological traitsassociated with improved drought resistance in winter wheat. *Field Crops Res.* **2007**, *103*, 11–14. [CrossRef]
5. Dodd, I.C.; Whalley, W.R.; Ober, E.S.; Parry, M.A.J. Genetic and management approaches to boost UK winter wheat yields by ameliorating water deficits. *J. Exp. Bot.* **2011**, *62*, 5241–5248. [CrossRef] [PubMed]
6. Lizana, C.; Wentworth, M.; Martinez, J.P.; Villegas, D.; Meneses, R.; Murchie, E.H.; Pastenes, C.; Lercari, B.; Vernieri, P.; Horton, P.; *et al.* Differential adaptation of two varieties of common bean to abiotic stress. I. Effects of drought on yield and photosynthesis. *J. Exp. Bot.* **2006**, *57*, 685–697. [CrossRef] [PubMed]
7. Martinez, J.P.; Silva, H.; Ledent, J.F.; Pinto, M. Effect of drought stress on the osmotic adjustment, cell wall elasticity and cell volume of six cultivars of common beans (*Phaseolus vulgaris* L.). *Eur. J. Agron.* **2007**, *6*, 30–38. [CrossRef]
8. Abberton, M.T.; Michaelson-Yeates, T.P.T.; Marshall, A.H.; Holdbrook-Smith, K.; Rhodes, I. Morphological characteristics of hybrids between white clover, *Trifolium. repens* L. and Caucasian clover, *Trifolium. ambiguum* M. Bieb. *Plant Breed.* **1998**, *117*, 494–496. [CrossRef]
9. Abberton, M.T.; Marshall, A.H. Progress in breeding perennial clovers for temperate agriculture. *J. Agric. Sci.* **2005**, *143*, 117–135. [CrossRef]
10. Barbour, M.; Caradus, J.R.; Woodfield, D.R.; Silvester, W.B. Water stress and water use efficiency of ten white clover cultivars. In *White Clover: New Zealand's Competitive Edge*; Woodfield, D.R., Ed.; Grassland Research and Practice Series No. 6; New Zealand Grassland Association: Palmerston North, New Zealand, 1996; pp. 159–162.
11. Brink, G.E.; Pederson, G.A. White clover response to a water application gradient. *Crop Sci.* **1998**, *38*, 771–775. [CrossRef]
12. Collins, R.P. The effect of drought stress and winter stress on the persistence of white clover. In *Lowland Grasslands of Europe: Utilization and Development*; Fisher, G., Frankow-Lindberg, B.E., Eds.; REUR Technical Series No. 64; FAO: Rome, Italy, 2002; pp. 17–32.
13. Annicchiarico, P.; Piano, E. Indirect selection for root development of white clover and implications for drought tolerance. *J. Agron. Crop Sci.* **2004**, *190*, 28–34. [CrossRef]
14. Caradus, J.R.; Woodfield, D.R. Genetic control of adaptive root characteristics in white clover. *Plant Soil* **1998**, *200*, 63–69. [CrossRef]
15. Hussain, S.W.; Williams, W.M.; Mercer, C.F.; White, D.W.R. Transfer of clover cyst nematode resistance from *Trifolium. nigrescens* Viv. to *T. repens* by interspecific hybridisation. *Theor. Appl. Genet.* **1997**, *95*, 1274–1281. [CrossRef]
16. Marshall, A.H.; Rascle, C.; Abberton, M.T.; Michaelson-Yeates, T.P.T.; Rhodes, I. Introgression as a route to improved drought tolerance in white clover (*Trifolium. repens* L.). *J. Agron. Crop Sci.* **2001**, *187*, 11–18. [CrossRef]
17. Williams, W.M.; Hussain, S.W. Development of a breeding strategy for interspecific hybrids between Caucasian clover and white clover. *NZ J. Agric. Res.* **2008**, *51*, 115–126. [CrossRef]
18. Williams, W.M. Trifolium interspecific hybridisation: Widening the white clover gene pool. *Crop Pasture Sci.* **2014**, *65*, 1091–1106. [CrossRef]
19. Nichols, S.N.; Hofman, R.W.; Williams, W.M. Drought resistance of Trifolium repens × *Trifolium uniflorum* interspecific hybrids. *Crop Pasture Sci.* **2014**, *65*, 911–921. [CrossRef]
20. Coolbear, P.; Hill, M.J.; Efendi, F. Relationships between vegetative and reproductive growth in a four year old stand of Caucasian clover (*Trifolium. ambiguum* M Bieb.) cv. Monaro. *Proc. Agron. Soc. N. Z.* **1994**, *24*, 77–82.
21. Taylor, N.L.; Smith, R.R. Kura clover (*Trifolium. ambiguum* M.B.) breeding, culture and utilization. *Adv. Agron.* **1998**, *63*, 153–178.
22. Black, A.D.; Moot, D.J.; Lucas, R.J. Development and growth characteristics of Caucasian and white clover seedlings, compared with perennial ryegrass. *Grass Forage Sci.* **2006**, *61*, 442–453. [CrossRef]
23. Fu, S.M.; Hill, M.J.; Hampton, J.G. Root system development in Caucasian clover cv. Monaro and its contribution to seed yield. *N. Z. J. Agric. Res.* **2001**, *44*, 23–29. [CrossRef]

24. Turner, L.B. The effect of water stress on the vegetative growth of white clover (*T. repens* L.): comparison of long-term water deficit and a short-term developing drought. *J. Exp. Bot.* **1991**, *42*, 311–316. [CrossRef]
25. Belaygue, C.; Wery, J.; Cowan, A.A.; Tardieu, F. Contribution of leaf expansion, rate of leaf appearance and stolon branching to growth of plant leaf area under water deficit in white clover. *Crop Sci.* **1996**, *36*, 1240–1246. [CrossRef]
26. Carlson, G.E. Growth of clover leaves developmental morphology at ten stages. *Crop Sci.* **1996**, *6*, 293–294. [CrossRef]
27. Hoagland, D.R.; Snyder, W.C. Nutrition of strawberry under controlled conditions: (a) Effects of deficiencies of boron and certain other elements: (b) Susceptibility to injury from sodium salts. *Proc. Am. Soc. Hortic. Sci.* **1933**, *30*, 288–296.
28. Payne, R.W.; Murray, D.A.; Harding, S.A.; Baird, D.B.; Soutar, D.M. *Introduction to GenStat® for Windows™*; VSN International: Hemel Hempstead, UK, 2010.
29. Humphreys, M.W.; Yadav, R.S.; Cairns, A.J.; Turner, L.B.; Humphreys, J.; Skøt, L. A changing climate for grassland research. *New Phytol.* **2006**, *169*, 9–26. [CrossRef] [PubMed]
30. Humphreys, M.O. Grass roots for improved soil structure and hydrology. *IBERS Knowl.-Based Innov.* **2011**, *2011*, 21–25.
31. Humphreys, M.O. Genetic improvement of forage crops—-Past, present and future. *J. Agric. Sci.* **2005**, *143*, 441–448. [CrossRef]
32. Turner, N.C.; Begg, J.E. Responses of pasture plants to water deficits. In *Plant Relations in Pastures*; Wilson, J.R., Ed.; CSIRO: Melbourne, Australia, 1978; pp. 50–66.
33. Widdup, K.H.; Knight, T.L.; Waters, C.J. Genetic variation for rate of establishment in Caucasian clover. *Proc. Agron. Soc. N. Z.* **1998**, *60*, 213–217.
34. Marshall, A.H.; Williams, T.A.; Abberton, M.T.; Michaelson-Yeates, T.P.T.; Powell, H.G. Dry matter production of white clover (*Trifolium. repens* L.), Caucasian clover (*T. ambiguum* M. Bieb) and their associated hybrids when grown with a grass companion over three harvest years. *Grass Forage Sci.* **2003**, *59*, 91–99.
35. Marshall, A.H.; Williams, T.A.; Abberton, M.T.; Michaelson-Yeates, T.P.T.; Olyott, P.; Powell, H.G. Forage quality of white clover (*Trifolium. repens* L.) × Caucasian clover (*T. ambiguum* M. Bieb) hybrids when grown with a grass companion over three harvest years. *Grass Forage Sci.* **2004**, *59*, 91–99. [CrossRef]
36. Blum, A. Drought resistance, water-use efficiency, and yield potential—-Are they compatible, dissonant or mutually exclusive? *Aust. J. Agric. Res.* **2005**, *56*, 1159–1168. [CrossRef]
37. Sponchiado, B.N.; White, J.W.; Castillo, J.A.; Jones, P.G. Root growth of four common bean cultivars in relation to drought tolerance in environments with contrasting soil types. *Exp. Agric.* **1989**, *25*, 249–257. [CrossRef]
38. Caradus , J.R. The structure and function of white clover root systems. *Adv. Agron.* **1990**, *43*, 1–46.
39. Ennos, R.A. The significance of genetic variation for root growth within a natural population of white clover (*Trifolium. repens* L.). *J. Ecol.* **1985**, *73*, 615–624. [CrossRef]
40. Clarke, J.M.; McCaig, T.N. Breeding for efficient root systems. In *Plant breeding—-Principles and Prospectsi*; Hayward, M.D., Bosemark, N.O., Romagosa, I., Eds.; Chapman & Hal: London, UK, 1993; pp. 485–499.
41. Wilson, D. Breeding for morphological and physiological traits. In *Plant Breeding II*; Frey, K.J, Ed.; Iowa State University Press: Ames, IA, USA, 1981; pp. 233–290.
42. Thomas, H. Effects of drought on growth and competitive ability of perennial ryegrass and white clover. *J. Appl. Ecol.* **1984**, *21*, 591–602. [CrossRef]
43. Stevenson, C.A.; Laidlaw, A.S. The effect of moisture stress on stolon and adventitious root development in white clover (*Trifolium. repens* L.). *Plant Soil* **1985**, *85*, 249–257. [CrossRef]
44. Fouldes, W. Response to soil moisture supply in three leguminous species. I. Growth, reproduction and mortality. *New Phytol.* **1978**, *80*, 535–545. [CrossRef]
45. Elgersma, A.; Fengrui, L. Effects of cultivar and cutting frequency on dynamics of stolon growth and leaf appearance rate in white clover grown in mixed swards. *Grass Forage Sci.* **1997**, *3*, 370–380. [CrossRef]

# 14

# Linking Management, Environment and Morphogenetic and Structural Components of a Sward for Simulating Tiller Density Dynamics in Bahiagrass (*Paspalum notatum*)

**Masahiko Hirata**

Department of Animal and Grassland Sciences, Faculty of Agriculture, University of Miyazaki, Miyazaki 889-2192, Japan; m.hirata@cc.miyazaki-u.ac.jp

Academic Editor: Cory Matthew

**Abstract:** A model which describes tiller density dynamics in bahiagrass (*Paspalum notatum* Flügge) swards has been developed. The model incorporates interrelationships between various morphogenetic and structural components of the sward and uses the inverse of the self-thinning rule as the standard relationship between tiller density and tiller weight (a density-size equilibrium) toward which tiller density progressively changes over time under varying nitrogen (N) rates, air temperature and season. Water and nutrient limitations were not considered except partial consideration of N. The model was calibrated against data from swards subjected to different N rates and cutting intensities, and further validated against data from a grazed sward and swards under different cutting intensities. As the calibration and validation results were satisfactory, the model was used as a tool to investigate the responses of tiller density to various combinations of defoliation frequencies and intensities. Simulations identified defoliation regimes required for stabilizing tiller density at an arbitrary target level, *i.e.*, sustainable use of the sward. For example, the model predicted that tiller density can be maintained at a medium level of about 4000 $m^{-2}$ under conditions ranging from weekly cuttings to an 8 cm height to 8-weekly cuttings to 4 cm. More intense defoliation is needed for higher target tiller density and *vice versa*.

**Keywords:** model; tiller density; tiller birth; tiller death; self-thinning rule; bahiagrass

---

## 1. Introduction

Grasslands are essential to human life. Our challenge is to sustain grasslands and make better use of them for agricultural production, conservation of the environment and wildlife, and other purposes (e.g., recreation and amenity). This requires understanding and predicting sward dynamics in grasslands in response to the environment (e.g., temperature and rainfall) and management (e.g., defoliation and fertilizer application).

Sward dynamics in grasslands can be mechanistically analyzed and understood by breaking down the sward into a set of morphogenetic and structural components [1,2]. This approach was taken initially for temperate forage species and later for tropical species. Based on the morphogenetic and structural mechanisms, persistence of grass (Poaceae) swards is dependent on the ability of the plant to maintain a high tiller density, which in turn depends on the longevity (rate of death) and recruitment (rate of appearance) of the tillers [3].

Bahiagrass (*Paspalum notatum* Flügge), a sod-forming, warm-season perennial, is widespread in the southern USA, and Central and South America [4]. It is also well adapted to the low-altitude regions of south-western Japan, and used for both grazing and hay [5]. This grass forms a highly persistent sward under a wide range of management [6–11]. Among tropical forage species, bahiagrass has been most detailedly studied in terms of the morphogenetic and structural components, providing a good deal of information for modeling sward dynamics of the grass [12–22].

In the present study, a model of tiller density dynamics in bahiagrass swards was built by integrating information selected from the literature. The framework of the model is a combination of interrelationships linking management, environment and morphogenetic and structural components of the sward. The model was calibrated and validated against data from swards subjected to various management conditions, and then used to explore defoliation management for sustainable use of bahiagrass swards. The aims of the study were to examine how the integrated interrelationships work as a whole and to characterize the model in comparison with previous models of tiller density dynamics in grasses. The structure and performance of the model have been partly described in Hirata [22,23].

## 2. The Model

The model simulates changes in tiller density by calculating tiller appearance and death, which are driven by variables relating to the sward, environment and management (Figure 1). Since the model forms a submodel of an integrated model of sward dynamics in grasslands, it requires daily herbage mass as an input from a submodel of herbage production and utilization. The model also needs mean daily air temperature, annual nitrogen (N) fertilizer rate and month of the year as inputs, and initial tiller density at the commencement of a simulation run. Water and nutrient limitations are not considered except partial consideration of N. Variables used in the model are given in Table 1.

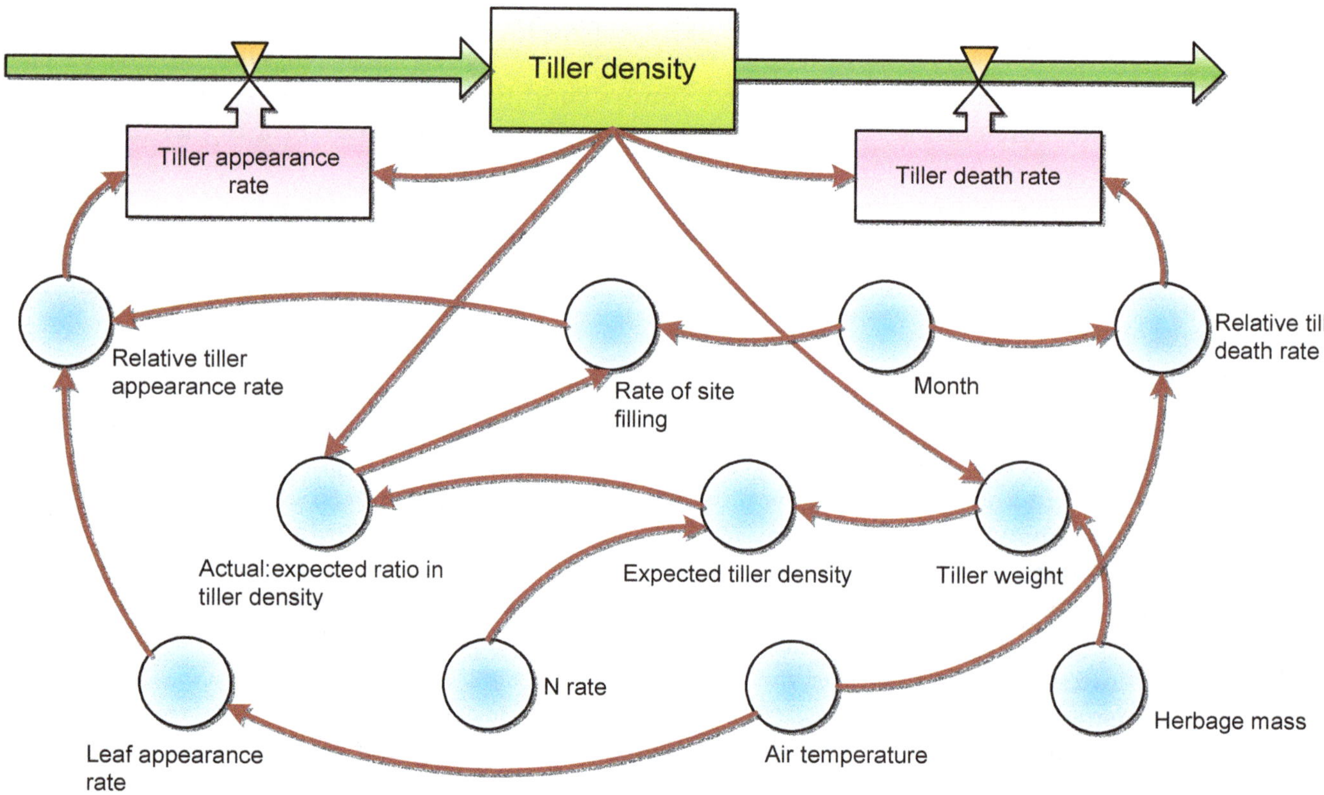


**Figure 1.** The framework of the model. Reproduced with permission from Hirata [23]; published by Wageningen Academic Publishers.

**Table 1.** Variables used in the model.

| Symbol | Description | Unit |
|---|---|---|
| $b_0$ | Intercept of the standard relationship between tiller density and tiller weight | log tillers m$^{-2}$ |
| $b_1$ | Slope of the standard relationship between tiller density and tiller weight | log tillers m$^{-2}$ (log mg DM tiller$^{-1}$)$^{-1}$ |
| $D$ | Tiller density | tillers m$^{-2}$ |
| $D_{A:E}$ | Actual: expected ratio in tiller density | fraction |
| $D_E$ | Expected tiller density | tillers m$^{-2}$ |
| $F_N$ | Annual nitrogen fertilizer rate | g m$^{-2}$ year$^{-1}$ |
| $F_S$ | Rate of site filling | tillers leaf$^{-1}$ |
| $M$ | Herbage mass | g DM m$^{-2}$ |
| $R_{leaf,app}$ | Leaf appearance rate | leaves tiller$^{-1}$ day$^{-1}$ |
| $R_{tiller,app}$ | Tiller appearance rate | tillers m$^{-2}$ day$^{-1}$ |
| $R'_{tiller,app}$ | Relative tiller appearance rate | tillers tiller$^{-1}$ day$^{-1}$ |
| $R_{tiller,death}$ | Tiller death rate | tillers m$^{-2}$ day$^{-1}$ |
| $R'_{tiller,death}$ | Relative tiller death rate | tillers tiller$^{-1}$ day$^{-1}$ |
| $t$ | Time | day |
| $T$ | Mean daily air temperature | °C |
| $W$ | Tiller weight | mg DM tiller$^{-1}$ |

### *2.1. Rate of Change in Tiller Density*

The daily rate of change in tiller density ($\Delta D/\Delta t$, tillers m$^{-2}$ day$^{-1}$) is expressed as the balance between tiller appearance rate ($R_{tiller,app}$, tillers m$^{-2}$ day$^{-1}$) and tiller death rate ($R_{tiller,death}$, tillers m$^{-2}$ day$^{-1}$):

$$\frac{\Delta D}{\Delta t} = R_{tiller,app} - R_{tiller,death} \quad (1)$$

The two rates are written as:

$$R_{tiller,app} = R'_{tiller,app} \times D \quad (2)$$

and

$$R_{tiller,death} = R'_{tiller,death} \times D \quad (3)$$

where $D$ is the tiller density (tillers m$^{-2}$), and $R'_{tiller,app}$ and $R'_{tiller,death}$ are the relative rates of tiller appearance and death (tillers tiller$^{-1}$ day$^{-1}$), respectively.

### *2.2. Tiller Appearance*

Relative tiller appearance rate is known to be the product of leaf appearance rate ($R_{leaf,app}$, leaves tiller$^{-1}$ day$^{-1}$) and the rate of site filling ($F_S$, tillers leaf$^{-1}$), *i.e.*, the rate at which axillary buds develop into tillers (visible without dissection) in relation to the rate at which leaf axils are formed [24,25]:

$$R'_{tiller,app} = F_S \times R_{leaf,app} \quad (4)$$

Leaf appearance rate is expressed as a threshold response function of the mean daily air temperature ($T$, °C) [14]:

$$\begin{aligned} R_{leaf,app} &= 0 && (\text{when } T \le 7.6) \\ &= 0.117 \times \frac{((T-7.6)/6.4)^{3.6}}{1+((T-7.6)/6.4)^{3.6}} && (\text{when } T > 7.6) \end{aligned} \quad (5)$$

This equation shows that leaves emerge when the temperature exceeds 7.6 °C and that the leaf appearance rate attains its half-maximal response when the temperature is 14.0 °C and approaches the maximal response (asymptote) of 0.117 leaves tiller$^{-1}$ day$^{-1}$ at higher temperatures (Figure 2).

The rate of site filling ($F_S$) is expressed as a function of the actual:expected ratio in tiller density ($D_{A:E}$), *i.e.*, the ratio of actual tiller density ($D$) to the expected tiller density ($D_E$, tillers m$^{-2}$), which was parameterized by revisiting the data given in Hirata and Pakiding [21] and Hirata [22]:

$$
\begin{aligned}
F_S &= \max[0,\ -0.120(D_{A:E} - 1.2)\ ] && \text{(for April)} \\
&= \max[0,\ -0.110(D_{A:E} - 1.2)\ ] && \text{(for May)} \\
&= \max[0,\ -0.100(D_{A:E} - 1.2)\ ] && \text{(for June)} \\
&= \max[0,\ -0.035(D_{A:E} - 1.2)\ ] && \text{(for July–November)} \\
&= 0 && \text{(for other months)}
\end{aligned} \tag{6}
$$

where

$$D_{A:E} = D / D_E \tag{7}$$

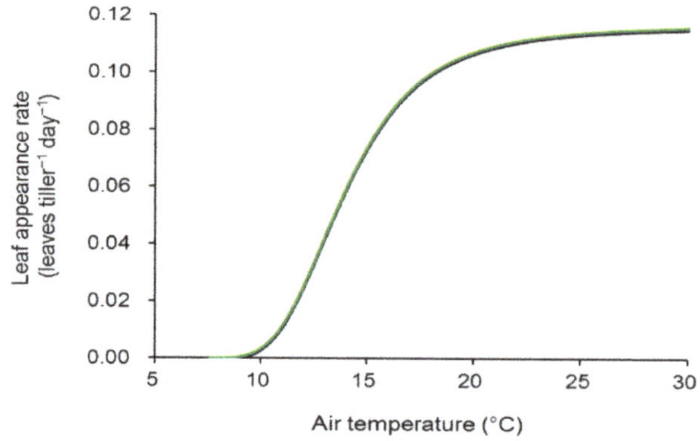


**Figure 2.** Relationship between leaf appearance rate and mean daily air temperature. Reproduced with permission from Pakiding and Hirata [14]; published by the Tropical Grassland Society of Australia.

Equation (6) shows that axillary buds have the potential of developing into tillers only in April–November and only when the actual tiller density is lower than 1.2 times the density expected from the standard relationship, with higher developmental rates at lower actual:expected ratios in tiller density and in April–June (April > May > June) than in July–November (Figure 3).

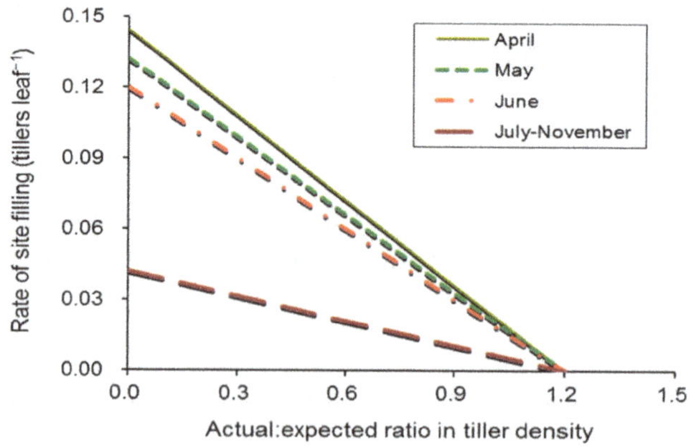


**Figure 3.** Relationships between rate of site filling and actual: expected ratio in tiller density.

The expected tiller density is calculated from tiller weight ($W$, mg DM tiller$^{-1}$), using the standard relationship between tiller density and tiller weight, *i.e.*, relationship in an almost stabilized sward (an equilibrium) under fixed management:

$$D_E = 10^{(b_0 - b_1 \log W)} \tag{8}$$

This standard density–weight relationship derives from the relationship, $\log D_E = b_0 - b_1 \log W$, a reverse form of the self-thinning rule [26] in terms of $x$- and $y$-variables, which makes it possible to estimate tiller density from tiller weight. The two parameters for the standard relationship ($b_0$ and $b_1$) are influenced by annual N fertilizer rate ($F_N$, g m$^{-2}$ year$^{-1}$) [21,22]:

$$b_0 = 4.355 + 0.0247(F_N - 5) \quad (9)$$

$$b_1 = -0.376 - 0.0092(F_N - 5) \quad (10)$$

Equations (8)–(10) show that the number of tillers carried on a unit land area increases as the tiller weight decreases, with a greater rate of increase at a higher N fertilizer rate (Figure 4).

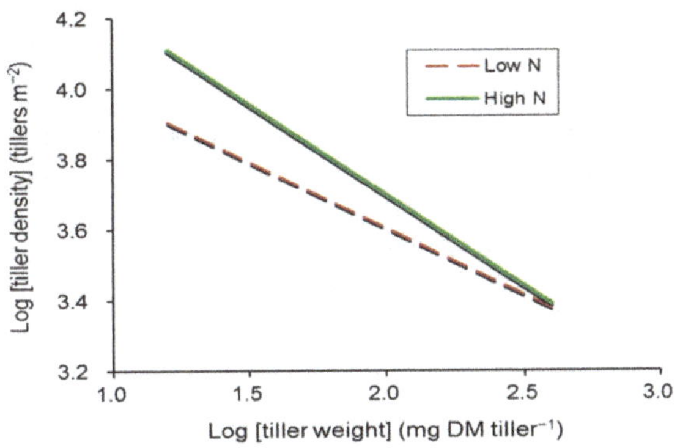


**Figure 4.** Standard relationships between tiller density and tiller weight. Low nitrogen = 5 g m$^{-2}$ year$^{-1}$, high nitrogen = 20 g m$^{-2}$ year$^{-1}$. Reproduced with permission from Hirata and Pakiding [21]; published by the Tropical Grassland Society of Australia.

Tiller weight is calculated as:

$$W = 1000 \times M/D \quad (11)$$

where $M$ is herbage mass (g DM m$^{-2}$) and 1000 is a unit conversion factor from g to mg.

### 2.3. Tiller Death

Relative tiller death rate is expressed as a function of the mean daily air temperature [22]:

$$\begin{aligned} R'_{\text{tiller,death}} &= 0.00084 \times \exp(0.031 \times T) \quad \text{(for spring–summer)} \\ &= 0.00020 \times \exp(0.083 \times T) \quad \text{(for autumn–winter)} \end{aligned} \quad (12)$$

This equation shows that the relative tiller death rate increases exponentially as the mean daily air temperature increases, maintaining higher values in spring–summer (March–August) than in autumn–winter (September–February) until the temperature reaches 27.9 °C (Figure 5).

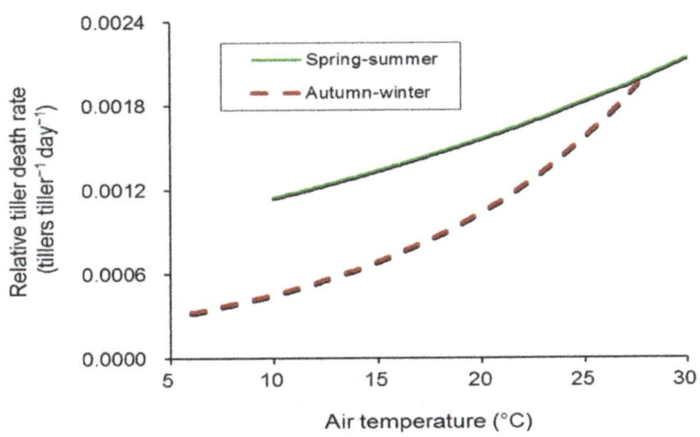


**Figure 5.** Relationship between relative tiller death rate and mean daily air temperature.

## 3. Model Performance

### 3.1. Calibration

The parameters for Equations (5), (6), (9), (10) and (12) were determined using a regression technique to calibrate the model. The performance of the model was first evaluated by simulating tiller density dynamics for the study from which most of the data used for calibrating the model were derived, *i.e.*, bahiagrass tiller density under different N fertilizer rates and cutting heights [18,21]. The simulations for the six experimental treatments (2 N rates $\times$ 3 cutting heights) were run for 1341 days ($t = 1, \cdots, 1341$) from 1 June 1996 to 1 February 2000. Daily herbage mass as an input (Figure 1) was estimated by interpolating data from monthly measurements. Mean daily air temperature was derived from daily records. Annual N rate was set at the actual dosage (5 and 20 g m$^{-2}$ year$^{-1}$ for low and high N treatments, respectively). Month of the year was calculated from the day number ($t$) and the initial date of simulation. Initial tiller density was set at the measured data.

Overall, the simulated tiller densities showed good agreement with the measured data, despite slight to moderate over- or under-prediction in some seasons in some treatments (Figure 6).

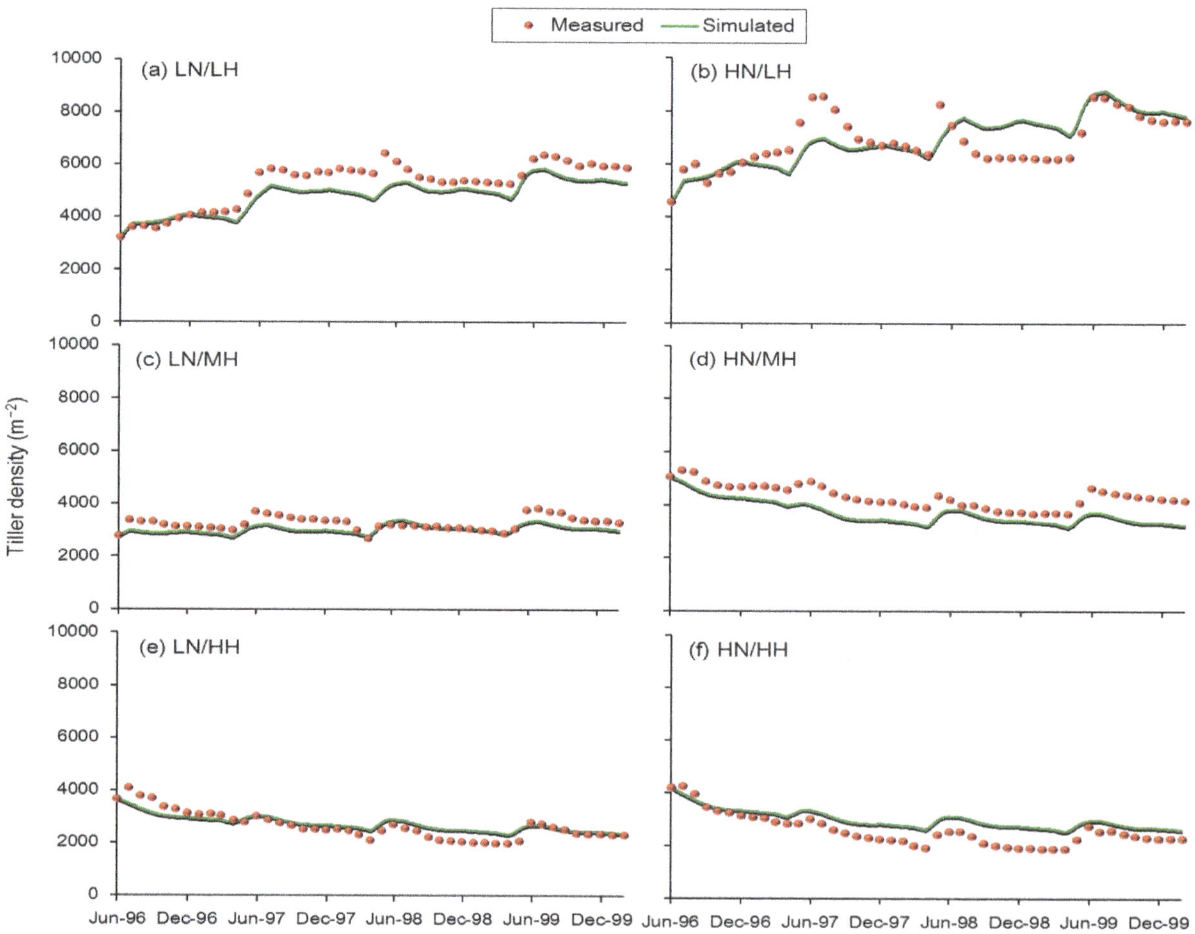


**Figure 6.** Measured (closed circle) and simulated (line) tiller densities in bahiagrass swards under different nitrogen rates and defoliation intensities (calibration results). Measured data derive from Pakiding and Hirata [18] and Hirata and Pakiding [21]. LN = low nitrogen (5 g m$^{-2}$ year$^{-1}$), HN = high nitrogen (20 g m$^{-2}$ year$^{-1}$), LH = low height (2 cm), MH = medium height (12 cm), HH = high height (22 cm) of cutting (heights above ground). Reproduced with permission from Hirata [22]; published by Research Signpost.

### 3.2. Validation

The model was then validated against independent data sets, *i.e.*, bahiagrass tiller density under cattle grazing [15] and under different cutting heights [10]. The simulations for the former study were run for 1462 days ($t = 1, \cdots, 1462$) from 18 May 1996 to 18 May 2000, and those for the latter study (5 cutting-height treatments) were run for 1092 days ($t = 1, \cdots, 1092$) from 30 May 1986 to 25 May 1989. Daily herbage mass was estimated by interpolating data from 2-weekly to seasonal measurements. Mean daily air temperature was derived from daily records. Annual N rate was set at the actual doses (4.5–9.7 g $m^{-2}$ $year^{-1}$ and 20 g $m^{-2}$ $year^{-1}$ for the former and latter studies, respectively). Month of the year was calculated from the day number ($t$) and the initial date of simulation. Initial tiller density used the measured data.

Under grazing, the simulated tiller densities showed good agreement with the measured data except some over-prediction in summer (June–August) 1997 (Figure 7). Under cutting, the simulated densities followed the measured values well over the course of 3 years, despite partial inability to follow the seasonal fluctuations within individual years (Figure 8). As a whole, the validation results were acceptable.

## 4. Use of the Model

As the calibration and validation results were successful, the model was used as a tool to investigate the responses of tiller density to various combinations of defoliation frequencies and intensities (heights above ground). The simulations were run for 3 years ($t = 1, \ldots, 1095$) from 1 May. Daily herbage mass was provided by a model of herbage production and utilization of a bahiagrass sward which was based on the data presented in Hirata [27,28], with no feedback from the tiller dynamics model. The mean daily air temperature ($T$, °C) was determined using the following equation, which approximates the long-term average of the annual cycle in Miyazaki (31°56′ N, 131°25′ E):

$$T = 17.5 + 10.5 \times \sin(2\pi(t + 9)/365) \quad (13)$$

Annual N rate was set at 10 g $m^{-2}$. Month of the year was calculated from the day number ($t$) and the initial date of simulation.

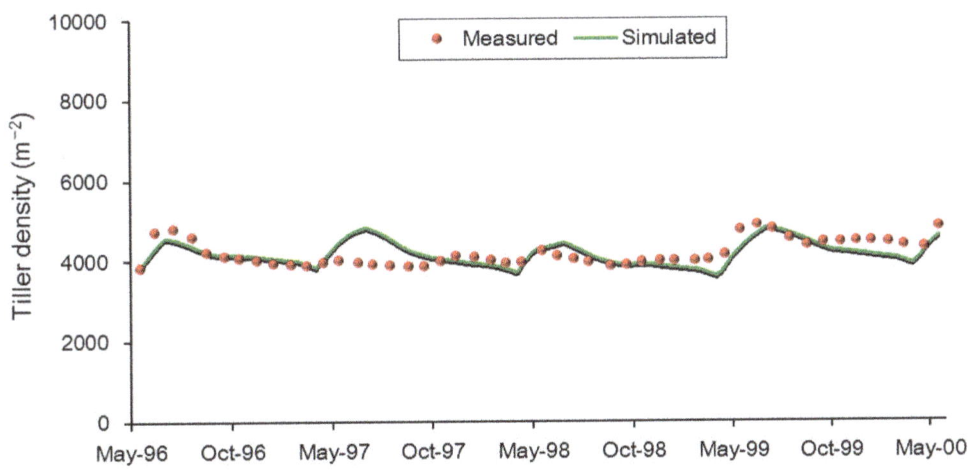


**Figure 7.** Measured (closed circle) and simulated (line) tiller densities in a bahiagrass pasture under cattle grazing (validation results). Measured data derive from Hirata and Pakiding [15].

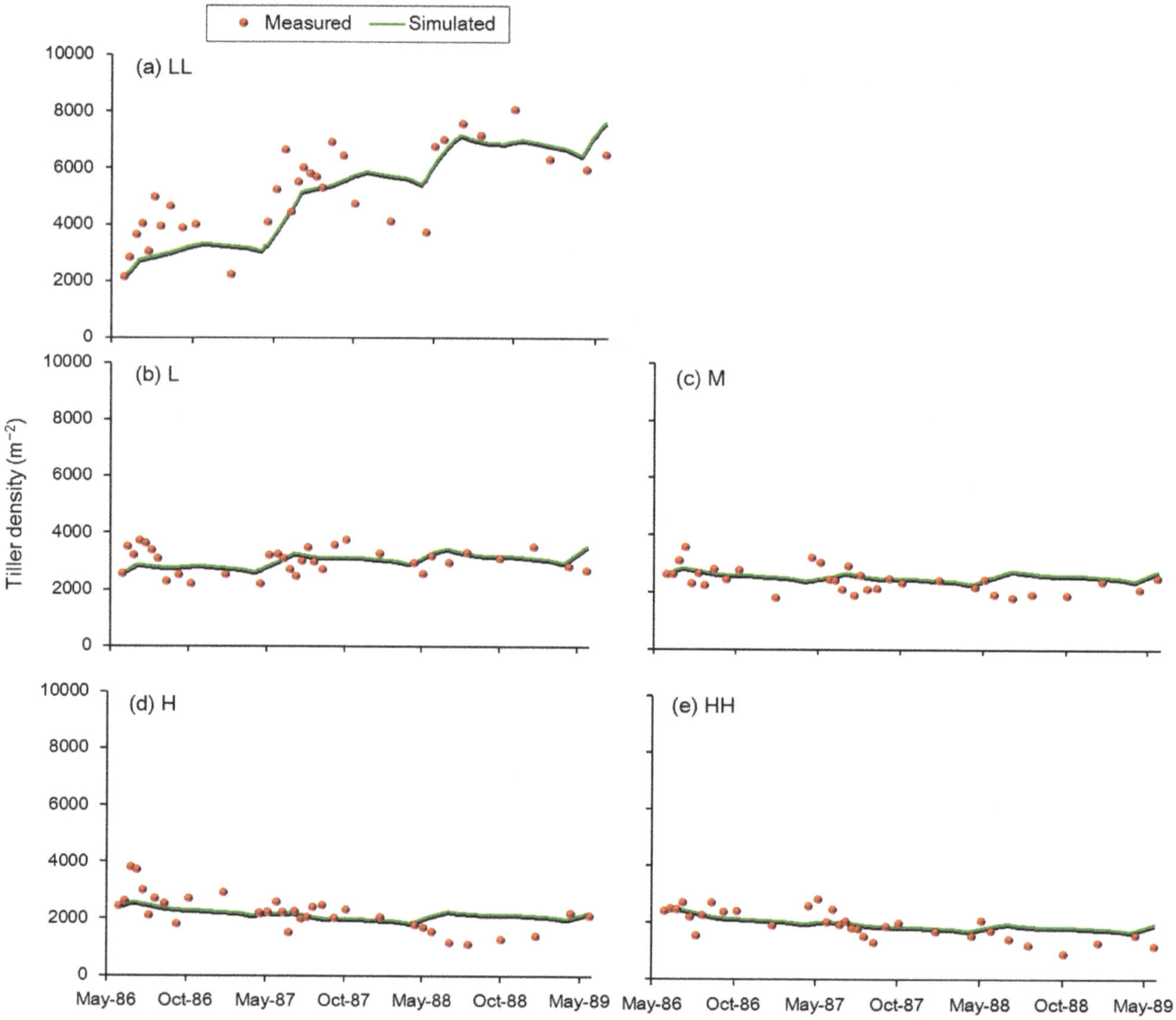


**Figure 8.** Measured (closed circle) and simulated (line) tiller densities in bahiagrass swards under different defoliation intensities (validation results). Measured data derive from Hirata [10]. LL = 2 cm, L = 7 cm, M = 12 cm, H = 17 cm, HH = 22 cm of cutting height above ground. Reproduced with permission from Hirata [22]; published by Research Signpost.

The simulations predicted gradual decrease in tiller density from the initial value of 4000 $m^{-2}$ under cutting to a medium (12 cm) or high (22 cm) height, irrespective of cutting intervals ranging between 2 and 6 weeks (Figure 9). By contrast, tiller density was predicted to increase with time under a low defoliation height (2 cm), with steeper increases under more frequent defoliation.

Further simulations identified defoliation regimes (frequency and height) required for stabilizing tiller density at a low (2500 $m^{-2}$), medium (4000 $m^{-2}$) or high (6000 $m^{-2}$) level over 3 years, *i.e.*, sustainable use of a sward (Figure 10). The model predicted that the high tiller density can be maintained under conditions ranging from weekly cuttings to a 4 cm height to 8-weekly cuttings to 2 cm, while the low density can be maintained under conditions ranging from weekly cuttings to 19 cm to 8-weekly cuttings to 12 cm. The model also showed that maintaining the medium tiller density requires defoliation with intermediate intensities ranging from 8 cm at weekly cuttings to 4 cm at 8-weekly cuttings. Defoliation height needed decreased as the target tiller density increased and *vice versa*.

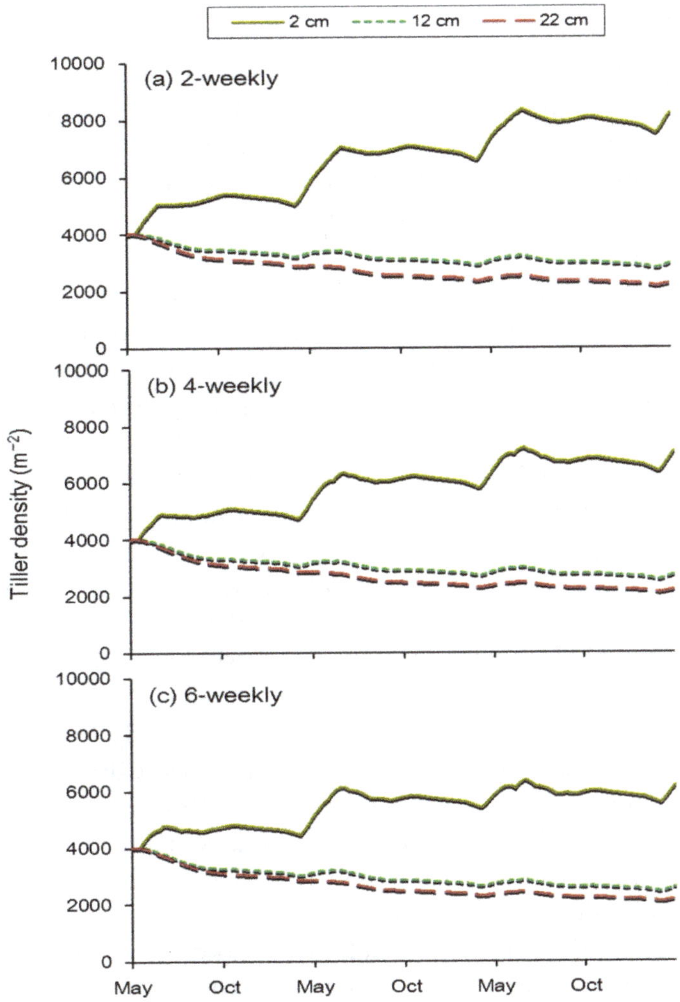


**Figure 9.** Predicted changes in tiller density in bahiagrass swards under various combinations of defoliation frequencies and intensities (heights above ground) at an annual N rate of 10 g $m^{-2}$. The initial tiller density (on 1 May) is 4000 $m^{-2}$.

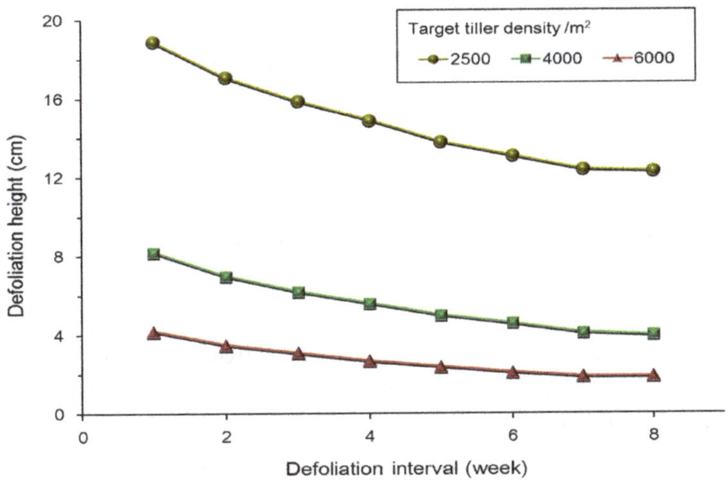


**Figure 10.** Predicted height (above ground) and interval of defoliation required for maintaining constant tiller density in bahiagrass swards at 2500 (○), 4000 (□) and 6000 (Δ) $m^{-2}$ at an annual N rate of 10 g $m^{-2}$.

## 5. Discussion

Maintaining plant population density is crucial to sustainable use of grasslands for agricultural production, conservation of the environment and wildlife, and other purposes such as recreation and amenity. It has been reported that tiller density in grass swards exhibits considerable response to the environment and management [10,18,21,22,29–31]. It is therefore important to develop a model which can predict tiller density dynamics under varying environmental and management conditions.

Previous models of tiller density dynamics in grass swards described the production of new tillers (tillering) in various ways. Some defined the tiller formation rate (per unit land area or per existing tiller) directly as a function of the nutrient state in the plants (amount and concentration of assimilate, carbon (C) and N) [32,33] and leaf area index (LAI) of the sward canopy (an index of self-shading) [33]. Others described the relative rate of tiller appearance mechanistically, as a product of the leaf appearance rate and the rate of site filling (Equation (4)), and expressed the former as a function of temperature [34,35] and the latter as a function of time after defoliation [34], LAI or light transmission at ground level [34,35] and availability (amount) of C and N within the plants [34,35]. These models described the death of tillers also in various ways, relating tiller mortality (relative rate of tiller death) to the developmental stage of tillers [32], LAI [33,34], thermal time [35], assimilate supply and C and N reserves [35] and self-thinning [35].

The current model differs from the previous models mainly in that it does not include the internal nutrient state of the plants or LAI as factors influencing tiller appearance and death (Figure 1), although the positive effect of plant N and negative effect of LAI (self-shading) on tillering can be achieved indirectly by the N-dependent standard relationships between tiller density and tiller weight (Figure 4 and Equations (8)–(10)). Because of the lack of the internal nutrient state, the present model cannot predict a decrease in tillering or an increase in tiller mortality caused by the nutrient limitations within the plants, which may take place when supply of nutrients via photosynthesis and uptake from the soil are restricted due to the deficiency in soil moisture and nutrients, *i.e.*, the model can be used when the growth of plants is not limited by either water or nutrients. The model is also unable to respond flexibly to N management with varying times and rates of split applications. Incorporating the plant nutrient state into the model and linking the model to a herbage production and utilization model simulating the internal nutrient state as well as herbage mass should thus broaden the management and environmental conditions to which the model can be applied.

Furthermore, the current model uses the standard density–weight relationship (reverse form of the self-thinning rule) only for controlling tiller appearance, leaving it unused for controlling tiller death (Figure 1). This may be a reason why the simulations were not able to follow some of the drastic decreases in tiller density which often followed an increase in mid-to-late spring and/or early-to-mid summer (Figures 6 and 8). Although a previous analysis reported a poor association between the relative tiller death rate and the actual:expected ratio in tiller density [21,22], their relationship may need to be reanalyzed in more detail to find a mechanism which can be incorporated into the model.

Despite the limitations discussed above, the present model can provide important information on how tiller population density changes in response to the environment and management (Figures 6–10) based on a simple and mechanistic structure. It can therefore be concluded that the model is of potential value as a prototype submodel for an integrated model of sward dynamics in grasslands. Refinement of the model needs to trade simplicity for complexity.

**Acknowledgments:** The author would like to thank Wempie Pakiding for his contribution to data collection in the field and Cory Matthew for his encouragement during manuscript preparation.

## References


1. Chapman, D.F.; Lemaire, G. Morphogenetic and structural determinants of plant regrowth after defoliation. In Proceedings of the XVII International Grassland Congress, Palmerston North, Hamilton, Lincoln and Rockhampton, New Zealand and Australia, 8–21 February 1993; pp. 95–104.
2. Lemaire, G.; Chapman, D. Tissue flows in grazed plant communities. In *The Ecology and Management of Grazing Systems*; Hodgson, J., Illius, A.W., Eds.; CAB International: Wallingford, UK, 1996; pp. 3–36.
3. Matthew, C.; Agnusdei, M.G.; Assuero, S.G.; Sbrissia, A.F.; Scheneiter, O.; da Silva, S.C. State of knowledge in tiller dynamics. In Proceedings of the 22nd International Grassland Congress, Sydney, Australia, 15–19 September 2013; New South Wales Department of Primary Industry: Orange, New South Wales, Australia, 2013; pp. 1041–1044.
4. Skerman, P.J.; Riveros, F. *Tropical Grasses*; FAO: Rome, Italy, 1989; pp. 571–575.
5. Hirata, M.; Ogawa, Y.; Koyama, N.; Shindo, K.; Sugimoto, Y.; Higashiyama, M.; Ogura, S.; Fukuyama, K. Productivity of bahiagrass pastures in south-western Japan: Synthesis of data from grazing trials. *J. Agron. Crop Sci.* **2006**, *192*, 79–91. [CrossRef]
6. Beaty, E.R.; Brown, R.H.; Morris, J.B. Response of Pensacola bahiagrass to intense clipping. In Proceedings of the XI International Grassland Congress, Surfers Paradise, Australia, 13–23 April 1970; University of Queensland Press: St. Lucia, Queensland, Australia, 1970; pp. 538–542.
7. Beaty, E.R.; Engel, J.L.; Powell, J.D. Yield, leaf growth, and tillering in bahiagrass by N rate and season. *Agron. J.* **1977**, *69*, 308–311. [CrossRef]
8. Stanley, R.L.; Beaty, E.R.; Powell, J.D. Forage yield and percent cell wall constituents of Pensacola bahiagrass as related to N fertilization and clipping height. *Agron. J.* **1977**, *69*, 501–504. [CrossRef]
9. Hirata, M.; Ueno, M. Response of bahiagrass (*Paspalum notatum* Flügge) sward to cutting height. 1. Dry weight of plant and litter. *J. Jpn. Grassl. Sci.* **1993**, *38*, 487–497.
10. Hirata, M. Response of bahiagrass (*Paspalum notatum* Flügge) sward to cutting height. 3. Density of tillers, stolons and primary roots. *J. Jpn. Grassl. Sci.* **1993**, *39*, 196–205.
11. Hirata, M. Response of bahiagrass (*Paspalum notatum* Flügge) sward to nitrogen fertilization rate and cutting interval. 1. Dry weight of plant and litter. *J. Jpn. Grassl. Sci.* **1994**, *40*, 313–324.
12. Pakiding, W.; Hirata, M. Tillering in a bahia grass (*Paspalum notatum*) pasture under cattle grazing: Results from the first two years. *Trop. Grassl.* **1999**, *33*, 170–176.
13. Hirata, M. Effects of nitrogen fertiliser rate and cutting height on leaf appearance and extension in bahia grass (*Paspalum notatum*) swards. *Trop. Grassl.* **2000**, *34*, 7–13.
14. Pakiding, W.; Hirata, M. Leaf appearance, death and detachment in a bahia grass (*Paspalum notatum*) pasture under cattle grazing. *Trop. Grassl.* **2001**, *35*, 114–123.
15. Hirata, M.; Pakiding, W. Tiller dynamics in a bahia grass (*Paspalum notatum*) pasture under cattle grazing. *Trop. Grassl.* **2001**, *35*, 151–160.
16. Hirata, M.; Pakiding, W. Dynamics in tiller weight and its association with herbage mass and tiller density in a bahia grass (*Paspalum notatum*) pasture under cattle grazing. *Trop. Grassl.* **2002**, *36*, 24–32.
17. Hirata, M.; Pakiding, W. Dynamics in lamina size in a bahia grass (*Paspalum notatum*) pasture under cattle grazing. *Trop. Grassl.* **2002**, *36*, 180–192.
18. Pakiding, W.; Hirata, M. Effects of nitrogen fertilizer rate and cutting height on tiller and leaf dynamics in bahiagrass (*Paspalum notatum* Flügge) swards: Tiller appearance and death. *Grassl. Sci.* **2003**, *49*, 193–202.
19. Pakiding, W.; Hirata, M. Effects of nitrogen fertilizer rate and cutting height on tiller and leaf dynamics in bahiagrass (*Paspalum notatum* Flügge) swards: Leaf appearance, death and detachment. *Grassl. Sci.* **2003**, *49*, 203–210.
20. Pakiding, W.; Hirata, M. Effects of nitrogen fertilizer rate and cutting height on tiller and leaf dynamics in bahiagrass (*Paspalum notatum* Flügge) swards: Leaf extension and mature leaf size. *Grassl. Sci.* **2003**, *49*, 211–216.
21. Hirata, M.; Pakiding, W. Tiller dynamics in bahia grass (*Paspalum notatum*): An analysis of responses to nitrogen fertiliser rate, defoliation intensity and season. *Trop. Grassl.* **2004**, *38*, 100–111.
22. Hirata, M. Canopy dynamics in bahia grass (*Paspalum notatum*) swards. In *Recent Research Developments in Crop Science*; Pandalai, S.G., Ed.; Research Signpost: Kerala, India, 2004; Volume 1, pp. 117–145.

23. Hirata, M. Modelling tiller density dynamics in a grass sward. In *XX International Grassland Congress: Offered Papers*; O'Mara, F.P., Wilkins, R.J., t'Mannetje, L., Lovett, D.K., Rogers, P.A.M., Boland, T.M., Eds.; Wageningen Academic Publishers: Wageningen, The Netherlands, 2005; p. 870.
24. Davies, A. Leaf tissue remaining after cutting and regrowth in perennial ryegrass. *J. Agric. Sci. Camb.* **1974**, *82*, 165–172. [CrossRef]
25. Thomas, H. Terminology and definitions in studies of grassland plants. *Grass Forage Sci.* **1980**, *35*, 13–23. [CrossRef]
26. Yoda, K.; Kira, T.; Ogawa, H.; Hozumi, K. Intraspecific competition among higher plants. XI. Self-thinning in overcrowded pure stands under cultivated and natural conditions. *J. Biol. Osaka City Univ.* **1963**, *14*, 107–129.
27. Hirata, M. Quantifying spatial heterogeneity in herbage mass and consumption in pastures. *J. Range Manag.* **2000**, *53*, 315–321. [CrossRef]
28. Hirata, M. Estimating herbage and leaf utilization in bahiagrass (*Paspalum notatum* Flügge) swards from height measurements. *Grassl. Sci.* **2002**, *48*, 105–109.
29. Korte, C.J. Tillering in 'Grasslands Nui' perennial ryegrass swards. 2. Seasonal pattern of tillering and age of flowering tillers with two mowing frequencies. *N. Z. J. Agric. Res.* **1986**, *29*, 629–638. [CrossRef]
30. Bullock, J.M.; Hill, B.C.; Silvertown, J. Tiller dynamics of two grasses—Response to grazing, density and weather. *J. Ecol.* **1994**, *82*, 331–340. [CrossRef]
31. Sbrissia, A.F.; da Silva, S.C.; Sarmento, D.O.L.; Molan, L.K.; Andrade, F.M.E.; Gonçalves, A.C.; Lupinacci, A.V. Tillering dynamics in palisadegrass swards continuously stocked by cattle. *Plant Ecol.* **2010**, *206*, 349–359. [CrossRef]
32. Dayan, E.; van Keulen, H.; Dovrat, A. Tiller dynamics and growth of Rhodes grass after defoliation: A model named TILDYN. *Agro-Ecosystems* **1981**, *7*, 101–112. [CrossRef]
33. Coughenour, M.B.; McNaughton, S.J.; Wallace, L.L. Simulation study of East-African perennial graminoid responses to defoliation. *Ecol. Model.* **1984**, *26*, 177–201. [CrossRef]
34. Schapendonk, A.H.C.M.; Stol, W.; van Kraalingen, D.W.G.; Bouman, B.A.M. LINGRA, a sink/source model to simulate grassland productivity in Europe. *Eur. J. Agron.* **1998**, *9*, 87–100. [CrossRef]
35. Soussana, J.F.; Oliveira Machado, A. Modelling the dynamics of temperate grasses and legumes in cut mixtures. In *Grassland Ecophysiology and Grazing Ecology*; Lemaire, G., Hodgson, J., de Moraes, A., de F. Carvalho, P.C., Nabinger, C., Eds.; CABI Publishing: Wallingford, UK, 2000; pp. 169–190.

# 15

# Carbon Assimilation, Biomass Partitioning and Productivity in Grasses

**Louis J. Irving**

Faculty of Life and Environmental Science, University of Tsukuba, 1-1-1 Tennodai, Tsukuba 305-8572, Japan; irving.louis.fb@u.tsukuba.ac.jp

Academic Editor: Cory Matthew

**Abstract:** Plant growth correlates with net carbon gain on a whole plant basis. Over the last several decades, the driving factors shaping plant morphology and performance have become increasingly clear. This review seeks to explore the importance of these factors for grass performance. Briefly, these fall into factors influencing photosynthetic rates directly, competition between plants in a canopy, and nutrient status and availability.

**Keywords:** carbon; nitrogen; biomass partitioning; shoot-root allocation

---

## 1. Introduction

Plant growth correlates with net carbon (C) gain on a whole plant basis [1]. Photosynthetic rates vary between species, but leaf nitrogen (N) content and light intensity are known to be major determinants [2]. The other main factor determining plant carbon gain is leaf area per plant, which is a function of mean leaf area and leaf number. In grasses, the production of secondary stems by tillering can greatly increase leaf number per plant, with tiller numbers appearing to be mediated by plant nutrient status, light availability and competition for light between plants in a stand.

A recent review [3] looked at the factors controlling the allocation of biomass to leaves, stems, and roots in a wide range of species. The most important factors were determined to be the quantity of light available (daily photon irradiance), nutrient and water availability, temperature, and plant age/size. However, two of these, water availability and temperature, tended to have large effects only at extremes—very low temperatures, and either strong water deficits or waterlogging conditions. Thus, for most plants, light intensity and nutrient availability seem to be the primary factors driving carbon acquisition and biomass allocation. While Poorter's review covered a huge number of studies, unfortunately they could not cover physiological mechanisms in great detail. The purpose of this review is to cover some of the mechanisms controlling plant performance, particularly in the areas of photon capture by shoots and resource acquisition by roots, with a focus on grasses.

Grasses (family Poaceae) are a plant group of unique importance in agriculture. Grasses contribute more than half the calories consumed worldwide—both directly, as cereals, and as forage grasses that form the basis for the production of meat and milk worldwide [4]. Thus, a complete understanding of the factors determining the productivity of grass plants is of fundamental importance in meeting the food needs of the growing human population. Cultivated and semi-natural grasslands cover an area estimated at 52 million km$^2$ [5], and account for approximately 15% of global primary productivity [4]. Grass architecture is based around a tiller axis comprised of a stack of phytomers with an apical meristem responsible for the production of new phytomers. Each phytomer has the capacity to produce one leaf, one tiller, and one or more roots. The tiller axis of vegetative grasses is generally small, and is often located around ground level, while the leaves grow from the base, as adaptations to prevent grazing damage to the meristems. Perennial ryegrass (*Lolium perenne* L.) has on average three leaves per tiller, with the initiation of new leaves coinciding with the senescence of the fourth leaf [6], while cereal grasses may have a much higher number of leaves.

Under competitive conditions, we expect plants to pursue a strategy where they use resources such as N and water to maximise growth and ultimately reproduction. Photoassimilates represent both the substrate for growth, and also energy storage for biochemical activities. The partitioning of these resources can have significant implications for plant productivity. For example, generally in grasses, around 80%–85% of plant biomass is partitioned to aboveground organs, such as leaves and stems, with 15%–20% allocated to roots. However, root mass allocation can be significantly affected by nutrient or light availability [3], with greater carbon allocation to the root system under low nutrient, high-light conditions. Even within the root system, the distribution of roots is frequently non-uniform, with localized root proliferation common in nutrient rich patches [7]. This proliferation produces root biomass in excess of that required to fully exploit the nutrient-rich patch, but is thought to increase nutrient capture rates in competitive environments, allowing such plants to outcompete those producing fewer roots, and thus come to dominate the ecosystem. It is worth stressing that the optimal biomass allocation will depend heavily on environmental constraints.

Thornley [8] developed an early model of vegetative plant growth and biomass allocation based upon carbon and nitrogen uptake and assimilation (Figure 1). Thornley's model holds that carbon fixed in the shoot is either used in the shoot or transported to the roots. Similarly, nitrogen absorbed by the roots can either be used directly *in situ*, or transported to the shoot. Although over the subsequent decades many more complex models have been developed [9–14], Thornley's model provides a clear, simple starting point from which to develop a more comprehensive understanding of biomass allocation in grasses. Within each compartment, carbon can be used either for maintenance or growth [15]. In the case of roots, carbon from the stem can be respired to provide energy for metabolic processes such as nutrient uptake and assimilation, or used in growth to explore the soil for further nutrients. Meanwhile, nitrogen in excess of that required by roots can be transported to the shoots for use there. Thus, in order to understand the partitioning of carbon within plants, we need to understand the factors which determine the size of the carbon/nitrogen pool and its usage in each organ. While quantitative descriptions of factors driving carbon allocation are becoming increasingly robust [3], there remains much to be done on understanding the underlying mechanisms in context. This paper aims to outline some of these processes.

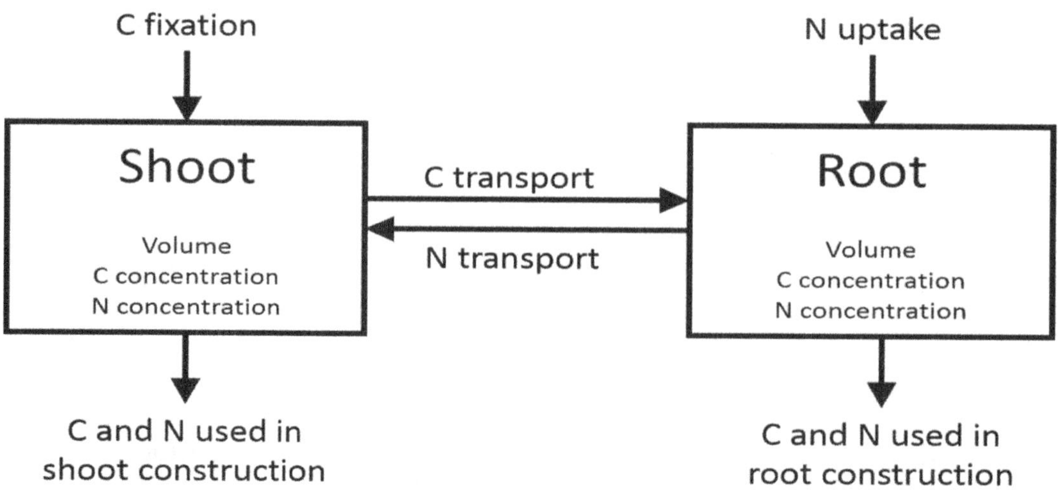


**Figure 1.** Two pool model of shoot and root growth. C and N status in each pool influences the volume of the pool. Adapted from [8].

## 2. Carbon Fixation

Given the primacy of photosynthesis in plant growth, it seems logical to first discuss factors influencing photosynthetic rate. However, given the size and complexity of the subject area, this is not intended to be a comprehensive analysis of photosynthesis, or of the factors controlling photosynthetic rates. Rather, the goal is to briefly explore a few of the more important factors, first on an individual plant basis, and then for competing plants in a later section.

At a biochemical level, the inefficiencies of C3 photosynthesis are well understood, and mainly center on the enzyme Rubisco [16]. The $CO_2$ fixing enzyme, ribulose-1,5-bisphosphate carboxylase/oxygenase (Rubisco; EC4.1.1.39), has a very low catalytic efficiency; thus requiring the investment of 20%–30% of leaf nitrogen in Rubisco [17]. Given its inefficiency, and the relatively high proportion of leaf nitrogen for which Rubisco accounts, a strong positive correlation between photosynthetic rates and leaf nitrogen concentration has been noted [18,19]. As we shall discuss, this requirement for high concentrations of Rubisco in photosynthesizing leaves limits productivity, and has profound implications for plant form and productivity.

Photorespiration is the process by which Rubisco fixes oxygen rather than $CO_2$, with the resulting production of $CO_2$ and ammonia, and represents a second major inefficiency of C3 photosynthesis—and thus plant productivity—with up to one-third of light energy intercepted by the plant used in photorespiration. Photosynthesis and photorespiration are competitive, with an increase in $CO_2$ or a decrease in $O_2$ having equivalent effects of increasing carbon fixation. Indeed, classical studies show that growing *Phaseolus vulgaris* L. at reduced oxygen levels led to a 40% increase in photosynthetic $CO_2$ uptake [20], while a 30% increase in grain yield was noted in rice plants grown at elevated $CO_2$ [21]. Thus, it is generally assumed that increasing atmospheric $CO_2$ levels will likewise lead to a decrease in photorespiration, and an increase in plant productivity. However, recent evidence suggests that malate formed during photorespiration is an important source of energy for the reduction of nitrate to nitrite—the first step in nitrate assimilation. Free air carbon dioxide enrichment (FACE) experiments suggest that future increases in atmospheric $CO_2$ levels may suppress photorespiration, leading to grain protein levels between 7% and 10% lower than current conditions in $C_3$ plants—with negative consequences for human nutrition and health globally [22,23]. However, this is strongly species-specific; in loblolly pine (*Pinus taeda* L.) saplings fed nitrate as their sole N source, $CO_2$ enrichment led to a suppression of growth compared to plants fed ammonium, while in the same study, neither elevated $CO_2$ levels nor nitrogen form had much of an effect on wheat (*Triticum aestivum* L.) growth [24]. While some studies have demonstrated acclimatory decreases in leaf photosynthesis as a result of elevated $CO_2$ levels [25], others have noted increases in leaf photosynthetic rates [26], which may be linked to decreased stomatal conductance and improved water use efficiency. Clearly there remain significant questions about the photosynthetic and growth responses in $C_3$ grasses, and the importance of elevated $CO_2$ to future productivity.

$C_4$ plants use a biochemical pump to increase the $CO_2$ concentration around Rubisco, thus suppressing photorespiration. $C_4$ Rubisco has carboxylation rates up to double those of $C_3$ plants [27] with $C_4$ plants achieving superior photosynthetic nitrogen use efficiencies ($CO_2$ fixed per unit nitrogen) due to their reduced N requirement for photosynthesis. Maize (*Zea mays* L.) and sorghum (*Sorghum bicolor* L.) are agronomically important $C_4$ grasses, with both species used both for grain and as silages for animal fodder. However, given that approximately 46% of the estimated 10,000 grass species use $C_4$ photosynthesis [28], there must be many other $C_4$ grasses which would be suitable targets for improvement for forage, despite their generally low nutritive value. $C_4$ photosynthesis has evolved at least 24 times in the grasses [29], and there are significant efforts currently to engineer $C_4$ rice (*Oryza sativa* L.) with the goal of significant increases in grain yield [30]. $C_4$ plants significantly increase the $CO_2$ concentration around Rubisco, which in principle would allow the use of more efficient forms of Rubisco, such as those originating in cyanobacteria [31]. Cyanobacterial Rubisco has catalytic rates approximately four-times those of $C_3$ plants, and might drive up carbon fixation rates and growth, or at least reduce N requirements for photosynthesis. This might be useful in crops destined for biofuel production, but the low tissue N levels would provide poor nutritive ability for animals. A further benefit of developing $C_4$ agricultural species might be an improvement in stress tolerance. In sorghum, drought induced yield declines were reduced in plants grown under elevated $CO_2$ conditions, suggesting that future increases in atmospheric $CO_2$ conditions may alleviate drought stress, even in $C_4$ plants [32].

Rubisco overexpressing (125% of WT) rice plants exhibited 32% and 15% increased biomass when grown at 280 and 400 ppm $CO_2$ respectively, but no difference at 1200 ppm. In the same study, Rubisco antisense plants (35% of WT) exhibited significantly lower biomass levels, irrespective of $CO_2$ level [33]. However, the reductions in relative growth rate in the antisense plants were small compared to the large decrease in leaf Rubisco levels. Clearly Rubisco concentration and growth are inherently linked to the atmospheric $CO_2$ levels. Increases in $CO_2$ levels have been shown to cause a decrease in leaf Rubisco levels in Arabidopsis [34,35] and rice [36], although Rubisco contents in pea (*Pisum sativum* L.), spinach (*Spinacia oleracea* L.), and parsley (*Petroselinum crispum* Mill.) (all $C_3$) are relatively unaffected by $CO_2$ levels [37]. It has been hypothesized that decreased N investment in Rubisco may lead to an increase in photosynthetic nitrogen use efficiency, and greater biomass production [16]. In line with this, wheat ($C_3$) plants grown under elevated $CO_2$ were taller, produced more tillers, had lower leaf area ratio (leaf area per unit shoot mass), and at high N availabilities, had relatively larger root systems than plants grown under control $CO_2$ levels [38]. $CO_2$ levels had no effect on maize ($C_4$) growth in the same experiment, perhaps suggesting that future increases in atmospheric $CO_2$ levels may have a smaller effect on $C_4$ plant morphology than for $C_3$ plants. If these findings can be applied to forage grasses, reduced leaf area and protein content would suggest that increasing $CO_2$ levels may lead to significant declines in forage quality in $C_3$ grasses.

## 3. Plant Morphology as a Partial Consequence of Photosynthetic N Requirement

Plant N concentration decreases allometrically with plant size [39]. The minimum N concentration which allows maximum growth is termed the "critical N concentration", with plants grown at sub-optimal N levels exhibiting proportionately decreased growth rates. Although there are clear differences in critical N concentrations between $C_3$ and $C_4$ plants due to the lower N requirements of $C_4$ plants, the relationship between plant size and N concentration can be found in both groups, with the primary factor appearing to be a decrease in "leafiness" through plant development [14]. As the plant grows, over time a greater proportion of biomass must be allocated to stems and other support material. Thus, the decrease in N concentration with age may reflect an increase in shaded or senescent leaves over time, or a shift in shoot to root ratio such that C fixation rates are higher than N uptake rates. Hikosaka and Terashima [40] demonstrated that the optimal allocation of N into photosynthetic components varies with light and nitrogen availability. Under high light conditions, high levels of Rubisco are required to maximise photosynthesis. However, as the leaf ages and transitions to a subordinate canopy position shaded by younger leaves, the optimal balance shifts towards the degradation of Rubisco and the remobilization of N to new leaves. These patterns are exacerbated in N-deficient plants.

Given that there is a light gradient down the canopy, with older leaves shaded by younger leaves, plant nitrogen use efficiency can be increased by translocating N in excess of photosynthetic requirements from older, shaded leaves, to younger leaves which have an optimal canopy position [41]. Grasses undergo progressive senescence where the older leaves senesce and are replaced by younger leaves on a continual basis. Photosynthetic rates at ambient $CO_2$ levels scale with leaf Rubisco content [42], which increases rapidly during leaf emergence, reaches a peak around full expansion, and declines through until senescence [43,44]. However, even under high exogenous supply, approximately half of the nitrogen for new leaf growth comes from the remobilization of leaf proteins including Rubisco from older leaves [45]. Thus, Rubisco represents a key juncture between carbon and nitrogen metabolic pathways. Rubisco retention in older leaves may increase the photosynthetic capacity of that leaf; however, particularly under N limiting conditions, carbon fixation might have been higher had the N been remobilized to younger leaves. In this case, maximising leaf photosynthesis may lead to a decrease in carbon fixation at the plant level. Thus, we expect a balance between the maintenance of leaf photosynthesis, and nitrogen remobilization to new tissues. Nitrogen remobilization is known to be important in supporting the development of new leaves—in *Poa trivialis* L. and *Panicum maximum* (Jacq.), despite removal of the exogenous N supply, new leaves continued to be produced using N remobilized from older organs [46]. Although some loss of N must be expected in senescent biomass, N remobilization is clearly important in determining plant productivity, with a recent paper demonstrating significantly reduced growth in mutant rice plants with reduced capacity for Rubisco degradation, particularly under N limiting conditions [47]. Thus, we can postulate an optimal leaf protein turnover rate which maximises canopy rather than leaf photosynthesis. Given the importance of $CO_2$ availability to biomass production in Rubisco overexpressing rice, this optimum is likely environmentally modulated, or contingent on other factors, such as plant size. Development of grasses using high catalytic efficiency Rubisco may reduce the need for internal recycling of nitrogen, leading to a more uniform N allocation between leaves.

A recent paper [48] demonstrated a negative correlation between the maximum leaf Rubisco concentration and plant dry mass in glasshouse grown *Lolium perenne* plants. A second paper, comparing two long-leaved with two short-leaved *L. perenne* genotypes, presented evidence that leaf Rubisco content correlated strongly with leaf elongation duration, tiller weight, and leaf elongation rate, with a weaker correlation noted between leaf Rubisco content and plant dry mass [49]. However, these correlations disappeared when Rubisco levels were considered in terms of concentration rather than content. This suggests that leaf Rubisco concentration is relatively invariant, with plant productivity more strongly controlled by leaf size. If we assume that the maximum Rubisco concentration in leaf cells is essentially a constant, and that leaves grow in proportion to the available carbon and nitrogen resources, we might expect leaf size to be contingent upon photosynthetic rates and nitrogen remobilization within the plant as a whole.

In *Fallopia japonica* (Houtt.), as plant nitrogen content increases a greater proportion of biomass is allocated to leaves, while the specific leaf weight (leaf mass per unit area) decreases [10]. Even assuming that the net assimilation rate is invariant with N availability, this would necessarily lead to an increase in whole plant carbon gain and relative growth rate. Plant biomass has been shown to correlate strongly with intercepted radiation, with the gradient contingent on N availability [50]. In a study using short and long-leaved varieties of tall fescue, MacAdam *et al.* [51] demonstrated that while cell length and the relative leaf elongation rate were apparently genetically controlled, environmental factors such as temperature and light availability, as well as N supply, were important in determining the rate of leaf expansion. Plants receiving high N supply produced a significantly greater number of mesophyll cells, relative to epidermal cells [51]. This may yield thicker leaves, and given the importance of mesophyll cells for photosynthesis, might partially explain higher rates of photosynthesis in plants grown under high N conditions.

Canopy photosynthesis optimization models predict that under low light availabilities, canopies should have a low leaf area index (LAI; leaf area per unit ground area) which minimizes the self-shading of leaves. Under high light conditions, a greater number of leaves, and higher LAI, are predicted. However, as pointed out by Anten [52], optimized canopy models do not take into account the fact that LAIs which are optimal for canopies may not be evolutionarily stable for the individuals within those canopies, since low LAI plants can be outcompeted by plants with higher LAI, which have higher individual photosynthetic carbon gain. Thus, the optimal LAI for the plant is higher than the optimal value for the canopy [53]. However, there must similarly be a maximum LAI for plants, beyond which the carbon cost of increasing LAI is greater than the photosynthetic benefit. At a given LAI, in the absence of N fertilization, canopy photosynthesis in field grown tall fescue (*Festuca arundinacea* Schreb.) was 30% reduced relative to well-fertilized plants. Since the measurements were controlled for LAI, these differences were attributed to a direct N effect on photosynthesis [54]. Grass growth is known to be strongly seasonal, and even N-deficient plants grown in spring or summer achieved significantly higher LAI than N-replete plants grown in autumn [55]. Presumably this results from a combination of low temperature, which limits leaf elongation rate [56], and reduced day-length and light intensity limiting carbon fixation. Taken together, these results suggest that nitrogen availability directly influences shoot growth via LAI, with photosynthetic declines resulting from N-deficiency important only under otherwise optimal conditions.

Along with absolute leaf area, specific leaf area (SLA; leaf area per unit dry mass) can also varyand is an important factor in determining plant photosynthetic performance. Anten and Hirose [57] demonstrated that in *Xanthium canadense* (Mill.), SLA decreased with increasing light intensity, independent of nutrient availability. Similar results were noted in *Lolium perenne* with comparatively low SLA (160 $cm^2/g$) at high light, and higher values (500 $cm^2/g$) at low light levels (RA Carnevalli, personal communication). Poorter and Evans [58] showed that plants with high SLA had similar photosynthetic rates to low SLA plants on an area basis, but higher rates on a unit mass basis. At lower light intensities, high SLA plants tended to have higher photosynthetic rates per unit nitrogen, due to a greater leaf area production.

## 4. Nutrient and Water Availability

Nitrogen deficiency is a major factor limiting the growth and yield of agronomically important plants, with inorganic N fertilizers used as a cost-effective method to increase crop yields globally [60]. Nutrient capture by roots is a significant cost for plants, both in terms of soil exploration by root production and the C costs of root activities, which includes both nutrient uptake and assimilation. However, plant N content has a primary role in determining whole plant photosynthetic potential, and thus photosynthesis and productivity, and C investment in roots can thus be seen as a method of maximising future photosynthesis. Given the necessary allometric relationships between leaf area and root biomass, correlations between plant N uptake and shoot growth have been confirmed [60].

Maximum plant growth rates will be achieved when the partitioning of biomass facilitates sufficient nutrient uptake rates to match the rate of carbon supply by photosynthesis [9]. Long-term shifts in canopy photosynthesis (e.g., due to seasonal changes in light availability) have been shown to strongly affect the availability of carbohydrates for the production of leaves, roots, and daughter tillers [61]. Thomas and Davies [62] demonstrated that while the shoot mass of *Lolium perenne* was relatively unaffected by shading, root mass was positively correlated with mean daily insolation. The ability to control biomass partitioning is typically considered to be an adaptive response either as a mechanism to maximise relative growth rates by increasing photosynthetic carbon supply, or as a result of a shift in the site of competition: at low N availabilities, plants compete for nitrogen via the roots, while at higher N levels, light capture becomes more important, and biomass is partitioned towards production of leaves. Conversely, Thomas and Sadras [63] have suggested that rather than being carbon limited, most plants have a carbon surplus, and they posit root growth as a mechanism to balance carbon influx with respiratory usage. However, as well as these evolutionary explanations, we should consider more physiological mechanisms for these phenomena. In nitrogen starved wheat plants, approximately equal proportions of the carbon transported to the roots are incorporated into biomass, respired, and translocated back to the shoot as amino acids [64]. As nitrogen is supplied to roots, we might assume that C demand will increase, as a result of the respiratory and C-skeleton costs of amino acid synthesis. Assuming C supply to the roots to be relatively invariant over the short term, this might lead to decreased allocation of carbon for root production. In line with this, Scheible *et al.* [65] noted that nitrate accumulation in tobacco shoots was accompanied by a decrease in root sugar content, and that root sugar content correlated with root growth. As exogenous N concentration increases further, despite the ability of plant root cells to store nitrate in the vacuole [66], given the finite size of root storage capacity, we start to see transport of nitrate to the shoots, where it would be assimilated and used directly [67].

Nitrogen uptake in plants scales with root length, and has a strong genetic component [68]. In nitrogen and phosphate deprived plants, roots proliferate in nitrate, ammonium or phosphate rich patches to maximise nutrient capture [69]. A reanalysis of Drew's data suggested that while all the available nitrate could have been captured without proliferation, it increased the rate of nitrate capture, which would be important for plants in competitive environments [7]. However, root proliferation may be more important for ions with low soil mobility, such as ammonium and phosphate. In Arabidopsis, nitrate was shown to increase the length but not the number of lateral roots [70]. However, application of other N sources did not stimulate lateral root elongation, suggesting that nitrate was acting as a signal molecule, and its effect on root elongation was not directly as a nutrient. Conversely, very high levels of N supply tended to retard root growth and branching [70]. Ammonium patches also cause root proliferation, although the roots are shorter and much more highly branched when provided ammonium compared with nitrate [71]. This indicates separate mechanisms for root proliferation by nitrate and ammonium, which may have a functional significance, since nitrate has a diffusibility around 10-times higher than ammonium in soil [72].

Nitrogen availability is known to affect the allocation of biomass to the shoots or roots. However, the N form is rarely considered. Zerihun *et al.* [73] noted that there are significantly different C costs associated with assimilating nitrate, ammonium, or glutamine, and that the physical location and time of assimilation could also have effects. For example, malate derived from photorespiration acts as a source of reducing power for nitrate, where assimilation occurs in the leaves during the light period, without significantly affecting carbon fixation rates [23]. N assimilation costs have been postulated to have significant effects on biomass partitioning, and indeed fast-growing grass species have been shown to have lower nutrient assimilation costs than slower growing grass species [74], although the data did not support this hypothesis in *Phaseolus vulgaris* [73]. The factors determining the relative cost of N assimilation remain poorly understood, and further research is required to explore the importance of C-costs of nutrient assimilation in determining plant growth rate and productivity. Work in tobacco has suggested a strong positive relationship between shoot to root ratio and leaf nitrate concentration [65], however that relationship broke down under phosphate deficiency or when plants were supplied ammonium, rather than nitrate. Andrews *et al.* [75] conversely argued that protein concentration, rather than leaf nitrate levels, was primarily important in shoot growth rates. Furthermore, as well as abiotic factors noted here, root mass allocation can also be affected by biotic factors—hemiparasitic plant infection has been shown to cause an increase in host plant root mass allocation, presumably caused by nutrient abstraction [76].

It is generally understood that the majority of grass plant roots occur in the surface layers of the soil [77]. A well-developed root system has the capacity to mitigate local variability in resource variability, providing a continuous supply of water and nutrients to leaves, despite the heterogeneous nature of the soil environment. Plants require large amounts of water for photosynthesis, since drought suppresses stomatal aperture leading to the depletion of intercellular $CO_2$ and increased photorespiration. Furthermore, carbohydrate accumulation in tissues during stress [78] presumably leads to a reduction in their availability for growth, and has been shown to cause feedback repression of photosynthesis in *Phaseolus vulgaris* [79]. Thus, plants growing in dry environments may suffer retarded growth as a result of insufficient water availability. Deep roots comprise approximately 30% of the grass root system, and help to maintain access to water even during drought conditions [80]. Huang and Fu [77] noted that in tall fescue, when the surface layer of soil was allowed to dry out but deeper soil had adequate water, carbon allocation to the shoot was decreased by more than 40%, while root production increased in both the upper, drier layer, and the lower layer. Presumably the functional significance of this is to maximise water uptake; however, the physiological mechanism by which it is achieved remains unclear.

## 5. Competition

The leaf canopy is a dynamic system with a constant turnover of leaves, influenced by environmental conditions. Within a dense stand where plants are competing strongly for light, vertical growth of the plant would be an optimal strategy. However, in more open canopies, vertical growth increases LAI and self-shading. Thus tillering is a preferred strategy, as this increases the number of growing leaves, presumably facilitating a greater increase in whole plant growth rate than could be achieved by an increase in leaf elongation rate alone. Plant C/N ratio is thought to be important in controlling tiller initiation, with nitrogen replete, carbon deficient *Carex rostrata* (Stokes.) plants producing significantly more daughter stems than carbon replete, nitrogen deficient plants [81]. Similarly, Davies [82] presented data showing a strong correlation between tiller number and nitrogen application in *Lolium perenne*, while Gautier *et al.* [83] demonstrated that shading caused a decrease in both leaf appearance rates and tiller formation. Conversely, as discussed previously, wheat plants grown under elevated $CO_2$ conditions produced more tillers than plants grown under control conditions [38]. The red/far-red ratio (R/FR) of light reaching the base of the canopy is an important shade signal for plants, and mediates a range of plant responses such as increased apical dominance, accelerated flowering and early seed production, and reduced tiller formation. The R/FR of daylight is around 1.2, but this decreases in deep canopies as the red light is absorbed by chlorophyll, making R/FR a signal for canopy closure [84]. A recent paper, which imposed increased R/FR on wheat cultivars demonstrated a 13%–23% increase in the number of tillers produced [85], while reduced tillering was found in wheat plants grown at high densities or under shade conditions where R/FR was decreased to around 0.4 [86]. Similarly, Gautier *et al.* [83] demonstrated a decrease in tiller formation in *Lolium perenne* plants grown under reduced R/FR. R/FR ratio has also been suggested to be a factor in tiller senescence, where decreases in ground-level R/FR coincided with a decline in tiller number [87], although whether this is a causal relationship or simply a result of canopy closure remains unclear. Irrespective of the mechanism, the shift in growth strategy from the production of many stems (tillering) to the concentration of resources in fewer, larger stems as competition for light increases appears common in grasses.

Within grass canopies, we expect a negative relationship between plant density and mass [88]. At low planting densities, the growth of new tillers fills the space in the canopy, while at high densities plant dry weight decreases, presumably due to competitive effects. Using *Lolium perenne* plants sown at a range of differing densities, Lonsdale and Watkinson [89] reported that plant shoot dry mass followed a $-3/2$ relationship with plant density under high light conditions. However, when the plants were put under heavy shade conditions (83% reduction in light intensity) or when the tiller masses became very large, this relationship shifted from $-3/2$ to approximately $-1$. It is thought that at high plant masses, the stand reaches a maximum yield with no further increase in stand mass possible [90]. The $-1$ slope suggests that at some point resource usage is maximised (e.g., light interception nears 100% for the canopy), leading to a zero-sum game, where an increase in resource use by one plant is necessarily at the expense of a competing plant. This raises the question of whether light attenuation places a maximum limit on the quantity of photosynthetically active biomass produced per unit area. More recent work suggests that rather than $-3/2$, the true relationship between plant mass and density follows a $-4/3$ slope [91].

In Matthew *et al.* [92], plots under low grazing intensity tended to have fewer, larger tillers, and a higher overall biomass. Plots grazed at higher intensities tended to have a greater number of smaller tillers. Grazing led to a more open canopy, presumably prompting tiller initiation and facilitating establishment. Under high light conditions, grazing is thought to have similar implications for plant morphology and yield as growth under low light conditions, due to the reduced leaf area for light interception. Clear seasonal differences in tiller density were apparent, with increased tiller numbers during the summer, suggesting that higher light levels in the summer may act as a substitute for a more open canopy in *Lolium perenne*. Interestingly, there was also seasonality in biomass per unit area for any given tiller density, which was explained by an increase in leaf area index. This suggests environmental factors, other than light, influence the maximum potential biomass. Temperature and light quality both seem to be viable candidates with low temperatures linked to decreased leaf appearance rates [93], although other factors such as water availability and nitrogen availability are also important.

Competition for light between plants promotes an increase in plant height. Taller plants must invest more carbon in support structures and less in leaves. Anten and Hirose [57] demonstrated than *Xanthium canadense* plants grown in stands invested a lower proportion of their mass in leaves than plants grown singly. Increased height allows increased access to light, but comes at the cost of carbon invested in stems. Thus, there must be an optimal plant height at which the C cost of additional stem production is greater than any increases in light interception and C fixation resulting from the increase in plant height. In stands where we have individuals of different species, or unrelated members of the same species, we would expect competitive pressures to drive individuals towards this, presumably environmentally influenced, maximum. However, the situation for clonal plants or tillers of an individual plant remains unclear. As covered earlier, the optimal LAI for an individual plant is higher than that for the canopy. Similarly, we might expect the optimal tiller number per plant to be higher than the optimal tiller density for the canopy. Clearly, excessive competition between multiple stems of a single plant would be disadvantageous for the plant, and we might expect to see a reduction of LAI towards the canopy optimum in clonal stands. On the other hand, limiting competitive effects might be difficult, as it would require the plant to be able to differentiate between tillers and other plants, and to communicate these differences internally. Little work has been conducted on the competitive pressures between tillers in a single plant.

## 6. Carbon Allocation

Photoassimilate is the major substrate for plant growth, thus it is impossible to understand plant performance without a thorough understanding of the factors determining assimilate transport and partitioning [94]. Phloem is the primary pathway for the transport of carbohydrates from the sites of assimilation (*i.e.*, mature leaves) to the sites of utilization (growth and respiration). However, the direct study of phloem is difficult, requiring the use of radioisotopes with rapid decay rates (e.g., $^{11}C$) and expensive equipment, such as positron imaging systems. In the majority of studies, the allocation of carbon to organs is studied using stable or radioisotopes with slower decay rates (e.g., $^{13}C$ or $^{14}C$), and although this is less direct it has allowed us to build models of carbon allocation within plants.

Carbon flux rates are generally assumed to be a function of the difference in concentration between C-sources such as photosynthesizing leaves, and sinks such as growing leaves, roots or seeds [11,13]. While source strength is fairly easily understood as deriving from photosynthetic C fixation, sink strength is a less well-defined concept. One way to think about it may be as the capacity of phloem companion cells in the proximity of a meristem to unload sugars. An upper limit may be established by the ability of cells to use carbon during division, expansion and differentiation [13]. $^{14}C$ labelling studies show the preferential allocation of carbon to young developing roots, with the oldest roots receiving little from the shoots [95]. Given that the older, established roots have a greater biomass and more apical meristems, we might postulate that they have a greater sink capacity than the younger roots. If C allocation is sink-driven, it would be difficult to understand this allocation. It has been hypothesized that the utilization of photoassimilates by young roots close to the shoot renders the supply insufficient for more distant sinks (C Matthew, personal communication), or it may be that transport resistance between the leaf and the root apical meristem is a key factor determining carbon allocation. Assuming that transport of assimilates through the phloem follows Poiseuille's law, the resistance of C flow down the root would be proportional to the root's length. Thus, the root tips of longer roots might be expected to receive less photoassimilate than younger, shorter roots, due to their length rather than their age or axial position. Phloem sap viscosity is known to be an important limitation to C export from photosynthesizing *Pinus sylvestris* L. leaves [96], suggesting a similar phenomenon might occur in grass roots, although this remains speculative.

In *Lolium perenne* and *Paspalum dilatatum* (Poir.), 48%–64% of the carbon incorporated into new tissues was assimilated within the preceding 24 h [97]. In contrast, for nitrogen, these figures were 3%–17%, with over 80% of N having been supplied by an intermediate storage pool. This intermediate storage pool undoubtedly refers to internal remobilization from senescing to newly growing organs.

Carbon older than 2–3 days contributed little to new tissue growth, and shaded plants tended to grow more slowly, further reinforcing the close linkage between photosynthesis and growth rate. Short-term labelling studies show that photoassimilates can travel 2–3 cm per minute in *Vicia faba* L. [98], potentially allowing C to reach leaf growth zones within 10 to 20 min from assimilation, and reinforcing the close linkage between photosynthesis and growth. Similarly to new tissues, when we look at root exudates, we find that approximately 58% of the C in *L. perenne* root exudates derives from recent assimilation [99].

Reciprocal translocation of C and N between tillers appears to be common, and is a potential mechanism by which mature tillers can support the growth and development of daughter tillers. Daughter tillers represent a potential C source for root growth and development [100]. However, the degree of resource sharing between tillers appears to vary between species and varieties, and presumably has an environmental component. In *Panicum maximum*, young daughter tillers have a tight vascular connection with their parent and a high degree of photoassimilate sharing is apparent, while secondary tillers (those connected to primary tillers) receive virtually no carbon from the parent stem, suggesting a low degree of vascular continuity [100]. In *Poa trivialis*, removal of an exogenous N supply effectively stopped N remobilization from mature leaves on the main stem to developing daughter tillers [46]. Clearly, tillering has important implications for plant productivity, although factors influencing the degree of resource sharing between the main stem and daughter tillers remain poorly understood. Furthermore, since each tiller can be split off and regenerate a whole plant, this raises a philosophical question about what constitutes the plant, and whether tillers which remain physically attached but have lost vascular connection can be meaningfully said to be the same plant.

It is generally understood that photosynthate partitioning is "local" in grasses, with upper leaves feeding the developing panicle in reproductive wheat plants, while lower leaves provide carbon for roots [101]. Carbon fixed by primary tillers was allocated first to the tiller itself, and secondly to main stem axial roots. However, carbon fixed by secondary tillers tended to be retained almost wholly in those tillers, with secondary tillers having little role in supporting main stem roots [101]. Work on the interaction between tillers of different ages within a plant are at a rudimentary level, and the balance between cooperation and competition between stems in a single plant as it develops remains completely unstudied, yet clearly of enormous importance. Osaki *et al.* [102] has pointed out the need for continued carbon investment in roots as the plant matures to maximise plant production. As grasses shift from vegetative to reproductive growth, we see a shift in the relative mass of leaves as carbon is allocated to stem elongation [3]. As the tillers increase in size, it seems likely that the shading of older leaves deeper in the canopy leads to a decline in carbon flux to the roots and therefore root activity, and perhaps the characteristic drying out we see in maturing cereal crops.

## 7. Conclusions

While much progress has been made over the last few decades in certain aspects, a great deal of work remains to be done before a comprehensive understanding of the factors controlling plant carbon allocation and plant productivity emerges. Key questions remain regarding the potential diminishment of $C_3$ grass nutritive value under elevated $CO_2$ conditions, and whether C4 plants will gain in importance as a target for forage plant improvement due to their potentially superior stress tolerance under high $CO_2$ conditions. Recent work in rice has unequivocally demonstrated the importance of Rubisco turnover for plant growth, and that Rubisco turnover rates appear to have a high degree of genetic variability. However, linking leaf behaviour to whole plant performance has yielded little so far, and a more comprehensive approach to the problem seems necessary. The factors controlling tiller formation remain unclear, with most studies in the area relying heavily on correlation and few causative links demonstrated. Both red/far-red ratio and carbon/nitrogen balance appear important, but there is little clarity on the relative importance of these factors. Similarly, the drivers of carbon allocation to differing classes of roots remains a key question, which will probably require a combined modelling/experimental approach. Finally, cooperation and competition between tillers in a single plant for N and C remains almost completely unexplored, and represents an important avenue for future research efforts. Although many questions remain, much work has been done since the heyday of forage research in the 1970s. We continue to hope that a more complete understanding of plant carbon metabolism will help us develop higher-yielding, fertilizer-efficient plants.

**Acknowledgments:** The author would like to thank the editor and two anonymous referees for their insightful comments.

## References

1. Kruger, E.L.; Volin, J.C. Reexamining the empirical relationship between plant growth and leaf photosynthesis. *Funct. Plant Biol.* **2006**, *33*, 421–429. [CrossRef]
2. Wright, I.J.; Reich, P.B.; Westoby, M.; Ackerly, D.D.; Baruch, Z.; Bongers, F.; Cavender-Bares, J.; Chapin, T.; Cornelissen, J.H.C.; Diemer, M.; *et al.* The worldwide leaf economics spectrum. *Nature* **2004**, *428*, 821–827. [CrossRef] [PubMed]
3. Poorter, H.; Niklas, K.J.; Reich, P.B.; Oleksyn, J.; Poot, P.; Mommer, L. Biomass allocation to leaves, stems and roots: Meta-analysis of interspecific variation and environmental control. *New Phytol.* **2012**, *193*, 30–50. [CrossRef] [PubMed]
4. Raven, J.A.; Thomas, H. Grasses. *Curr. Biol.* **2010**, *20*, R837–R839. [CrossRef] [PubMed]
5. Rehuel, D.; de Cauwer, B.; Cougnon, M. The role of forage crops in multifunctional agriculture. In *Fodder Crops and Amenity Grasses*; Boller, B., Posselt, U.K., Veronesi, F., Eds.; Springer-Verlag: New York, NY, USA, 2010; pp. 1–12.
6. Fulkerson, W.J.; Donaghy, D.J. Plant-soluble carbohydrate reserves and senescence-key criteria for developing an effective grazing management system for ryegrass-based pastures: A review. *Aust. J. Exp. Agric.* **2001**, *41*, 261–275. [CrossRef]
7. Robinson, D. Resource capture by localized root proliferation: Why do plants bother? *Ann. Bot.* **1996**, *77*, 179–185. [CrossRef]
8. Thornley, J.H.M. A balanced quantitative model for root:shoot ratios in vegetative plants. *Ann. Bot.* **1972**, *36*, 431–441.
9. Robinson, D. Compensatory changes in the partitioning of dry matter in relation to nitrogen uptake and optimal variations in growth. *Ann. Bot.* **1986**, *58*, 841–848.
10. Hirose, T. A vegetative plant growth model: Adaptive significance of phenotypic plasticity in matter partitioning. *Funct. Ecol.* **1987**, *1*, 195–202. [CrossRef]
11. Durand, J.-L.; Varlet-Grancher, C.; Lemaire, G.; Gastal, F.; Moulia, B. Carbon partitioning in forage crops. *Acta Biotheor.* **1991**, *39*, 213–224. [CrossRef]
12. Sheehy, J.E.; Gastal, F.; Mitchell, P.L.; Durand, J.-L.; Lemaire, G.; Woodward, F.I. A nitrogen-led model of grass growth. *Ann. Bot.* **1996**, *77*, 165–177. [CrossRef]
13. Lemaire, G.; Millard, P. An ecophysiological approach to modelling resource fluxes in competing plants. *J. Exp. Bot.* **1999**, *50*, 15–28. [CrossRef]
14. Gastal, F.; Lemaire, G.; Durand, J.-L.; Louarn, G. Quantifying crop responses to nitrogen and avenues to improve nitrogen-use efficiency. In *Crop Physiology, Applications for Genetic Improvement and Agronomy*, 2nd ed.; Sadras, V.O., Calderini, D., Eds.; Academic Press/Elsevier: Oxford, UK, 2015; pp. 159–206.
15. Amthor, J.S. The role of maintenance respiration in plant growth. *Plant Cell Environ.* **1984**, *7*, 561–569.
16. Yamori, W.; Irving, L.J.; Adachi, S.; Busch, F.A. Strategies for optimizing photosynthesis with biotechnology to improve crop yield. *Handb. Photosynth.* **2016**, in press.
17. Galmés, J.; Kapralov, M.V.; Andralojc, P.J.; Conesa, M.À.; Keys, A.J.; Parry, M.A.J.; Flexas, J. Expanding knowledge of the Rubisco kinetics variability in plant species: Environmental and evolutionary trends. *Plant Cell Environ.* **2014**, *37*, 1989–2001. [CrossRef] [PubMed]
18. Evans, J.R. Photosynthesis and nitrogen relationships in leaves of $C_3$ plants. *Oecologia* **1989**, *78*, 9–19. [CrossRef]
19. Makino, A.; Sato, T.; Nakano, H.; Mae, T. Leaf photosynthesis, plant growth and nitrogen allocation in rice under different irradiances. *Planta* **1997**, *203*, 390–398. [CrossRef]

20. Parkinson, K.J.; Penman, H.L.; Tregunna, E.B. Growth of plants in different oxygen concentrations. *J. Exp. Bot.* **1974**, *25*, 132–145. [CrossRef]
21. Yoshida, S. Effects of $CO_2$ enrichment at different stages of panicle development on yield components and yield of rice (*Oryza sativa* L.). *Soil Sci. Plant Nutr.* **1973**, *19*, 311–316. [CrossRef]
22. Myers, S.S.; Zanobetti, A.; Kloog, I.; Huybers, P.; Leakey, A.D.B.; Bloom, A.J.; Carlisle, E.; Dietterich, L.H.; Fitzgerald, G.; Hasegawa, T.; *et al.* Increasing $CO_2$ threatens human nutrition. *Nature* **2014**, *510*, 139–142. [CrossRef] [PubMed]
23. Bloom, A.J. The increasing importance of distinguishing among plant nitrogen sources. *Curr. Opin. Plant Biol.* **2015**, *25*, 10–16. [CrossRef] [PubMed]
24. Bloom, A.J.; Asensio, J.S.R.; Randall, L.; Rachmilevitch, S.; Cousins, A.B.; Carlisle, E.A. $CO_2$ enrichment inhibits shoot nitrate assimilation in $C_3$ but not $C_4$ plants and slows growth under nitrate in C3 plants. *Ecology* **2012**, *93*, 355–367. [CrossRef] [PubMed]
25. Sicher, R.C.; Bunce, J.A. Relationship of photosynthetic acclimation to changes of Rubisco activity in field grown winter wheat and barley during growth in elevated carbon dioxide. *Photosynth. Res.* **1997**, *52*, 27–38. [CrossRef]
26. Garcia, R.L.; Long, S.P.; Wall, G.W.; Osborne, C.P.; Kimball, B.A.; Nie, G.Y.; Pinter, P.J., Jr.; LaMorte, R.L.; Wechsung, F. Photosynthesis and conductance of spring-wheat leaves: Field response to continuous free-air atmospheric $CO_2$ enrichment. *Plant Cell Environ.* **1998**, *21*, 659–669. [CrossRef]
27. Ghannoum, O.; Evans, J.R.; Chow, W.S.; Andrews, T.J.; Conroy, J.P.; von Caemmerer, S. Faster Rubisco is the key to superior nitrogen use efficiency in NADP-Malic enzyme relative to NAD-Malic enzyme $C_4$ grasses. *Plant Physiol.* **2005**, *137*, 638–650. [CrossRef] [PubMed]
28. Sage, R.F.; Li, M.; Monson, R.K. The taxonomic distribution of $C_4$ photosynthesis. In *$C_4$ Plant Biology*; Sage, R.F., Monson, R.K., Eds.; Academic Press: San Diego, USA, 1999; pp. 551–584.
29. Studer, R.A.; Christin, P.A.; Williams, M.A.; Orengo, C.A. Stability-activity tradeoffs constrain the adaptive evolution of RubisCO. *Proc. Natl. Acad. Sci. USA* **2014**, *111*, 2223–2228. [CrossRef] [PubMed]
30. Von Caemmerer, S.; Quick, W.P.; Furbank, R.T. The development of $C_4$ rice: Current progress and future challenges. *Science* **2012**, *336*, 1671–1672. [CrossRef] [PubMed]
31. Lin, M.T.; Occhialini, A.; Andralojc, P.J.; Parry, M.A.J.; Hanson, M.R. A faster Rubisco with potential to increase photosynthesis in crops. *Nature* **2014**, *513*, 547–550. [CrossRef] [PubMed]
32. Conley, M.M.; Kimball, B.A.; Brooks, T.J.; Pinter, P.J., Jr.; Hunsaker, D.J.; Wall, G.W.; Adam, N.R.; LaMorte, R.L.; Matthias, A.D.; Thompson, T.L.; *et al.* $CO_2$ enrichment increases water-use efficiency in sorghum. *New Phytol.* **2001**, *151*, 407–412. [CrossRef]
33. Sudo, E.; Suzuki, Y.; Makino, A. Whole plant growth and N utilization in transgenic rice plants with increased or decreased Rubisco content under different $CO_2$ partial pressures. *Plant Cell Physiol.* **2014**, *55*, 1905–1911. [CrossRef] [PubMed]
34. Cheng, S.H.; Moore, B.D.; Seemann, J.R. Effects of short and long-term elevated $CO_2$ on the expression of ribulose-1,5-bisphosphate carboxylase/oxygenase genes and carbohydrate accumulation in leaves and *Arabidopsis thaliana* (L.) Heynh. *Plant Physiol.* **1998**, *116*, 715–723. [CrossRef] [PubMed]
35. Moore, B.D.; Cheng, S.H.; Sims, D.; Seemann, J.R. The biochemical and molecular basis for photosynthetic acclimation to elevated atmospheric $CO_2$. *Plant Cell Environ.* **1999**, *22*, 567–582. [CrossRef]
36. Seneweera, S.; Makino, A.; Hirotsu, N.; Norton, R.; Suzuki, Y. New insight into the photosynthetic acclimation to elevated $CO_2$: The role of leaf nitrogen and ribulose-1,5-bisphosphate carboxylase/oxygenase content in rice leaves. *Environ. Exp. Bot.* **2011**, *71*, 128–136. [CrossRef]
37. Moore, B.D.; Cheng, S.H.; Rice, J.; Seemann, J.R. Sucrose cycling, Rubisco expression, and prediction of photosynthetic acclimation to elevated atmospheric $CO_2$. *Plant Cell Environ.* **1998**, *21*, 905–915. [CrossRef]
38. Hocking, P.J.; Meyer, C.P. Effects of $CO_2$ enrichment and nitrogen stress on growth, and partitioning of dry matter and nitrogen in wheat and maize. *Aust. J. Plant Physiol.* **1991**, *18*, 339–356. [CrossRef]
39. Gastal, F.; Lemaire, G. N uptake and distribution in crops: An agronomical and ecophysiological perspective. *J. Exp. Bot.* **2002**, *53*, 789–799. [CrossRef] [PubMed]
40. Hikosaka, K.; Terashima, I. A model of the acclimation of photosynthesis in the leaves of $C_3$ plants to sun and shade with respect to nitrogen use. *Plant Cell Environ.* **1995**, *18*, 605–618. [CrossRef]
41. Anten, N.P.R.; Schieving, F.; Werger, M.J.A. Patterns of light and nitrogen distribution in relation to whole canopy carbon gain in $C_3$ and $C_4$ mono- and dicotyledonous species. *Oecologia* **1995**, *101*, 504–513. [CrossRef]

42. Makino, A. Rubisco and nitrogen relations in rice: Leaf photosynthesis and plant growth. *Soil Sci. Plant Nutr.* **2003**, *49*, 319–327. [CrossRef]
43. Mae, T.; Makino, A.; Ohira, A. Changes in the amounts of ribulose bisphosphate carboxylase synthesized and degraded during the life-span of rice life (*Oryza sativa* L.). *Plant Physiol.* **1983**, *24*, 1079–1086.
44. Irving, L.J.; Robinson, D. A dynamic model of Rubisco turnover in cereal leaves. *New Phytol.* **2006**, *169*, 493–504. [CrossRef] [PubMed]
45. Mae, T.; Ohira, K. The relationship of nitrogen related to leaf growth and senescence in rice plants (*Oryza sativa* L.). *Plant Cell Physiol.* **1981**, *22*, 1067–1074.
46. Santos, P.M.; Thornton, B.; Corsi, M. Nitrogen dynamics in the intact grasses *Poa trivialis* and *Panicum maximum* receiving contrasting supplies of nitrogen. *J. Exp. Bot.* **2002**, *53*, 2167–2176. [CrossRef] [PubMed]
47. Wada, S.; Hayashida, Y.; Izumi, M.; Kurusu, T.; Hanamata, S.; Kanno, K.; Kojima, S.; Yamaya, T.; Kuchitsu, K.; Makino, A.; *et al.* Autophagy Supports Biomass Production and Nitrogen Use Efficiency at the Vegetative Stage in Rice. *Plant Physiol.* **2015**, *168*, 60–73. [CrossRef] [PubMed]
48. Khaembah, E.N.; Irving, L.J.; Thom, E.R.; Faville, M.J.; Easton, H.S.; Matthew, C. Leaf Rubisco turnover in a perennial ryegrass (*Lolium perenne* L.) mapping population: Genetic variation, identification of associated QTL, and correlation with plant morphology and yield. *J. Exp. Bot.* **2013**, *64*, 1305–1316. [CrossRef] [PubMed]
49. Khaembah, E.N.; Gastal, F.; Carre, S.; Irving, L.J.; Barre, P.; Matthew, C. Morphology and Rubisco turnover characteristics of perennial ryegrass breeding populations after two and four cycles of divergent selection for long or short leaf length. *Crop Pasture Sci.* **2013**, *64*, 687–695. [CrossRef]
50. Bélanger, G.; Gastal, F.; Lemaire, G. Growth analysis of a tall fescue sward fertilised with different rates of nitrogen. *Crop Sci.* **1992**, *6*, 1371–1376. [CrossRef]
51. MacAdam, J.W.; Volenec, J.J.; Nelson, C.J. Effects of nitrogen on mesophyll cell division and epidermal cell elongation in tall fescue leaf blades. *Plant Physiol.* **1989**, *89*, 549–556. [CrossRef] [PubMed]
52. Anten, N.P.R. Optimal photosynthetic characteristics of individual plants in vegetation stands and implications for species coexistence. *Ann. Bot.* **2005**, *95*, 495–506. [CrossRef] [PubMed]
53. Anten, N.P.R. Evolutionarily stable leaf area production in plant populations. *J. Theor. Biol.* **2002**, *217*, 15–32. [CrossRef] [PubMed]
54. Gastal, F.; Bélanger, G. The effects of nitrogen fertilization and the growing season on photosynthesis of field-grown tall fescue (*Festuca arundinacea* Shreb.) canopies. *Ann. Bot.* **1993**, *72*, 401–408. [CrossRef]
55. Bélanger, G.; Gastal, F.; Warembourg, F.R. Carbon balance of tall fescue (*Festuca arundinacea* Shreb.): Effects of nitrogen fertilization and the growing season. *Ann. Bot.* **1994**, *74*, 653–659. [CrossRef]
56. Gastal, F.; Bélanger, G.; Lemaire, G. A model of the leaf expansion rate of tall fescue in response to nitrogen and temperature. *Ann. Bot.* **1992**, *70*, 437–442.
57. Anten, N.P.R.; Hirose, T. Biomass allocation and light partitioning among dominant and subordinate individuals in *Xanthium canadense* stands. *Ann. Bot.* **1998**, *82*, 665–673. [CrossRef]
58. Poorter, H.; Evans, J.R. Photosynthetic nitrogen-use efficiency of species that differ inherently in specific leaf area. *Oecologia* **1998**, *116*, 26–37. [CrossRef]
59. Andrews, M.; Raven, J.A.; Lea, P.J. Do plants need nitrate? The mechanisms by which nitrogen form affects plants. *Ann. Appl. Biol.* **2013**, *163*, 174–199. [CrossRef]
60. Lemaire, G.; van Oosterom, E.; Sheehy, J.; Jeuffroy, M.H.; Massignam, A.; Rossato, L. Is crop N demand more closely related to dry matter accumulation or leaf area expansion during vegetative growth? *Field Crops Res.* **2007**, *100*, 91–106. [CrossRef]
61. Hikosaka, K. Leaf canopy as a dynamic system: Ecophysiology and optimality in leaf turnover. *Ann. Bot.* **2005**, *95*, 521–533. [CrossRef] [PubMed]
62. Thomas, H.; Davies, A. Effect of shading on the regrowth of *Lolium perenne* swards in the field. *Ann. Bot.* **1977**, *42*, 705–715. [CrossRef]
63. Thomas, H.; Sadras, V.O. The capture and gratuitous disposal of resources by plants. *Funct. Ecol.* **2001**, *15*, 3–12.
64. Lambers, H.; Simpson, R.J.; Beilharz, V.C.; Dalling, M.J. Translocation and utilization of carbon in wheat (*Triticum aestivum*). *Physiol. Plant.* **1982**, *56*, 18–22. [CrossRef]
65. Scheible, W.R.; Lauerer, M.; Schulze, E.D.; Caboche, M.; Stitt, M. Accumulation of nitrate in the shoot acts as a signal to regulate shoot–root allocation in tobacco. *Plant J.* **1997**, *11*, 671–691. [CrossRef]

66. Miller, A.J.; Fan, X.; Orsel, M.; Smith, S.J.; Wells, D.M. Nitrate transport and signalling. *J. Exp. Bot.* **2007**, *58*, 2297–2306. [CrossRef] [PubMed]
67. Andrews, M.; Morton, J.D.; Lieffering, M.; Bisset, L. The partitioning of nitrate assimilation between root and shoot of a range of temperate cereals and pasture grasses. *Ann. Bot.* **1992**, *70*, 271–276.
68. Laperche, A.; Devienne-Barret, F.; Maury, O.; le Gouis, J.; Ney, B. A simplified conceptual model of carbon/nitrogen functioning for QTL analysis of winter wheat adaptation to nitrogen deficiency. *Theor. Appl. Genet.* **2006**, *113*, 1131–1146. [CrossRef] [PubMed]
69. Drew, M.C. Comparison of the effects of a localized supply of phosphate, nitrate, ammonium and potassium on the growth of the seminal root system, and the shoot, in barley. *New Phytol.* **1975**, *75*, 479–490. [CrossRef]
70. Zhang, H.; Jennings, A.; Barlow, P.W.; Forde, B.G. Dual pathways for regulation of root branching by nitrate. *Proc. Natl. Acad. Sci. USA* **1999**, *96*, 6529–6534. [CrossRef] [PubMed]
71. Lima, J.E.; Kojima, S.; Takahashi, H.; von Wirén, N. Ammonium triggers lateral root branching in *Arabidopsis* in AMMONIUM TRANSPORTER1;3-dependent manner. *Plant Cell* **2010**, *22*, 3621–3633. [CrossRef] [PubMed]
72. Raven, J.A.; Wollenweber, B.; Handley, L.L. A comparison of ammonium and nitrate as nitrogen sources for photolithotrophs. *New Phytol.* **1992**, *121*, 19–32. [CrossRef]
73. Zerihun, A.; McKenzie, B.A.; Morton, J.D. Photosynthate costs associated with the utilization of different nitrogen-forms: Influence on the carbon balance of plants and shoot-root biomass partitioning. *New Phytol.* **1998**, *138*, 1–11. [CrossRef]
74. Scheurwater, I.; Cornelissen, C.; Dictus, F.; Welschien, R.; Lambers, H. Why do fast- and slow-growing grass species differ so little in their rate of root respiration, considering the large differences in the rate of growth and ion uptake? *Plant Cell Environ.* **1998**, *21*, 995–1005. [CrossRef]
75. Andrews, M.; Raven, J.A.; Lea, P.J.; Sprent, J.I. A role for shoot protein in shoot—Root dry matter allocation in higher plants. *Ann. Bot.* **2006**, *97*, 3–10. [CrossRef] [PubMed]
76. Irving, L.J.; Cameron, D.D. You are what you eat: Interactions between root parasitic plants and their hosts. *Adv. Bot. Res.* **2009**, *50*, 87–138.
77. Huang, B.; Fu, J. Photosynthesis, respiration, and carbon allocation of two cool-season perennial grasses in response to surface soil drying. *Plant Soil* **2000**, *227*, 17–26. [CrossRef]
78. Hare, P.D.; Cress, W.A.; van Staden, J. Dissecting the roles of osmolyte accumulation during stress. *Plant Cell Environ.* **1998**, *21*, 535–553. [CrossRef]
79. Araya, T.; Noguchi, K.; Terashima, I. Effect of carbohydrate accumulation on photosynthesis differ between sink and source leaves of *Phaseolus vulgaris* L. *Plant Cell Physiol.* **2006**, *47*, 644–652. [CrossRef] [PubMed]
80. Hamblin, A.; Tennant, D. Root length density and water uptake in cereals and grain legumes: How well are they correlated? *Aust. J. Agric. Res.* **1987**, *38*, 513–527. [CrossRef]
81. Saarinen, T.; Haansuu, P. Shoot density of *Carex rostrate* Stokes in relation to internal carbon:nitrogen balance. *Oecologia* **2000**, *122*, 29–35. [CrossRef]
82. Davies, A. Changes in growth rate and morphology of perennial ryegrass swards at high and low nitrogen levels. *J. Agric. Sci.* **1971**, *77*, 123–134. [CrossRef]
83. Gautier, H.; Varlet-Grancher, C.; Hazard, L. Tillering responses to the light environment and to defoliation in populations of perennial ryegrass (*Lolium perenne* L.) selected for contrasting leaf length. *Ann. Bot.* **1999**, *83*, 423–429. [CrossRef]
84. Smith, H. Phytochromes and light signal perception by plants—An emerging synthesis. *Nature* **2000**, *407*, 585–591. [CrossRef] [PubMed]
85. Toyota, M.; Tatewaki, M.; Morokuma, M.; Kusutani, A. Tillering responses to high red/far red ratio of four Japanese wheat cultivars. *Plant Prod. Sci.* **2014**, *17*, 124–130. [CrossRef]
86. Evers, J.B.; Vos, J.; Andrieu, B.; Struik, P.C. Cessation of tillering in spring wheat in relation to light interception and red: Far-red ratio. *Ann. Bot.* **2006**, *97*, 649–658. [CrossRef] [PubMed]
87. Sparkes, D.L.; Holme, S.J.; Gaju, O. Does light quality initiate tiller death in wheat? *Eur. J. Agron.* **2006**, *24*, 212–217. [CrossRef]
88. Simons, R.G.; Davies, A.; Troughton, A. The effect of spacing on the growth of two genotypes of perennial ryegrass. *J. Agric. Res.* **1973**, *80*, 495–502. [CrossRef]
89. Lonsdale, W.M.; Watkinson, A.R. Light and self-thinning. *New Phytol.* **1982**, *90*, 431–445. [CrossRef]
90. Lonsdale, W.M. The self-thinning rule: Dead or alive? *Ecology* **1990**, *71*, 1373–1388. [CrossRef]

91. Enquist, B.J.; Brown, J.H.; West, G.B. Allometric scaling of plant energetics and population density. *Nature* **1998**, *395*, 163–165. [CrossRef]
92. Matthew, C.; Lemaire, G.; Sackville-Hamilton, N.R.; Hernandez-Garay, A. A modified self-thinning equation to describe size/density relationships for defoliated swards. *Ann. Bot.* **1995**, *76*, 579–587. [CrossRef]
93. Davies, A.; Thomas, H. Rates of leaf and tiller production in young spaced perennial ryegrass plants in relation to soil temperature and solar radiation. *Ann. Bot.* **1983**, *57*, 591–597.
94. Minchin, P.E.H.; Lacointe, A. New understanding on phloem physiology and possible consequences for modelling long-distance carbon transport. *New Phytol.* **2005**, *166*, 771–779. [CrossRef] [PubMed]
95. Matthew, C.; Kemball, W.D. Allocation of carbon-14 to roots of different ages in perennial ryegrass (*Lolium perenne* L.). In Proceedings of the XVIII International Grassland Congress, Calgary: Association Management Center, Section 7, Calgary, Canada, 8–19 June, 1997; pp. 1–2.
96. Nikinmaa, E.; Hölttä, T.; Hari, P.; Kolari, P.; Mäkelä, A.; Sevanto, S.; Vesala, T. Assimilate transport in phloem sets conditions for leaf gas exchange. *Plant Cell Environ.* **2013**, *36*, 655–669. [CrossRef] [PubMed]
97. Lattanzi, F.A.; Schnyder, H.; Thornton, B. The sources of carbon and nitrogen supplying leaf growth. Assessment of the role of stores with compartmental models. *Plant Physiol.* **2005**, *137*, 383–395. [CrossRef] [PubMed]
98. Matsuhashi, S.; Fujimaki, S.; Kawachi, N.; Sakamoto, K.; Ishioka, N.S.; Kume, T. Quantitative modeling of photoassimilate flow in an intact plant using the positron emitting tracer imaging system (PETIS). *Soil Sci. Plant Nutr.* **2005**, *51*, 417–423. [CrossRef]
99. Thornton, B.; Paterson, E.; Midwood, A.J.; Sim, A.; Pratt, S.M. Contribution of current carbon assimilation in supplying root exudates of *Lolium perenne* measured using steady-state $^{13}C$ labelling. *Physiol. Plant.* **2004**, *120*, 434–441. [CrossRef] [PubMed]
100. Carvalho, D.D.; Irving, L.J.; Carnevalli, R.A.; Hodgson, J.; Matthew, C. Distribution of current photosynthate in two Guinea grass (*Panicum maximum* Jacq.) cultivars. *J. Exp. Bot.* **2006**, *57*, 2015–2024. [CrossRef] [PubMed]
101. Osaki, M.; Shinano, T.; Yamada, M.; Yamada, S. Function of node unit in photosynthate distribution to root in higher plants. *Photosynthetica* **2004**, *42*, 123–131. [CrossRef]
102. Osaki, M.; Morikawa, K.; Matsumoto, M.; Shinano, T.; Iyoda, M.; Tadano, T. Productivity of high yielding crops. III. Accumulation of ribulose-1,5-bisphosphate carboxylase/oxygenase and chlorophyll in relation to productivity of high yielding crops. *Soil Sci. Plant Nutr.* **1993**, *39*, 399–408. [CrossRef]

# 16

# Ecophysiology of $C_4$ Forage Grasses—Understanding Plant Growth for Optimising their use and Management

**Sila Carneiro da Silva [1,*], André Fischer Sbrissia [2] and Lilian Elgalise Techio Pereira [3]**

1 Animal Science Department, University of São Paulo, E.S.A. "Luiz de Queiroz", Av. Pádua Dias, 11, CEP 13418-900, Piracicaba, SP, Brazil
2 Centro de Ciências Agroveterinárias, Santa Catarina State University, Avenida Camões, 2090, CEP 88520-000, Lages, SC, Brazil; andre.sbrissia@udesc.br
3 Animal Science Department, University of São Paulo, Faculdade de Engenharia de Alimentos e Zootecnia, Av. Duque de Caxias Norte, 225, CEP 13635-900 Pirassununga, SP, Brazil; ltechio@usp.br
* Author to whom correspondence should be addressed; siladasilva@usp.br

Academic Editor: Cory Matthew

**Abstract:** Grazing management has been the focus of the research with forage plants in Brazil for many years. Only in the last two decades, however, significant changes and advances have occurred regarding the understanding of the key factors and processes that determine adequate use of tropical forage plants in pastures. The objective of this review is to provide an historical overview of the research with forage plants and grasslands in Brazil, highlighting advances, trends, and results, as well as to describe the current state of the art and identify future perspectives and challenges. The information is presented in a systematic manner, favoring an integrated view of the different trends and research philosophies. A critical appraisal is given of the need for revision and change of paradigms as a means of improving and consolidating the knowledge on animal production from pastures. Such analysis idealizes efficient, sound and sustainable grazing management practices necessary to realize the existing potential for animal production in the tropics.

**Keywords:** tropical pastures; ecophysiology; morphogenesis; grazing management

---

## 1. Introduction

Brazil is a South American country situated between latitudes 5°16′20″ S and 33°45′03″ S, therefore part of the tropical/sub-tropical region of the world. Animal husbandry and production systems are mainly based on grazed pastures and harvested forages. The country has around 196 million ha of permanent pastures [1], of which around 51% are comprised of cultivated (introduced) species [2], particularly of the genera *Brachiaria* and *Panicum*, all $C_4$ grasses with high herbage dry matter yield potential. These are very different from most temperate forage grasses, showing a wide range of plant morphology and structure, varying from prostrate/semi-prostrate to tall-tufted erect growing plants [3]. Despite the potential for generating high animal productivity [4] (25–30,000 kg milk/ha per year and 1000–1600 kg body weight gain/ha per year), national average productivity is low (around 800–1000 kg milk/ha per year and 60–100 kg body weight gain/ha per year) [5], the consequence of inadequate empirical management practices. For that reason, focus has historically been given to research on improving the agronomic, nutritional, and productive traits of forage plants and grazing management. Effective progress, however, became more evident when studies started to adopt a more ecological approach and were planned taking into account functional responses related to plant growth, form, and function, favouring the understanding of the associated processes and providing the necessary knowledge basis for planning sound and efficient management practices. In this context, the inclusion of evaluations regarding plant ecophysiology in the experimental protocols during the early 1990s played an important role and represented a turning point in grassland research in the tropics.

Against that background, the objective of this review is to provide an historical overview of the research with tropical forage grasses in Brazil, highlight advances and identify future perspectives and challenges. In spite of the Brazil focus, the findings should be applicable to the same climate worldwide.

## 2. Forage Grass Research—A Historical Overview

The research with forage plants and pastures in countries with developed animal husbandry systems started in the early 1900s. Among the scientific publications that came out at that time, some became important contributions and references for all the progress and achievements obtained by grassland scientists, even at the present time. Graber [6], cited by Volenec *et al.* [7], was one of the first to report that the concentration of total non-structural carbohydrates (TNC) in roots decreased soon after defoliation during the regrowth phase of alfalfa (*Medicago sativa* L.). Watson [8], cited by Black [9], demonstrated that a measure of the size of the plant photosynthetic apparatus was important in making comparisons of crop yield and productivity, and proposed the concept of leaf area index (LAI). Brougham [10–16] demonstrated the importance of LAI for understanding the relationship between canopy light interception (LI) and herbage accumulation and the interaction between defoliation frequency and severity.

Brougham [10] fitted a sigmoid curve to the variation in herbage mass during regrowth of perennial ryegrass (*Lolium perenne* L.), white clover (*Trifolium repens* L.) and red clover (*Trifolium pratense* L.) mixed swards (variation in herbage mass with time), as well as describing the asymptotic relationship between LAI and LI [12,17]. The author demonstrated that plant growth was a function of canopy light interception and LAI, and that the rate of herbage dry matter accumulation reached a maximum constant value when there was enough foliage to intercept almost all the incident light. In general, three distinct phases may be identified in a regrowth curve. The first, soon after defoliation, is characterized by an exponential increase in herbage mass with associated increase in rate of herbage accumulation. This phase is highly influenced by plant organic reserves, climatic and edaphic conditions, and residual leaf area after cutting or grazing [12]. The second is characterized by constant rates of herbage accumulation (linear increase in sward herbage mass). During this phase, intra and inter specific competition between plants become increasingly intense, particularly when the sward is close to maximum canopy light interception. During the third phase there is a reduction in herbage accumulation rate, the consequence of a proportionally larger increase in leaf senescence relative to leaf growth caused by leaves having reached their leaf lifespan and severe shading at the bottom of the sward [18].

These studies provided the basis for developing grazing management strategies based on the concepts of LAI and TNC accumulation and mobilization [19–21], despite the difficulties of measuring them. During the 1960s, Smith [22] carried out a series of experiments with alfalfa in the USA with the objective of demonstrating the importance of plant organic reserves and LAI for adequately managing pastures subjected to intermittent defoliation. As a result, the residual LAI and the organic reserve concentration remained as important considerations to be taken into account in planning and idealizing grazing management practices and strategies. Alcock [23] proposed three simple concepts for explaining plant responses to defoliation: (1) total availability and reutilization of organic reserves, (2) root growth, and (3) leaf area development and canopy light interception. The role of organic reserves had already been recognized as important for a long time [24], but with no applied results, since no grazing management strategy was produced based on it, except for harvesting alfalfa. The same happened with root growth. On the other hand, the LAI and LI concepts were used and studied in a series of experiments whose results helped to establish a strong knowledge basis for understanding the process of herbage accumulation of forage plants subjected to defoliation (cutting or grazing) regimes. Brown & Blaser [25] considered the use of LAI to define management practices an oversimplification of the process, arguing that on tall swards there were usually few leaves close to ground level and that lenient grazing would be necessary to ensure enough residual LAI to allow maximum canopy light interception. The authors also argued that on tall swards the photosynthetic efficiency of the leaves positioned at the base of the sward would be lower than that of leaves positioned at the top, resulting in low harvest efficiency and high risk of reduction in tiller population density.

In Brazil, during the first Symposium on Pasture Management, in 1973, it was recognized that there was a complex interplay among LAI, tillering, and TNC reserves in the regrowth of forage plants, and that the knowledge of those relationships in temperate species should be valid for tropical forages as well [26,27]. Jacques [27], however, argued that "despite its importance, the LAI was not enough to plan and sustain adequate management practices", and highlighted the distinct interests and approaches of the predominant "research schools" at that time (North American and British).

Several of the existing grazing management guidelines have been based on the argument that pasture species should be used under intermittent defoliation regimes such as rotational stocking, generating a series of successive regrowth cycles (sigmoid curves), in order to better use the growth characteristics of forage plants. Under these circumstances, defoliations (grazings or harvests) should happen at the end of the linear growth phase (phase 2) so that maximum average rate of herbage dry matter accumulation could be obtained. Herbage nutritive value under those conditions, however, was usually low and it could be valuable to interrupt regrowth at an earlier stage with the objective of harvesting better quality herbage [28]. Under continuous stocking, in cases when pastures are maintained at a steady state condition characterized by constant levels of sward LAI, height, or herbage mass, the idea of maintaining LAI to ensure 95% LI would not be valid, since leaf senescence is proportional to leaf growth, resulting in zero or even negative values of net herbage accumulation [29]. In this case, the recommendation would be to manage swards at lower heights and lower LAI relative to intermittent defoliation regimes as a means of ensuring higher rates of herbage accumulation and harvest efficiency of the produced herbage [30].

From the 1960s onwards the "North-American" school started to be influenced by Gerald Mott's work, with the introduction of the put and take stocking method for adjusting stocking rate in grazing experiments [31]. The method consisted of using fixed groups of animals, called testers, that would represent the experimental unit for measuring animal responses, and another group that would act as stocking rate regulators, which were added to or removed from paddocks as a means of adjusting defoliation intensity based on an arbitrary criterion, originally suggested by Mott as grazing pressure. Grazing pressure was defined as the relationship between animal live weight and herbage mass per unit area [31,32]. By derivation, Mott [31] defined another term, carrying capacity, or the maximum stocking rate that would achieve a target level of animal performance in a specified grazing system. The put-and-take stocking method was more intensively used for research purposes with little use in practical situations, the consequence of the difficulties of its implementation in farm conditions. These concepts were used for many years by several researchers, but, according to Maraschin [33], without fully understanding the underlying principles involved. According to the same author, for efficient use of the herbage produced, stocking rate should always be defined in relation to the carrying capacity of pastures, a condition associated with the optimum grazing pressure. The goal under this approach would be to find an equilibrium between animal performance and animal production per unit area as a means of generating higher economical returns of pasture utilization, but without taking into account aspects of plant ecophysiology and the ecology of grassland ecosystems.

The "North American" school had a strong influence on the research with forage plants and pastures in Brazil for a long time. As a consequence, the emphasis was on describing the growth curve of pastures after cutting or grazing, their seasonality of herbage production, and the morphological and chemical composition of the produced herbage, with no particular attention to dynamic aspects related to plant population and competition for light normally associated with plant recovery after defoliation. Stocking rate and grazing pressure were considered key features and started to be used as control variables (treatments) in grazing experiments. This is corroborated by the large number of conferences on the topic at the Symposium on Pasture Management at ESALQ in Brazil [34–51] and research abstracts published in the Annual Meeting Proceedings of the Brazilian Society of Animal Science [52]. After some time, grazing pressure evolved to the concept of herbage allowance (the amount of herbage on offer per animal), and evidences [53,54] showed that individual herbage intake was maximized when herbage allowance corresponded to three–four times the daily requirement of dry matter [55]. In tropical pastures, after successive grazings, this could result in excessively high levels of herbage on offer characterized by large quantities of dead material and stems, low nutritive value of the herbage, and reduced intake by the grazing animals [56–58]. Mott [59], aware of that limitation, warned that the conversion of the primary production in animal product in tropical pastures would be considerably different from that in temperate pastures. As a consequence, he suggested that management practices should aim to provide the maximum amount of live tissues with high digestibility to animals, particularly leaves, as a means of increasing intake and performance. This idea led to the concept of green dry matter allowance [60], which evolved to leaf dry matter allowance. These practices were effective in providing a good fit between herbage allowance and animal performance data (as originally reported by Mott [31]) in relation to the traditional asymptotic curve [61], but magnified the problem of excessive herbage mass with high proportions of stems and dead material, with negative implications on herbage and animal production.

Although the conflict between the necessary sward conditions for generating high rates of herbage accumulation and those for achieving maximum intake and performance was evident in the literature since the early 1970s [56–58], harvest efficiency was frequently compromised as a means of favoring high levels of animal performance [31], resulting in low productivity and large dry matter losses to senescence, death, and decay [62]. Better understanding and more sound grazing management practices started to emerge during the 1980s when the results from more detailed studies evaluating the dynamics of plant population and growth became available [18]. Bircham & Hodgson [63] were the first to describe the dynamics of the herbage accumulation in continuously stocked perennial ryegrass swards, and showed that net herbage accumulation was the result of the balance between two concomitant and antagonistic processes—growth and senescence. These respond differently to agronomic and management practices. Consequently, evaluation of herbage accumulation only, *i.e.*, without taking into account the independent processes of growth and senescence, could result in imperfect understanding of patterns of plant and pasture response to defoliation [64]. The results also demonstrated that the herbage accumulation process could be adjusted through manipulation of sward structural characteristics such as LAI, height, and herbage mass, allowing for the development of grazing management targets that could be used to guide and control the grazing process in farm conditions. In this context of strict control of sward structure, the need for multi-year experiments became relatively less important when compared, for example, with experimental protocols that used stocking rate or grazing pressure as control variables, since the factors causing the treatment x year interactions were the same as those causing the treatment x environment and/or the treatment x season of the year interactions.

According to Hodgson [65], effective understanding of how plants and animals respond to variations in sward conditions, and consequently to management, could only be achieved in grazing experiments with rigid control of sward structural characteristics at a given state (continuous stocking management with variable stocking rate) or following a pre-specified pattern of variation (pre and post-grazing conditions of an intermittent stocking management). Korte *et al.* [66], studied perennial ryegrass subjected to cutting regimes characterized by two frequencies and two severities of defoliation, and planned their experimental treatments based on the findings of Brougham during the 1950s, using the 95% LI as the reference condition for initiating defoliation. The authors concluded that during the vegetative growth stage the 95% LI criterion could be used to define the best moment for initiating defoliation and, relative to longer defoliation intervals, would result in greater herbage production with higher proportion of leaves and lower proportion of dead material. That would indicate the ideal harvest point during regrowth (determinant of cutting and/or grazing interval), a condition that would be associated with the end of the linear phase of the sigmoidal growth curve described by Brougham [10]. These findings indicated convergence of the available knowledge and corroborated the central role of LAI as a determinant of plant responses to grazing, highlighting the need to study and understand aspects related to sward structure, light use, and the balance between growth and senescence as a means of planning and defining efficient grazing management strategies [67]. Chapman & Lemaire [68] reinforced the importance of LAI as a determinant of forage plant responses, and demonstrated that it was the result of the combined expression of the morphogenetic and structural characteristics of plants in a given environment. The paper established a reference point because it integrated the understanding from morphogenetic and ecophysiological studies from experimentation with forage plants, providing the necessary knowledge base for understanding the ecological and functional responses of plants and animals in grazing systems [69].

In Brazil, the first papers with information on the morphogenesis and ecophysiology of tropical grasses were published by Pinto *et al.* [70,71]. They reported results on leaf and stem accumulation, average tiller weight, proportion of vegetative and reproductive tillers, and rates of leaf appearance and leaf and stem elongation for andropogon (*Andropogon gayanus* Kunth), guinea (*Panicum maximum* Jacq.), and setaria (*Setaria sphacelata* Schumach.) grasses subjected to two levels of nitrogen fertilization. Gomide [72], in a review paper, showed results from several experiments on plant morphogenesis carried out in Brazil until that date. He indicated that the stabilization of the number of leaves per tiller and of tillers per plant could be a possible indicator for orienting and controlling the grazing management of tropical forage plants. On the other hand, Lemaire [73] argued that the morphogenesis of the main tropical forage species would have to be thoroughly studied in order to provide the necessary conditions for understanding plant responses to changes in management and environment. That introduced a change in research paradigm and to how the experimentation with forage plants would be conducted in the country from then onwards. In this new context, sustainability became an important feature, emphasizing pasture stability and productivity as the main goals to be achieved by the idealized management practices, highlighting the importance of rationalization and integration of the existing knowledge and results [74].

The inclusion of ecophysiology principles and a more integrative approach into the experimental protocols used changed the research with forage plants in Brazil, with key words like plant growth and development, herbage intake, ingestive behaviour, utilization, and conversion starting to be slowly but irreversibly incorporated into the vocabulary of researchers. As a result, knowledge and understanding of plant structural and morphogenetic characteristics became an important tool for determining the adequate sward conditions (height, herbage and leaf mass, LAI *etc.*) to ensure efficient and sustainable animal production from pastures.

## 3. The Ecophysiology of Tropical Grasses and Grazing Management

Among the commonly most used forage grasses in cultivated grasslands in Brazil are those from the *Brachiaria* (Syn. *Urochloa*), *Panicum*, *Pennisetum*, and *Cynodon* genera. Although they all have an effective vegetative perennation mechanism based on clonal growth (e.g., tillering—vegetative growth resulting in the natural production of potentially autonomous daughter plants—ramets), there is a wide range in plant architecture and growth habit which determines the functional relationships among individuals and the adaptive plasticity of their morphological structures.

*Cynodon* species and cultivars, as well as their intra and interspecific hybrids, are creeping-type grasses spread either by stolons, rhizomes or both. *Brachiaria*, *Panicum*, and *Pennisetum purpureum* Schum. are tussock-type grasses (also known as caespitose, bunchgrass, tufted grass, or phalanx) characterized by a compact pattern of tiller organization in an erect and clumped growth form and representing the main type of grass plants used in cultivated pastures in Brazil. For this group, there is a wide range of plant size and architectural configurations, which should be considered in order to understand their vegetative perennation mechanisms (e.g., tillering) and adaptation to grazing. Among these tall-tufted growing grasses there are species and cultivars of *Pennisetum purpureum*

Schum. and *Panicum maximum* Jacq. with a very high dry matter production potential, which are especially demanding in terms of soil fertility and fertilization. The *Brachiaria* genus has a large variety of morphological types and growth habits varying from small tussock forming plants such as *Brachiaria brizantha Hochst* ex A. Rich Stapf. and *Brachiaria ruziziensis* R. Germ & Evrard, creeping plants such as *Brachiaria humidicola* (Rendle) Schweick, and *Brachiaria purpurascens* Henr. Blumea (Syn. *Brachiaria mutica* (Forsk) Stapf.), to intermediate morphological types characterized by an initial prostrate growth that evolves to an erect form such as *Brachiaria decumbens* Stapf. and *Brachiaria hybrid* cvs Mulato I (*B. ruziziensis* × *B. brizantha*) and Mulato II (*B. ruziziensis* × *B. decumbens* × *B. brizantha*).

According to Cruz & Boval [75], the large phenotypic variability in forage grasses characterized by the wide range of morphological types reveals the need for particular strategies for different species to control canopy development and biomass production. In this context, plant growth and development, leaf turnover, and population dynamics must be analyzed as integrated physiological and adaptive processes that may determine significant changes in the morphological composition of the produced herbage, sward structure, and spreading and colonization pattern. As a result, the recent research with tropical forage grasses has been focused on identifying grazing management strategies that harmonize with and optimize the natural growth cycle of plants, favoring their growth and production. During the last two decades significant progress was made [76] and management targets were generated as a means of transferring this knowledge to practical use [3].

### 3.1. Morphological Plant Types and Growth

Plant responses to grazing have the objective of maximizing sward leaf area index (LAI) in order to optimize assimilate production and energy supply to plant growth. They are usually integrated and occur at two levels of complexity—the turnover of leaves in individual tillers and the turnover of tillers in tiller population [77]. The LAI is determined by three components: tiller population density, number of leaves per tiller, and leaf size (leaf lamina area). In grasses, leaf size is mainly a function of lamina length, which is controlled by defoliation height, and the number of leaves per tiller is relatively stable, leaving the tiller population density as the component where changes in LAI can be readily expressed [78]. Tiller population density is the result of the balance between tiller appearance and death, which characterizes the tiller population dynamics that influences tiller population stability [79]. These, in turn, determine modifications in the demographic profile of the tiller population, altering the age profile of tillers, sward structure (canopy architecture), photosynthetic efficiency of sward leaf area, and the persistence of the tiller population [80].

The integration of plant responses determining plant growth in grazed communities was described by Chapman & Lemaire [68] and related to plant morphogenesis. Although originally described for temperate grasses, the approach developed by the authors provided a solid basis for understanding the growth patterns of tropical grasses. During the vegetative growth stage, the morphogenesis of temperate grasses is characterized by leaf appearance, leaf elongation, and leaf lifespan. Since only leaves are produced as above-ground organs, those responses were identified as the major morphogenetic characteristics determining plant and sward structure. Cruz & Boval [75] studied a group of temperate and tropical grasses and proposed the existence of two main morphogenetic types of grasses: tufted and stoloniferous species. According to the authors, while in tufted grasses leaf appearance and final leaf length are related to the length of the sheath tube, in stoloniferous grasses they are related to stolon elongation. In stoloniferous plants, grazing management strategies characterized by lenient defoliation or by high nitrogen availability favor the accumulation of stolons. As a result, leaf appearance rate is increased and final leaf length becomes shorter towards the apex of the stolons. The permanent internode elongation in tropical stoloniferous plants and the stem elongation only during the reproductive stages of growth in temperate grasses differentiate the two morphogenetic types [75]. Although the authors indicated that the growth pattern of tropical tufted grasses would be similar to that of temperate grasses, Cowan & Lowe [81] argued that in tropical tufted grasses stem elongation during the vegetative growth stage was not negligible as for temperate grasses, leading Hodgson and

Da Silva [82] to consider stem elongation as an important additional morphogenetic characteristic determining plant responses to grazing for tufted tropical grasses. These are plants with production of both leaves and stems as above-ground organs, a condition that would characterize them as an intermediary plant type. The existence of this intermediary morphogenetic type was initially indicated by Cruz & Boval [75] and recently discussed by Pereira *et al.* [83] for *Brachiaria brizantha* cv. Marandu. The evidence available describes a significant and regular stem elongation process that occurs during the vegetative growth stage of tropical tufted grasses [84] subjected to lenient and/or intermittent grazing which is independent of floral induction.

The tufted growth pattern and its implications to sward structure were originally described for *Panicum maximum* cv. Mombaça under intermittent stocking management [85]. During the early stages of regrowth the main morphological component accumulated is leaf. As LAI increases, competition for light within the sward canopy increases and plants change their growth pattern as a means of optimising light capture through stem elongation. The shift in growth pattern occurs when canopy light interception reaches and exceeds 95% [84]. According to Ballaré [86], during the early stages of regrowth, when LAI is small, the amount of photosynthetic radiation received by the leaves is not affected by neighboring plants, a condition that favors investment in leaf production. However, before any reduction in light availability occurs, small variations in light quality, caused by the preferential absorption of the blue and red wavelengths, and reflection of the far red, can provide plants with information regarding their surroundings [87]. The signaling mechanisms related to changes in the light environment within the canopy allow plants to redirect growth to more favorable patches (areas with better light availability), being common in morphological modifications of tropical grasses associated with shade avoidance responses (in contrast to shade acclimation or tolerance). As a result, the interval between successive defoliations determines the amplitude of the plastic responses plants have to develop [88]. When managed with long regrowth intervals that allow canopy light interception to exceed 95% (e.g., maximum light interception—$LI_{max}$), sward herbage mass is greater but with a higher proportion of stem and dead material relative to leaf, resulting in greater total dry matter production, but smaller leaf dry matter yield [84] and lower nutritive value of the produced herbage [89]. In addition, long grazing intervals characterized by the pre-grazing target of $LI_{max}$ result in a greater proportion of the produced herbage being lost to the soil as a consequence of physical damage and/or rejection, decreasing the efficiency of the grazing process [85,90,91]. The investment in stem elongation results in taller swards with an increased proportion of stems in the upper layers of their vertical profile [92], increasing the difficulty of grazing and in maintaining the post-grazing management targets [93,94]. As a result, the benefit of greater total herbage accumulation is offset by the greater grazing losses when long regrowth intervals are used, indicating that, although herbage accumulation per grazing cycle is smaller when managing with shorter grazing intervals ($LI_{95\%}$), the more frequent defoliation results in larger number of grazing cycles and total herbage accumulation (with higher leaf proportion) than when managing with longer grazing intervals ($LI_{max}$) [84,91,94,95].

The same pattern of response was later described for *Panicum maximum* cv. Tanzânia [90,92], *Brachiaria brizantha* cv. Xaraés [96,97], *Brachiaria brizantha* cv. Marandu [98], *Pennisetum purpureum* cv. Cameroon [99], *Brachiaria decumbens* cv. Basilisk [100], mulato brachiariagrass (*Brachiaria ruziziensis* x *Brachiaria brizantha* cv. Marandu) [101], *Panicum maximum* cv. Aruana [93], and *Pennisetum purpureum* cv. Napier [94,95], showing consistency within a wide range of morphological types and a strong light effect determining plant growth. More recently, in a study where a forage grass (*Brachiaria brizantha* cv. Piatã) was grown in monoculture or in association with trees (*Eucalyptus* sp.) planted at different densities (181 and 718 trees/ha) and managed under rotational grazing, the importance of light as the determining factor of plant growth and responses was corroborated and expanded, since plant morphogenetic and population dynamic responses were strongly influenced by the amount of photosynthetic radiation available [102]. Under those conditions, a wide range of light availability was generated through regrowth interval, tree density, distance from the tree rows, and season of the year (in this case representing the varying solar angle from summer to winter), illustrating the key role played by light availability within the sward in determining plant growth and herbage production. Further, the evidence suggests that, among the environmental factors, light would determine the potential of herbage production (ceiling LAI), and temperature and water availability (as well as nutrients) would determine if that potential could be realised and, if so, how fast the responses and processes would happen.

### 3.2. Population Dynamics and Stability

Although adjustments in morphogenetic responses of individual tillers are an important way to maximize leaf area, Matthew *et al.* [78] argued that tiller population density is the main component of sward leaf area when leaf size increase is restricted by defoliation. In plant communities, the beginning of regrowth is characterized by increases in LAI arising from both new tiller production and leaf growth on the existing tillers. However, as regrowth progresses and LAI increases, competition for light within the sward canopy increases, tiller recruitment ceases, and population density starts to decrease. From this point onwards further increase in tiller size and LAI result in reduction in tiller population, characterizing the size-density compensation or self-thinning mechanism described by Matthew *et al.* [103] for temperate grasses and later corroborated by Sbrissia *et al.* [104,105], Sbrissia & Da Silva [106], and Calsina *et al.* [107] for tropical grasses.

According to Sbrissia & Da Silva [106], this mechanism ensures that sward LAI remains relatively stable for a wide range of defoliation regimes, highlighting the importance of the vegetative mechanism of plant perennation (tillering) for developing grazing management strategies for tropical grasses. In stoloniferous species like *Cynodon* spp. and aerial-tiller producing grasses such as most tropical grasses, the physiological integration of plants seems to play an important role in adaptation and response to defoliation. Interconnected ramets of clonal plants, although potentially independent, can specialize functionally in performing a limited number of tasks, such as the uptake of resources from above and below ground, carbohydrate storage, vegetative spread, and sexual reproduction [108]. Such specialization and cooperation is comparable to a division of labor in economic systems or in colonies of social animals. The ecological significance of labor division in clonal plants may be found in the increased efficiency of entire clones for exploiting their environments. Preliminary evidences of this pattern of labor division through specialization in development of particular phytomers [109] were reported by Sbrissia *et al.* [104] for Coastcross bermudagrass (*Cynodon* spp.). According to the authors, a new cluster of tillers was formed along a stolon at regular intervals of three phytomers (node with roots and a daughter tiller followed by two successive leaves—Figure 1), suggesting that the first leaf would support root development, the second would support tiller development and the third would support stolon internode elongation. The clonal integration, associated with the labor division, explains the low values of the individual tiller leaf area-to-volume ratio (R value; [110]) recorded for two *Cynodon* spp. cultivars (Coastcross and Tifton-85) (13.1–17.1; [104,105]). Low R values imply that tillers present a small leaf area in relation to their volume, suggesting a reduced ability to use the incident light and, therefore, could be an indicative of reduced competitive ability. However, a plant with low R value could maintain its competitiveness through a higher level of clonal integration, since the R value of an object which is a cluster of similar modules is much greater than the R value of an isolated module [103]. For Tifton-85, assuming a hypothetical clonal integration of four tillers, R values increased (26.2 to 34.8; [105]) and became closer to 50, the reported value for perennial ryegrass [111], leading authors to infer that the natural high competitive ability of *Cynodon* spp. is the result of some degree of clonal integration, since R values of individual tillers are low. This is in line with the argument of De Kroon *et al.* [112], according to whom whole-plant plasticity is the sum of all environmentally induced modular responses plus all interaction effects that are due to communication and behavioral integration of modules.

In tall, tufted, tussock-forming grasses such as elephant grass (*Pennisetum purpureum*) and Mombaça and Tanzania guinea (*Panicum maximum*) grass, tillering is an important component of tussock growth and expansion which determines the efficiency of soil surface occupation through variation in the frequencies of colonized and bare ground areas in the pasture. In this plant type, the proportion of areas with tussocks or bare ground associated with tussock size represent an important indicator of grazing management effect on pasture persistence and productivity. Modifications of the horizontal sward structure in response to grazing were originally reported for rotationally managed Mombaça guinea grass by Lopes [113] and Montagner [114], and demonstrated that in areas with deficient grazing management or not in regular use, the stability of the tiller population was lower, tussock perimeter varied within a wider range (very small to very large tussocks), the proportion of bare ground was higher, and tiller population density was lower relative to pastures managed in a regular and controlled manner using adequate targets to control grazing [80]. Under those conditions, adjustments in foliage angle were the strategy used by plants to optimize light capture, resulting in 95% canopy light interception at lower sward height (60 cm; [113]). However, during the second year of the experiment, after one year adapting to regular and controlled grazing management, the tiller population density and the proportion of areas with tussocks increased, the proportion of bare ground decreased, the variation in tussock perimeter decreased (very small and very large tussocks disappeared) and the frequency of average size tussocks increased, resulting in 95% canopy light interception at 90 cm [114], the same value reported by Canevalli *et al.* [85] for Mombaça guinea grass. These results showed adaptation of plant population to defoliation regimes used and the importance of tiller and tussock population and size for the necessary adjustments in sward LAI. They also highlight the need to carefully plan experimental treatments and experiments, since there may be the need to allow for sward adaptation to defoliation regimes before starting the measurement period. A more uniform ground cover with smaller tussocks, greater tiller population density, and short distances between tussocks results in more efficient sward carbon acquisition [115] and may also minimize weed encroachment [116].

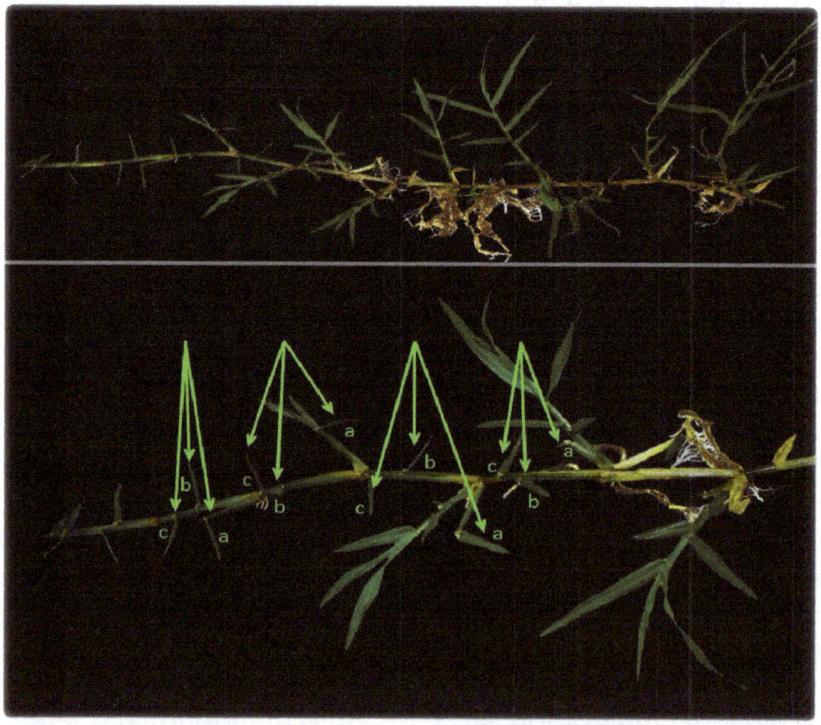

**Figure 1.** Triplet structure of *Cynodon* spp. cv. Tifton-85 showing cluster formation at regular intervals of three phytomers. Letters a, b and c indicate leaf 1, leaf 2, and leaf 3 of each cluster, respectively.

More recently, new evidence on modification of sward horizontal structure in adaptation to grazing management was reported by Pereira *et al.* [94,95] for rotationally managed Napier elephant grass (Figure 2). The authors showed that grazing management interfered with the vegetative mechanism used by plants for perennation by altering the number and the proportion of aerial and basal tillers in tiller population, with consequences on sward horizontal structure and herbage accumulation. Swards under more frequent defoliation (managed with the $LI_{95\%}$ pre-grazing target) showed a larger contribution of basal tillers, while those subjected to less frequent defoliation (managed with the $LI_{Max}$ pre-grazing target) showed a larger number of aerial tillers per support unit (basal tillers + decapitated tillers), an analogous growth strategy to that associated with the increase in tiller number per integrated physiological unit (IPU) described by Derner *et al.* [117] for the tussock-forming grass *Schizachyrium scoparium* (Michx) Nash. For similar tussock perimeters, swards managed with the $LI_{95\%}$ target showed higher tiller population stability index, higher frequency of tussocks and lower frequency of bare ground relative to those managed with the $LI_{Max}$ target. According to the authors, basal tillers would be responsible for increasing the basal area of tussocks and reducing the sharing of assimilates and nutrients among tillers within an IPU, promoting more uniform distribution of tillers within tussocks, and of tussocks in the area, a condition that would favor carbon acquisition [115] and could increase the growth potential of swards. These differences in patterns of plant distribution between targets of pre-grazing LI show that the duration of the regrowth period (e.g., grazing interval) strongly affects tiller replacement. Similarly, management strategies that favor aerial tillering in tall-tufted tussock forming grasses could adversely affect the dynamics of tussock growth and expansion and the capacity of plants to exploit soil surface, as well as increase the risk of erosion and presence of weeds.

**Figure 2.** Tussock size and distribution in *Pennisetum purpureum* cv. Napier paddocks subjected to strategies of rotational stocking management characterized by pre-grazing targets of 95% and maximum canopy light interception during regrowth ($LI_{95\%}$ and $LI_{Max}$) and post-grazing targets of 35 and 45 cm.

In addition to changes in sward structure and in the spreading and colonization ability of plants, tiller appearance and death also interfere with the age profile of the tiller population [80]. Younger tillers (< 2 months old) have higher leaf appearance and elongation rates than old tillers (> 4 months old) [118,119], highlighting the importance of grazing management as a means of manipulating the age profile of the tiller population. More frequent and/or more intense grazing for intermittent or continuous stocking management, respectively, results in higher intensity of defoliation and, consequently, higher rates of tiller death and appearance (higher turnover in tiller population). Similarly, agronomic practices such as fertilization and irrigation accelerate plant growth and increase the turnover of leaves and tillers, resulting in a younger profile of the tiller population [120]). This was demonstrated by Paiva *et al.* [121] for marandu palisade grass receiving contrasting rates of nitrogen fertilizer. Tiller age has a significant impact on leaf turnover and elongation, with higher values recorded for young relative to old tillers. Young tillers are also more responsive to nitrogen fertilization [119,121], increasing the benefits of agronomic practices and fertilization and also favoring the production of high nutritive value herbage [89], an important condition for ensuring high animal performance.

Seasonal variation in the availability of climatic conditions affects both the number of tillers produced and the longevity of the tiller generations. Periods of high availability of climatic growth factors (e.g., temperature and rainfall), as in late spring and summer, are characterized by high rates of tiller appearance and death and, therefore, short-lived generations. However, the balance between tiller appearance and death is usually positive and results in increased tiller population density. During autumn/winter and early spring, tiller survival increases, although not enough to offset the reduction in tiller appearance, and tiller population density decreases [120,122]. Such variations in tillering dynamics represent a natural cycle of plant growth throughout the year, and occur regardless of the grazing strategy, grazing method, and fertilization level [120].

Similar to the pre-grazing management targets, definition of post-grazing targets is also important and dependent on the plant's resistance and adaptation to grazing, since it interferes with how quick sward leaf area is restored. In general, more severe grazing results in lower sward residual leaf area and herbage mass soon after grazing and, consequently, lower canopy light interception and longer grazing intervals than under more lenient grazing [90,92]. However, these characteristics may be offset by the corresponding higher rates of herbage accumulation, greater herbage removal and higher grazing efficiency in more severely grazed pastures, within certain limits. For Mombaça guinea grass [85], Andropogon grass (*Andropogon gayanus* cv. Planaltina [123]), Xaraés palisade grass (*Brachiaria brizantha* cv. Xaraés [96]), and Napier elephant grass [94,95,124] under rotational grazing, similar values of total herbage accumulation have been reported for grazing severities equivalent to a removal of 40%–60% of the pre-grazing height when associated with grazing at $LI_{95\%}$. On the other hand, definition of post-grazing targets also influences the short term rate of herbage intake of grazing animals by interfering with their ability to graze [3]. Carvalho [125] and Carvalho *et al.* [126] described how bite size, biting rate, and intake rate varied from the beginning to the end of grazing of rotationally managed *Cynodon* spp. and *Sorghum bicolor* L. During the first stages of the grazing process the short-term rate of intake remained stable, starting to decrease linearly after 40%–50% removal of the initial height [127]. For *Cynodon* spp. the decrease occurs at a faster rate because succeeding layers of herbage are more restrictive to bite formation than for *Sorghum bicolor* [128]. Similar results were reported for *Brachiaria brizantha* cv. Marandu [129] and for Mulato brachiariagrass hybrid [130]. As grazing progresses and herbage mass decreases towards the end of grazing, the proportions of stem and dead material increase in the consumed herbage as the consequence of the changing plant-part composition of the sward vertical profile [129–131]. This pattern of variation in the short-term rate of intake in relation to the decreasing sward height during grazing was formally described by Fonseca *et al.* [127,132] and corroborated by Mezzalira *et al.* [128], indicating the potential for manipulating pre- and post-grazing conditions as a means of optimising herbage production and intake in rotationally managed pastures. In general, more frequent defoliations (at 95% canopy light interception) than traditionally used, associated with moderate grazing severity (post-grazing heights around 50% of the pre-grazing height), result in greater leaf dry matter production [84,93,96,124], higher nutritive value [89,90] and intake rate of animals [127–129,132], augmenting animal performance and productivity [133], indicating that rational intensification of grassland use could be an effective way of ensuring sustainability of tropical pastoral systems of animal production. These results are in line with the findings of Knoke *et al.* [134] and provide feasible options for recovering degraded areas, reducing pressure on forest lands and releasing additional area for cropping and food production.

Under grazing, organic reserves are also an important feature to be considered, since they may represent an important source of energy during the early stages of regrowth, particularly when grazing is severe (low stubble heights), favoring regrowth and the competitive ability of plants. According to Matthew *et al.* [103], some grasses may lack organic reserves for bud release to increase tiller population density, a condition that could lead to failure in replacing dead tillers and result in pasture degradation. Despite the importance of stored reserves, the number of grazing studies with tropical grasses evaluating organic reserves is not large. Carvalho *et al.* [135] evaluated the concentration and the amount of total non-structural carbohydrates (TNC) in three *Cynodon* spp. cultivars (Tifton-85, Florakirk, and Coastcross) subjected to a range of grazing intensities under continuous stocking represented by the management heights of 5, 10, 15, and 20 cm. There was no reduction in either concentration or pool of TNC in shoots and roots of any of the grasses as a consequence of more severe grazing. Similar results were reported by Da Silva *et al.* [136] for Marandu palisade grass in an analogous experiment where management heights were 10, 20, 30, and 40 cm. Both experiments showed significant variation in the pool of TNC throughout the year, indicating a strong seasonal effect similar to other species [137]. Recorded values were greater during winter and early spring for both roots and shoots, and there was a reduction in the TNC pool in shoots during periods of active plant growth (late spring and summer). The TNC pool was larger in shoots than in roots, regardless of grazing intensity and season of the year [136], consistent with the findings of White [138], who indicated that the major storage areas of carbohydrate reserves in perennial grasses are usually the lower regions of the stems (stem bases), stolons, crowns, and rhizomes instead of roots.

The results from those experiments indicate the existence of a trade-off between leaf area quantity and quality in swards subjected to varying levels of grazing intensity under continuous stocking management. Under lenient grazing, LAI is high but photosynthetic potential of leaves is low, a consequence of the low light availability within the sward canopy. On the other hand, under severe grazing, swards are capable of compensating the smaller LAI through higher rates of tiller appearance and larger population of younger tillers [122], with greater growth potential than older tillers [121]. Since under continuous stocking management a relatively constant proportion of the sward leaf area is removed [139], the remaining leaf area may be sufficient for supplying assimilates, a condition in which the quality of the sward leaf area is crucial for maintaining plant growth [136]. This highlights the importance of the turnover in tiller population and the tillering process in tropical forage grasses, since the ability to replace tillers ensures rapid restoration of photosynthetic tissues and plant growth.

Under intermittent stocking, part of the initial growth of orchardgrass (*Dactylis glomerata* L.) [140] and perennial ryegrass (*Lolium perenne* L.) [140,141] after defoliation is sustained by assimilates supplied by the organic reserves during the first days of regrowth. However, Ward & Blaser [140] demonstrated that both the reserve carbohydrates and the remaining leaf area (current assimilates supplier) are involved in restoring the carbon balance of plants throughout regrowth. Leaf area restoration and regrowth depend not only on carbohydrate reserves but also on nitrogen reserves [141,142]. Carbon stored as carbohydrate is usually used following defoliation, but nitrogen usually comes from internal remobilization and recycling due to protein turnover associated with leaf senescence [142].

Although the importance of determining limits of grazing severity (management height) for ensuring leaf area restoration, plant growth and persistence is well recognized [143], understanding how defoliation frequency affects carbon and nitrogen reserves in tropical perennial forage grasses is less clear. Turner *et al.* [144] showed that frequent defoliation of orchardgrass (*Dactylis glomerata* L.) resulted in reduced water-soluble carbohydrate assimilation and, therefore, leaf, root, and tiller dry matter accumulation during subsequent periods of regrowth. Lestienne *et al.* [145] showed that defoliation frequency did not substantially affect N uptake, mobilization, and allocation between roots and adult and growing leaves on a plant basis, although tiller number per plant was largely increased under repeated defoliation in perennial ryegrass. In rotationally grazed tropical grasses, it is expected that defoliation severity has a relatively greater impact on mobilization and use of plant organic reserves than on temperate grasses, since it is the defoliation frequency that has been shown to be more related to the turnover of leaves in individual tillers and of tillers in tiller population. However, as discussed previously, some species have shown similar herbage accumulation within a range of grazing severity levels. As a result, field experiments are needed to evaluate whether the high population of young tillers is capable of restoring organic reserve levels during regrowth and to better understand the mechanisms involved with the use of organic reserves by tropical grasses.

In general, defoliation severities equivalent to 40%–60% removal of the initial (pre-grazing) height are within the limits of grazing resistance and use of plants, and ensure favorable conditions for high rates of herbage intake and animal performance when associated with the right pre-grazing management targets. In this context, defoliation frequency is relatively more important than defoliation severity for controlling stem elongation, the main morphological component that determines degeneration of sward structure [84] and imposes restrictions to grazing [3], particularly for tall-tufted grass species such as *Panicum* sp. and *Pennisetum* sp.

## 4. Future Perspectives and Challenges

The research with forage tropical grasses has shown significant progress since the late 1990s and early 2000s, a consequence of conceptual changes in approach and experimentation strategies. Such change provided the basis for relatively fast progress towards a knowledge of plant functional mechanisms comparable to that already available for temperate grasses. Although part of the same botanical family, temperate and tropical grasses have different photosynthetic pathways, a condition that has a direct influence on some important growth processes. For example, from an ontogenetic point of view, the results have demonstrated that, despite having a similar modular organization (sequential appearance of phytomers on the same tiller), there may be some differences in the dynamics of phytomer appearance and expansion between $C_3$ and $C_4$ grasses. As with $C_3$ grasses, $C_4$ grasses have, for a given species, a relatively constant number of live leaves per tiller. However, it is common for the majority of $C_4$ grasses for the concomitant expansion of two or more leaves and only one senescing leaf on the same tiller. As a result, the senescence rate is greater in a $C_4$ grass leaf relative to a $C_3$, suggesting that the former have a faster mobilization of nutrients from senescing leaves for supporting their growth. Another example is related to the stem elongation (stem + pseudostem) process. In contrast with $C_3$ grasses, $C_4$ grasses invest in stem elongation during vegetative growth. From an ecological point of view, such a strategy is interesting because it allows tillers to place their leaves in the upper strata of the sward, increasing the range of LAI traversed from the residue after grazing to the light interception capacity at the next grazing and consequently increasing their competitive ability and productivity. If on the one hand there is consistent evidence of a strong correlation between competition for light and high rates of stem elongation in $C_4$ grasses in the literature, on the other hand there is still the need for a better understanding about a possible ontogenetic programming of plants for producing stems (stem + pseudostem). That is because even with relatively high light availability (low light competition), tillers of tropical grasses can elongate the stem for maintaining the upright form. If the reasons for increasing stem elongation in relatively dense swards (high LAI) seem to be well explained by the amount and quality of the incident light, there is still no explanation for how and why stem elongation occurs in situations where light is not a limiting factor. Nevertheless, these two examples (leaf ontogenesis and stem elongation) are indicators of a likely genetic programming of $C_4$ grasses that is still not fully understood.

In relation to studies on a plant population level, several experiments have tried to describe the demographic pattern and tillering dynamics as well as the tiller size-density compensation mechanism for tropical forage grasses. In general, the results generated for some important genera such as *Panicum*, *Brachiaria* and *Pennisetum* have shown that tillering occurs at varying rates throughout the year, with a high turnover in tiller population during periods of high availability of resources (especially water) and suitable conditions (temperature) determined by climatic factors (late spring and summer) and a relative transitory instability in tiller population during periods of limited growth (winter and early spring). In spite of that, some studies have demonstrated that there seems to be an interesting interaction between defoliation severity and tiller population stability. For example, there is evidence that maintenance of ideal targets of grazing management for ensuring animal performance (relatively moderate defoliation, regardless of the grazing method used) result in high tiller survival during periods of intense vegetative growth (late spring and summer), favoring population stability without the need for intense tillering. Conversely, as tillers of most grass species have a relatively short period of life (less than six months, on average), there is high mortality during periods of adverse climatic conditions, causing instability in tiller population during autumn and winter. Therefore, it seems reasonable to hypothesize that strategic reductions in grazing management target heights during the autumn/winter period could stimulate tillering at the beginning of the following growing season, increasing population stability and getting pastures back into production earlier in the season. However, the impact of such a management strategy on the dynamics of the tiller population in the long term is still not clear. This highlights the importance of multi-year experiments depending on the question being asked.

Still on a plant population level, it is also important to point out the similarities in the tiller size-density compensation mechanisms found between $C_3$ and $C_4$ forage grasses characterized by the inverse relationship between tiller weight and population density on grazed swards, particularly those managed under different intensities of continuous stocking. Despite that, there is still limited information regarding the dynamics of that mechanism during regrowth of intermittently grazed pastures. A better understanding of the process could have important implications. For example, it is normally accepted in the scientific literature that after severe cutting or grazing the restoration of sward leaf area occurs, mainly, through the recruitment of new tillers from the plant base. As a result, an interesting exercise would be to determine if there would be a maximum residual LAI that would minimize tiller recruitment after grazing, increasing population stability and reducing grazing interval. However, since tillers have a genetically programmed lifespan, studies of this nature would have to consider the trade-offs between the two processes and evaluate whether management strategies that favor tiller survival could hinder plant persistence by reducing their ability to renew tiller population in subsequent grazing cycles.

Finally, several experiments with temperate grasses have shown that mixtures of plants from different species can increase (or at least maintain) the productive capacity of pastures relative to monocultures [146]. That possibility has not yet been seriously considered for tropical grasses. $C_4$ plants have a wide functional and morphological diversity (e.g., stoloniferous, prostrate, tall-tufted plants), a condition that allows for species combinations and coexistence between them (as corroborated by several experiments on native multi-specific pastures, e.g., [147,148]). For example, *Cynodon* species (stoloniferous/rhizomatous plants) are functionally different from *Pennisetum* species (tall-tufted, tussock forming plants). While the former direct a large proportion of their resources to producing structures like stolons (with large capacity for storing reserves), allowing plants to forage for light and resources horizontally, the latter can invest in structures like aerial tillers capable of occupying higher positions in the vertical strata of the sward, increasing LAI, and increasing light interception. Such functional and morphological diversity of tropical grasses indicates the possibility of coexistence of different plant types in the same area, given their complementary growth habits. This would allow for biodiversity in tropical ecosystems, and could represent a new window of opportunities towards attaining sustainable systems of animal production from pastures in the tropics. It seems that the question is not *whether* it can be done, but *how* to do it. Multi-specific pastures are normally associated with increased carbon sequestration, higher levels of soil organic matter, less evaporation with possible reduction in water stress to plants, increased meso and macrofauna biodiversity in the soil, and increased nutrient cycling [149]. There is no doubt that such benefits would boost multi-functionality and services provided by pastoral ecosystems in the tropics and could have an important economical, environmental, and social impact.

**Acknowledgments:** Thanks are due to Domicio do Nascimento Jr. for his contribution with the historical overview presented and to Carlos Guilherme Silveira Pedreira for his help in reviewing this manuscript, to the Brazilian National Council for Scientific and Technological Development (CNPq) for the sponsorship provided.

## References

1. FAO—Food and Agriculture Organization of the United Nations. Available online: http://faostat3.fao.org/browse/area/21/E (acessed on 8 May 2015).
2. IBGE—Instituto Brasileiro de Geografia e Estatística. *Censo Agropecuário 2006*; IBGE: Rio de Janeiro, Brazil, 2006.
3. Da Silva, S.C.; Carvalho, P.C.F. Foraging behaviour and herbage intake in the favourable tropics/subtropics. In *Grassland: A Global Resource*; McGilloway, D.A., Ed.; Wageningen Academic Publishers: Wageningen, The Netherlands, 2005; pp. 81–96.
4. Corsi, M.; Martha, G.B., Jr.; Nascimento, D., Jr.; Balsalobre, M.A.A. Impact of grazing management on productivity of tropical grassland. In Proceedings of the XIX International Grassland Congress, São Pedro, Brazil, 11–21 February 2001; Gomide, J.A., Mattos, W.R.S., da Silva, S.C., Eds.; FEALQ: São Pedro, Brazil, 2001; pp. 801–806.
5. Nascimento, D., Jr.; Barbosa, R.A.; Marcelino, K.R.A.; Garcez Neto, A.F.; Difante, G.S.; Lopes, B.A. A produção animal em pastagens no Brasil: Uso do conhecimento técnico e resultados. In *XX Simpósio Sobre Manejo da Pastagem*; Peixoto, A.M., Moura, J.C., da Silva, S.C., Faria, V.P., Eds.; FEALQ: Piracicaba, São Paulo, Brazil, 2003; pp. 1–82.
6. Graber, L.F.; Nelson, N.T.; Luekel, W.A.; Albert, W.B. *Organic Food Reserves in Relation to the Growth of Alfalfa and Other Perennial Herbaceous Plants*; Research Bulletin 80; Agricultural Experiment Station of the University of Wisconsin: Madison, WI, USA, 1927.
7. Volenec, J.J.; Ourry, A.; Joern, B.C. A role for nitrogen reserves in forage regrowth and stress tolerance. *Physiol. Plantarum* **1996**, *97*, 185–193. [CrossRef]
8. Watson, D.J. Comparative physiological studies on the growth of field crops. I. Variation in net assimilation rate and leaf area between species and varieties, and within and between years. *Ann. Bot.* **1947**, *11*, 41–76.
9. Black, J.N. The interrelationship of solar radiation and leaf area index in determining the rate of dry matter production of swards of subterranean clover (*Trifolium subterraneum* L.). *Aust. J. Agric. Res.* **1963**, *14*, 20–38. [CrossRef]
10. Brougham, R.W. A study in rate of pasture growth. *Aust. J. Agric. Res.* **1955**, *6*, 804–812. [CrossRef]
11. Brougham, R.W. Effects of intensity of defoliation on regrowth of pastures. *Aust. J. Agric. Res.* **1956**, *7*, 377–387. [CrossRef]
12. Brougham, R.W. Pasture growth rate studies in relation to grazing management. *Proc. New Zeal. Soc. An.* **1957**, *17*, 46–55.
13. Brougham, R.W. Interception of light by the foliage of pure and mixed stands of pasture plants. *Aust. J. Agric. Res.* **1958**, *9*, 39–52. [CrossRef]
14. Brougham, R.W. The effects of season and weather on the growth rate of a ryegrass clover pasture. *N. Z. J. Agric. Res.* **1959**, *2*, 283–296. [CrossRef]
15. Brougham, R. The effects of frequent hard grazing at different times of the year on the productivity and species yields of a grass-clover pasture. *N. Z. J. Agric. Res.* **1960**, *3*, 125–136. [CrossRef]
16. Brougham, R.W. The leaf growth of *Trifolium repens* as influenced by seasonal changes in the light environment. *J. Ecol.* **1962**, *50*, 449–459. [CrossRef]
17. Warren-Wilson, J. Influence of spatial arrangement of foliage area on light interception and pasture growth. In Proceedings of the VIII International Grassland Congress, Reading, England, 11–21 July 1960; Alden Press: Berkshire, UK, 1961; pp. 275–279.
18. Hodgson, J.; Bircham, J.S.; Grant, S.A.; King, J. The influence of cutting and grazing management on herbage growth and utilization. In *Plant Physiology and Herbage Production*; Wright, C.E., Ed.; Occasional Symposium, British Grassland Society: Nottingham, UK, 1981; Volume 13, pp. 51–62.
19. Hyder, D.N.; Sneva, F.A. Growth and carbohydrate trends in crested wheatgrass. *J. Range Manag.* **1959**, *12*, 271–276. [CrossRef]

20. Ryle, G.J.; Powell, C.E. Defoliation and regrowth in the graminaceous plant: The role of current assimilate. *Ann. Bot.* **1975**, *39*, 297–310.
21. Richards, J.H.; Caldwell, M.M. Soluble carbohydrates, concurrent photosynthesis and efficiency in regrowth following defoliation: A field study with *Agropyron* species. *J. Appl. Ecol.* **1985**, *22*, 907–920. [CrossRef]
22. Smith, D. Carbohydrate root reserves in alfalfa, red clover, and birdsfoot trefoil under several management schedules. *Crop Sci.* **1962**, *2*, 75–78.
23. Alcock, M.B. The physiological significance of defoliation to the subsequent regrowth of grass-clover mixtures and cereals. In *Grazing in Terrestrial and Marine Environments*; Crisp, D.J., Ed.; Blackwell Publications: Oxford, UK, 1964; pp. 25–41.
24. Weinmann, H. Total available carbohydrates in grasses and legumes. *Herb. Abstr.* **1961**, *31*, 225–261.
25. Brown, R.H.; Blaser, R.E. Leaf area index in pasture growth. *Herb. Abstr.* **1968**, *38*, 1–9.
26. Gomide, J.A. Fisiologia do crescimento livre de plantas forrageiras. In *I Simpósio sobre Manejo da Pastagem*; Faria, V.P., Moura, J.C., Eds.; FEALQ: Piracicaba, São Paulo, Brazil, 1973; pp. 83–89.
27. Jacques, A.V.A. Fisiologia do crescimento de plantas forrageiras: Área foliar e reservas orgânicas. In *I Simpósio sobre Manejo da Pastagem*; Faria, V.P., Moura, J.C., Eds.; FEALQ: Piracicaba, São Paulo, Brazil, 1973; pp. 95–101.
28. Rodrigues, L.R.A.; Rodrigues, T.J.D. Ecofisiologia de plantas forrageiras. In *Ecofisiologia da produção agrícola*; Castro, P.R.C., Ferreira, S.O., Yamada, T., Eds.; Potafós: Piracicaba, São Paulo, Brazil, 1987; pp. 203–230.
29. Parsons, A.J.; Penning, P.D. The effect of the duration of regrowth on photosynthesis, leaf death and the average rate of growth in a rotationally grazed sward. *Grass Forage Sci.* **1988**, *43*, 15–27. [CrossRef]
30. Parsons, A.J.; Leafe, E.L.; Collett, B.; Penning, P.D.; Lewis, J. The physiology of grass production under grazing. II. Photosynthesis, crop growth and animal intake of continuously-grazed swards. *J. Appl. Ecol.* **1983**, *20*, 127–139. [CrossRef]
31. Mott, G.O. Grazing pressure and the measurement of pastures production. In Proceedings of the VIII International Grassland Congress, Reading, England, 11–21 July 1960; pp. 606–611.
32. Mott, G.O. Evaluating forage production. In *Forages*, 3rd ed.; Heath, M.E., Metcalfe, D.S., Barnes, R.F., Eds.; The Iowa University Press: Ames, IA, USA, 1973; pp. 126–135.
33. Maraschin, G.E. Relembrando o passado, entendendo o presente e planejando o futuro. Uma herança em forrageiras e um legado em pastagens. In *XXXVII Reunião Anual da Sociedade Brasileira de Zootecnia*; Nascimento, D., Jr., Lopes, P.S., Pereira, J.C., Eds.; UFV: Viçosa, Minas Gerais, Brazil, 2000; pp. 113–180.
34. Maraschin, G.E. Pastejo rotacionado. In *III Simpósio Sobre Manejo da Pastagem*; Peixoto, A.M., Moura, J.C., Furlan, R.S., Faria, V.P., Eds.; FEALQ: Piracicaba, São Paulo, Brazil, 1976; pp. 253–282.
35. Maraschin, G.E. Manejo de plantas forrageiras dos gêneros *Digitaria*, *Cynodon* e *Chloris*. In *IX Simpósio sobre Manejo da Pastagem*; Peixoto, A.M., Moura, J.C., Faria, V.P., Eds.; FEALQ: Piracicaba, São Paulo, Brazil, 1988; pp. 109–140. (In Portuguese)
36. Maraschin, G.E. Produção de carne a pasto. In *XIII Simpósio Sobre Manejo da Pastagem*; Peixoto, A.M., Moura, J.C., Faria, V.P., Eds.; FEALQ: Piracicaba, São Paulo, Brazil, 1997; pp. 243–276. (In Portuguese)
37. Corsi, M. Espécies forrageiras para pastagem. In *III Simpósio sobre Manejo da Pastagem*; Peixoto, A.M., Moura, J.C., Furlan, R.S., Faria, V.P., Eds.; FEALQ: Piracicaba, São Paulo, Brazil, 1976; pp. 5–44. (In Portuguese)
38. Corsi, M. Parâmetros para intensificar o uso das pastagens. In *VI Simpósio sobre Manejo da Pastagem*; Peixoto, A.M., Moura, J.C., Faria, V.P., Eds.; FEALQ: Piracicaba, São Paulo, Brazil, 1980; pp. 214–263. (In Portuguese)
39. Corsi, M. Manejo de plantas forrageiras do gênero *Panicum*. In *IX Simpósio sobre Manejo da Pastagem*; Peixoto, A.M., Moura, J.C., Faria, V.P., Eds.; FEALQ: Piracicaba, São Paulo, Brazil, 1988; pp. 57–76. (In Portuguese)
40. Corsi, M.; Martha, G.B., Jr. Manutenção da fertilidade do solo em sistemas intensivos de pastejo rotacionado. In *XIV Simpósio sobre Manejo da Pastagem*; Peixoto, A.M., Moura, J.C., Faria, V.P., Eds.; FEALQ: Piracicaba, São Paulo, Brazil, 1997; pp. 161–192. (In Portuguese)
41. Barreto, I.L. Pastejo contínuo. In *III Simpósio Sobre Manejo da Pastagem*; Peixoto, A.M., Moura, J.C., Furlan, R.S., Faria, V.P., Eds.; FEALQ: Piracicaba, São Paulo, Brazil, 1976; pp. 219–251. (In Portuguese)

42. Blaser, R.E. Pasture-animal management to evaluate plants and to develop forage systemas. In *IX Simpósio Sobre Manejo da Pastagem*; Peixoto, A.M., Moura, J.C., Faria, V.P., Eds.; FEALQ: Piracicaba, São Paulo, Brazil, 1988; pp. 1–40.
43. Leite, G.G. Manejo de plantas forrageiras dos gêneros *Andropogon*, *Hyparrhenia* e *Setaria*. In *IX Simpósio sobre Manejo da Pastagem*; Peixoto, A.M., Moura, J.C., Faria, V.P., Eds.; FEALQ: Piracicaba, São Paulo, Brazil, 1988; pp. 185–218. (In Portuguese)
44. Leite, G.G.; Euclides, V.B.P. Utilização de pastagens de *Brachiaria* spp. In *XI Simpósio sobre Manejo da Pastagem*; Peixoto, A.M., Moura, J.C., Faria, V.P., Eds.; FEALQ: Piracicaba, São Paulo, Brazil, 1994; pp. 267–298. (In Portuguese)
45. Simão Neto, M. Sistemas de pastejo. 2. In *VIII Simpósio sobre Manejo da Pastagem*; Peixoto, A.M., Moura, J.C., Faria, V.P., Eds.; FEALQ: Piracicaba, São Paulo, Brazil, 1986; pp. 291–307. (In Portuguese)
46. Hillesheim, A. Manejo do gênero *Pennisetum* sob pastejo. In *IX Simpósio sobre Manejo da Pastagem*; Peixoto, A.M., Moura, J.C., Faria, V.P., Eds.; FEALQ: Piracicaba, São Paulo, Brazil, 1988; pp. 77–108. (In Portuguese)
47. Rodrigues, L.R.A.; Reis, R.A. Bases para o estabelecimento do manejo de capins do gênero *Panicum*. In *XII Simpósio Sobre Manejo da Pastagem*; Peixoto, A.M., Moura, J.C., Faria, V.P., Eds.; FEALQ: Piracicaba, São Paulo, Brazil, 1995; pp. 197–218. (In Portuguese)
48. Rodrigues, L.R.A.; Reis, R.A. Conceituação e modalidades de sistemas intensivos de pastejo rotacionado. In *XIV Simpósio Sobre Manejo da Pastagem*; Peixoto, A.M., Moura, J.C., Faria, V.P., Eds.; FEALQ: Piracicaba, São Paulo, Brazil, 1997; pp. 1–24. (In Portuguese)
49. Gomide, J.A. Sistemas de manejo de gramíneas do gênero *Melinis*. In *IX Simpósio sobre Manejo da Pastagem*; Peixoto, A.M., Moura, J.C., Faria, V.P., Eds.; FEALQ: Piracicaba, São Paulo, Brazil, 1988; pp. 41–56. (In Portuguese)
50. Euclides, V.P.B. Valor alimentício de espécies forrageiras do gênero Panicum. In *XII Simpósio sobre Manejo da Pastagem*; Peixoto, A.M., Moura, J.C., Faria, V.P., Eds.; FEALQ: Piracicaba, São Paulo, Brazil, 1995; pp. 245–274. (In Portuguese)
51. Zimmer, A.H.; Euclides, V.B.P.; Macedo, M.C.M. Manejo de plantas forrageiras do gênero *Brachiaria*. In *IX Simpósio sobre Manejo da Pastagem*; Peixoto, A.M., Moura, J.C., Faria, V.P., Eds.; FEALQ: Piracicaba, São Paulo, Brazil, 1988; pp. 141–184. (In Portuguese)
52. Faria, V.P.; Pedreira, C.G.S.; Santos, F.A.P. Evolução do uso de pastagens para bovinos. In *XIII Simpósio sobre Manejo da Pastagem*; Peixoto, A.M., Moura, J.C., Faria, V.P., Eds.; FEALQ: Piracicaba, São Paulo, Brazil, 1996; pp. 1–14. (In Portuguese)
53. Gibb, M.J.; Treacher, T.T. The effect of herbage allowance on herbage intake and performance of ewes and their twin lambs grazing perennial ryegrass. *J. Agric. Sci.* **1978**, *90*, 139–147. [CrossRef]
54. Jamieson, W.S.; Hodgson, J. The effect of daily herbage allowance and sward characteristics upon the ingestive behaviour and herbage intake of calves under strip-grazing management. *Grass Forage Sci.* **1979**, *34*, 261–271. [CrossRef]
55. Hodgson, J. The influence of grazing pressure and stocking rate on herbage intake and animal performance. In *Pasture Utilisation by the Grazing Animal*; Hodgson, J., Jackson, D., Eds.; Occasional Symposium, British Grassland Society: Hurley, UK, 1976; pp. 93–103.
56. Stobbs, T.H. The effect of plant structure on the intake of tropical pasture. I. Variation in the bite size of grazing cattle. *Aust. J. Agric. Res.* **1973**, *24*, 809–819. [CrossRef]
57. Stobbs, T.H. The effect of plant structure on the intake of tropical pasture. II. Differences in sward structure, nutritive value, and bite size of animals grazing Setaria anceps and Chloris gayana at various stages of growth. *Aust. J. Agric. Res.* **1973**, *24*, 821–829. [CrossRef]
58. Chacon, E.; Stobbs, T.H. Influence of progressive defoliation of grass sward on the eating behaviour of cattle. *Aust. J. Agric. Res.* **1976**, *27*, 702–727. [CrossRef]
59. Mott, G.O. Potential productivity of temperate and tropical grassland systems. In Proceedings of the XIV International Grassland Congress, Lexington, KY, USA, 15–24 June 1981; Smith, J.A., Hays, V.W., Eds.; Westview Press: Boulder, CO, USA, 1983; pp. 35–41.
60. Bircham, J.S.; Sheath, G.W. Pasture utilisation in hill country: 2. A general model describing pasture mass and intake under sheep and cattle grazing. *N. Z. J. Agric. Res.* **1986**, *29*, 639–648. [CrossRef]

61. Poppi, D.P.; Hughes, T.P.; L'Huillier, P.J. Intake of pasture by grazing ruminants. In *Feeding Livestock on Pasture*; Nicol, A.M., Ed.; Occasional Publication, New Zealand Society of Animal Production: Hamilton, New Zealand, 1987; pp. 55–63.
62. Leafe, E.L.; Parsons, A.J. Physiology of growth of a grazed sward. In Proceedings of the XIV International Grassland Congress, Lexington, KY, USA, 15–24 June 1981; Westview Press: Boulder, CO, USA, 1983; pp. 403–406.
63. Bircham, J.S.; Hodgson, J. The influence of sward condition on rates of herbage growth and senescence in mixed sward under continuous stocking management. *Grass Forage Sci.* **1983**, *38*, 323–331. [CrossRef]
64. Hodgson, J. *Grazing Management: Science into Practice*; Longman Scientific and Technical: Harlow, Essex, UK, 1990.
65. Hodgson, J. The control of herbage intake in the grazing ruminant. *Proc. Nutr. Soc.* **1985**, *44*, 339–346. [CrossRef] [PubMed]
66. Korte, C.J.; Watkin, B.R.; Harris, W. Use of residual leaf area index and light interception as criteria for spring-grazing management of a ryegrass-dominant pasture. *N. Z. J. Agric. Res.* **1982**, *25*, 309–319. [CrossRef]
67. Parsons, A.J. The effects of season and management on the growth of grass swards. In *The Grass Crop: The Physiological Basis of Production*; Jones, M.B., Lazenby, A., Eds.; Chapman & Hall: London, UK, 1988; pp. 129–177.
68. Chapman, D.F.; Lemaire, G. Morphogenetic and structural determinants of regrowth after defoliation. In Proceedings of the XVII International Grassland Congress, Palmerston North, New Zealand, 8–21 February 1993; pp. 95–104.
69. Da Silva, S.C.; Pedreira, C.G.S. Princípios de ecologia aplicados ao manejo da pastagem. In *III Simpósio Sobre Ecossistema de Pastagens*; Favoretto, V., Rodrigues, L.R.A., Rodrigues, T.J.D., Eds.; FUNEP: Jaboticabal, São Paulo, Brazil, 1997; pp. 1–62. (In Portuguese)
70. Pinto, J.C.; Gomide, J.A.; Maestri, M. Produção de matéria seca e relação folha/caule de gramíneas forrageiras tropicais, cultivadas em vasos, com duas doses de nitrogênio. *Rev. Bras. Zootecn.* **1994**, *23*, 313–326. (In Portuguese)
71. Pinto, J.C.; Gomide, J.A.; Maestri, M. Crescimento de folhas de gramíneas forrageiras tropicais, cultivadas em vasos, com duas doses de nitrogênio. *Rev. Bras. Zootecn.* **1994**, *23*, 327–332. (In Portuguese)
72. Gomide, J.A. Morfogênese e análise de crescimento de gramíneas tropicais. In *Simpósio Internacional Sobre Produção Animal em Pastejo*; Pereira, O.G., Obeid, J.A., Nascimento, D., Jr., Fonseca, D.M., Eds.; Universidade Federal de Viçosa: Viçosa, Minas Gerais, Brazil, 1997; pp. 411–430. (In Portuguese)
73. Lemaire, G. The physiology of grass growth under grazing: Tissue turnover. In *Simpósio Internacional Sobre Produção Animal em Pastejo*; Pereira, O.G., Obeid, J.A., Nascimento, D., Jr., Fonseca, D.M., Eds.; Universidade Federal de Viçosa: Viçosa, Minas Gerais, Brazil, 1997; pp. 115–144.
74. Hodgson, J.; da Silva, S.C. Sustainability of grazing systems: Goals, concepts and methods. In *Grassland Ecophysiology and Grazing Ecology*; Lemaire, G., Hodgson, J., Moraes, A., Nabinger, C., Carvalho, P.C.F., Eds.; CAB International: Wallingford, UK, 2000; pp. 1–14.
75. Cruz, P.; Boval, M. Effect of nitrogen on some morphogenetic traits of temperate and tropical perennial forage grasses. In *Grassland Ecophysiology and Grazing Ecology*; Lemaire, G., Hodgson, J., Moraes, A., Nabinger, C., Carvalho, P.C.F., Eds.; CAB International: Wallingford, UK, 2000; pp. 151–168.
76. Da Silva, S.C.; Nascimento, D., Jr. Research advances in tropical grasses on pasture: Morphophysiological characteristics and grazing management. *Rev. Bras. Zootecn.* **2007**, *36*, 121–138.
77. Da Silva, S.C. Understanding the dynamics of herbage accumulation in tropical grass species: the basis for planning efficient grazing management practices. In *II Symposium on Grassland Ecophysiology and Grazing Ecology*; [CD-ROM]; Pizarro, E., Carvalho, P.C.F., da Silva, S.C., Eds.; UFPR: Curitiba, Paraná, Brazil, 2004.
78. Matthew, C.; Assuero, S.G.; Black, C.K.; Sackville-Hamilton, N.R. Tiller dynamics in grazed swards. In *Grassland Ecophysiology and Grazing Ecology*; Lemaire, G., Hodgson, J., Moraes, A., Nabinger, C., Carvalho, P.C.F., Eds.; CAB International: Wallingford, UK, 2000; pp. 127–150.
79. Bahmani, I.; Thom, E.R.; Matthew, C.; Hooper, R.J.; Lemaire, G. Tiller dynamics of perennial ryegrass cultivars derived from different New Zealand ecotypes: Effects of cultivar, season, nitrogen fertilizer, and irrigation. *Aust. J. Agric. Res.* **2003**, *54*, 803–817. [CrossRef]

80. Da Silva, S.C.; Nascimento, D., Jr.; Sbrissia, A.F.; Pereira, L.E.T. Dinâmica de população de plantas forrageiras em pastagens. In *IV Simpósio sobre Manejo Estratégico da Pastagem*; Pereira, O.G., Obeid, J.A., Nascimento, D., Jr., Fonseca, D.M., Eds.; Editora UFV: Viçosa, Minas Gerais, Brazil, 2008; pp. 75–100.
81. Cowan, R.T.; Lowe, K.F. Tropical and subtropical grass management and quality. In *Grass for Dairy Cattle*; Cherney, J.H., Cherney, D.J.R., Eds.; CABI Publishing: Wallingford, UK, 1998; pp. 101–136.
82. Hodgson, J.; da Silva, S.C. Options in tropical pasture management. In Proceedings of the Annual Meeting of the Brazilian Animal Science Society, Recife, Brazil, 29 July–1 August 2002; SBZ: Recife, Brazil, 2002; Volume 39, pp. 180–202.
83. Pereira, L.E.T.; Paiva, A.J.; Guarda, V.D.A.; Pereira, P.M.; Caminha, F.O.; da Silva, S.C. Herbage utilisation efficiency of continuously stocked marandu palisadegrass subjected to nitrogen fertilisation. *Sci. Agric.* **2015**, *72*, 114–123.
84. Da Silva, S.C.; Bueno, A.A.O.; Carnevalli, R.A.; Uebele, M.C.; Bueno, F.O.; Hodgson, J.; Matthew, C.; Arnold, J.C.; Morais, J.P.G. Sward structural characteristics and herbage accumulation of *Panicum maximum* cv. Mombaça subject to rotational stocking managements. *Sci. Agric.* **2009**, *66*, 8–19. [CrossRef]
85. Carnevalli, R.A.; Da Silva, S.C.; Bueno, A.A.O.; Uebele, M.C.; Bueno, F.O.; Hodgson, J.; Silva, G.N.; Morais, J.P.G. Herbage production and grazing losses in *Panicum maximum* cv. Mombaça under four grazing management. *Trop. Grassl.* **2006**, *40*, 165–176.
86. Ballaré, C.L. Keeping up with the neighbours: Phytochrome sensing and other signalling mechanisms. *Trends Plant Sci.* **1999**, *4*, 97–102. [CrossRef]
87. Murphy, J.S.; Briske, D.D. Regulation of tillering by apical dominance—Chronology, interpretive value, and current perspectives. *J. Range Manag.* **1992**, *45*, 419–429. [CrossRef]
88. Lemaire, G. Ecophysiology of grasslands: Dynamic aspects of forage plant population in grazed swards. In Proceedings of the XIX International Grassland Congress, São Pedro, Brazil, 11–21 February 2001; Gomide, J.A., Mattos, W.R.S., da Silva, S.C., Eds.; FEALQ: São Pedro, Brazil, 2001; pp. 29–38.
89. Santos, P.M.; Corsi, M.; Pedreira, C.G.S.; Lima, C.G. Tiller cohort development and digestibility in Tanzania guinea grass (*Panicum maximum* cv. Tanzânia) under three levels of grazing intensity. *Trop. Grassl.* **2006**, *40*, 84–93.
90. Difante, G.S.; Nascimento, D., Jr.; Euclides, V.P.B.; da Silva, S.C.; Barbosa, R.A.; Gonçalves, W.V. Sward structure and nutritive value of Tanzânia guineagrass subject to rotational stocking managements. *Rev. Bras. Zootecn.* **2009**, *38*, 9–19. [CrossRef]
91. Silveira, M.C.T.; da Silva, S.C.; Souza, S.J., Jr.; Barbero, L.M.; Rodrigues, C.S.; Limão, V.A.; Pena, K.S.; Nascimento, D., Jr. Herbage accumulation and grazing losses on Mulato grass subjected to strategies of rotational stocking management. *Sci. Agric.* **2013**, *70*, 242–249. [CrossRef]
92. Barbosa, R.A.; Nascimento, D., Jr.; Euclides, V.P.B.; Da Silva, S.C.; Zimmer, A.H.; Torres, R.A.A., Jr. Capim Tanzânia submetido a combinações entre intensidade e frequência de pastejo. *Pesqui. Agropecu. Bras.* **2007**, *42*, 329–340. [CrossRef]
93. Zanini, G.D.; Santos, G.T.; Sbrissia, A.F. Frequencies and intensities of defoliation in Aruana guineagrass swards: Morphogenetic and structural characteristics. *Rev. Bras. Zootecn.* **2012**, *41*, 1848–1857. [CrossRef]
94. Pereira, L.E.T.; Paiva, A.J.; Geremia, E.V.; Da Silva, S.C. Components of herbage accumulation in elephant grass cvar Napier subjected to strategies of intermittent stocking management. *J. Agr. Sci.* **2014**, *152*, 954–966. [CrossRef]
95. Pereira, L.E.T.; Paiva, A.J.; Geremia, E.V.; da Silva, S.C. Grazing management and tussock distribution in elephant grass. *Grass Forage Sci.* **2014**, *70*, 1–12. [CrossRef]
96. Pedreira, B.C.; Pedreira, C.G.S.; da Silva, S.C. Estrutura do dossel e acúmulo de forragem de *Brachiaria brizantha* cultivar Xaraés em resposta a estratégias de pastejo. *Pesqui. Agropecu. Bras.* **2007**, *42*, 281–287. (In Portuguese) [CrossRef]
97. Pedreira, B.C.; Pedreira, C.G.S.; da Silva, S.C. Acúmulo de forragem durante a rebrotação de capim-xaraés submetido a três estratégias de desfolhação. *Rev. Bras. Zootecn.* **2009**, *38*, 618–625. (In Portuguese) [CrossRef]
98. Giacomini, A.A.; da Silva, S.C.; Sarmento, D.O.L.; Zeferino, C.V.; Souza, S.J., Jr.; Trindade, J.K.; Guarda, V.D.A.; Nascimento, D., Jr. Growth of marandu palisadegrass subjected to strategies of intermittent stocking. *Sci. Agric.* **2009**, *66*, 733–741.

99. Voltolini, T.T.; Santos, F.A.P.; Martinez, J.C.; Clarindo, R.L.; Penati, M.A.; Imaizumi, H. Características produtivas e qualitativas do capim-elefante pastejado em intervalo fixo ou variável de acordo com a interceptação da radiação fotossinteticamente ativa. *Rev. Bras. Zootecn.* **2010**, *39*, 1002–1010. (In Portuguese) [CrossRef]
100. Portela, J.N. Intensidade e Frequência de Desfolhação como Definidores da Estrutura do Dossel, da Morfogênese e do Valor Nutritivo de *Brachiaria Decumbens* Stapf. cv. Basilisk sob Lotação Intermitente. Ph.D. Thesis, Escola Superior de Agricultura "Luiz de Queiroz"—Universidade de São Paulo, Piracicaba, Brazil, 19 November 2010.
101. Barbero, L.M. Respostas Morfogênicas e Caracteristicas Estruturais do Capim-Mulato Submetido a Estratégias de Pastejo Rotativo. Ph.D. Thesis, Escola Superior de Agricultura "Luiz de Queiroz", Universidade de São Paulo, Piracicaba, Brazil, 28 March 2011.
102. Crestani, S. Respostas Morfogênicas e Dinâmica da População de Perfilhos e Touceiras em *Brachiaria brizantha* cv. Piatã Submetida a Regimes de Sombra em Área de Integração Lavoura-Pecuária-Floresta. Ph.D. Thesis, Escola Superior de Agricultura "Luiz de Queiroz", Universidade de São Paulo, Piracicaba, Brazil, 27 February 2015.
103. Matthew, C.; Lemaire, G.; Hamilton, N.R.S.; Hernández-Garay, A. A modified self-thinning equation to describe size/density relationships for defoliated swards. *Ann. Bot.* **1995**, *76*, 579–587. [CrossRef]
104. Sbrissia, A.F.; da Silva, S.C.; Carvalho, C.A.B.; Carnevalli, R.A.; Pinto, L.F.M.; Fagundes, J.L.; Pedreira, C.G.S. Tiller size/population density compensation in Coastcross grazed swards. *Sci. Agric.* **2001**, *58*, 655–665. [CrossRef]
105. Sbrissia, A.; da Silva, S.; Matthew, C.; Carvalho, C.A.B.; Carnevalli, R.A.; Pinto, L.F.M.; Fagundes, J.L.; Pedreira, C.G.S. Tiller size/density compensation in grazed Tifton 85 bermudagrass swards. *Pesqui. Agropecu. Bras.* **2003**, *38*, 1459–1468. [CrossRef]
106. Sbrissia, A.F.; Da Silva, S.C. Compensação tamanho/densidade populacional de perfilhos em pastos de capim-marandu. *Rev. Bras. Zootecn.* **2008**, *37*, 35–47. (In Portuguese) [CrossRef]
107. Calsina, L.M.; AgnusdeI, M.G.; Assuero, S.G.; Pérez, H. Size⁄density compensation in *Chloris gayana* Kunth cv. Fine Cut subjected to different defoliation regimes. *Grass Forage Sci.* **2012**, *67*, 255–262. [CrossRef]
108. Stuefer, J.F. Two types of division of labour in clonal plants: Benefits, costs and constraints. *Perspect. Plant Ecol. Evol. Syst.* **1998**, *1*, 47–60. [CrossRef]
109. Matthew, C.; van Loo, E.N.; Thom, E.R.; Dawson, L.A.; Care, D.A. Understanding shoot and root development. In Proceedings of the XIX International Grassland Congress, São Pedro, Brazil, 11–21 February 2001; Gomide, J.A., Mattos, W.R.S., da Silva, S.C., Eds.; FEALQ: São Pedro, Brazil, 2001; pp. 19–27.
110. Sackville Hamilton, N.R.; Matthew, C.; Lemaire, G. In defence of the -3/2 boundary rule: A re-evaluation of self-thinning concepts and status. *Ann. Bot.* **1995**, *76*, 569–577. [CrossRef]
111. Hernández-Garay, A.; Matthew, C.; Hodgson, J. Tiller size-density compensation in ryegrass miniature swards subject to differing defoliation heights and a proposed productivity index. *Grass Forage Sci.* **1999**, *54*, 347–356. [CrossRef]
112. De Kroon, H.; Huber, H.; Stuefer, J.F.; van Groenendael, J.M. A modular concept of phenotypic plasticity in plants. *New Phytol.* **2005**, *166*, 73–82. [CrossRef] [PubMed]
113. Lopes, B.A. Características Morfofisiológicas e Acúmulo de Forragem em Capim-Mombaça Submetido a Regimes de Desfolhação. Ph.D. Thesis, Universidade Federal de Viçosa, Viçosa, Brazil, 3 May 2006.
114. Montagner, D.B. Morfogênese e Acúmulo de Forragem em Capim-Mombaça Submetido a Intensidades de Pastejo Rotativo. Ph.D. Thesis, Universidade Federal de Viçosa, Viçosa, Brazil, 17 August 2007.
115. Ryel, R.J.; Beyschlag, W.; Caldwell, M.M. Light field heterogeneity among tussock grasses: Theoretical considerations of light harvesting and seedling establishment in tussocks and uniform tiller distributions. *Oecologia* **1994**, *98*, 241–246. [CrossRef]
116. Castillo, J.M.; Rubio-Casal, A.E.; Luque, T.; Figueroa, M.E.; Jimnenez-Nieva, F.J. Intratussock tiller distribution and biomass of *Spartina densiflora* Brongn. in a invaded salt marsh. *Lagascalia* **2003**, *23*, 61–73.
117. Derner, J.D.; Briske, D.D.; Polley, H.W. Tiller organization within the tussock grass *Schizachyrium scoparium*: A field assessment of competition-cooperation tradeoffs. *Botany* **2012**, *90*, 669–677. [CrossRef]
118. Carvalho, D.D.; Matthew, C.; Hodgson, J. Effect of aging in tillers of *Panicum maximum* on leaf elongation rate. In Proceedings of the XIX International Grassland Congress, São Pedro, Brazil, 11–21 February 2001; Gomide, J.A., Mattos, W.R.S., da Silva, S.C., Eds.; FEALQ: São Pedro, Brazil, 2001; pp. 41–42.

119. Paiva, A.J.; Pereira, L.E.T.; da Silva, S.C.; Dias, R.A.P. Identification of tiller age categories based on morphogenetic responses of continuously stocked marandu palisade grass fertilised with nitrogen. *Cienc. Rural* **2015**, *45*, 867–870. [CrossRef]
120. Caminha, F.O.; da Silva, S.C.; Paiva, A.J.; Pereira, L.E.T.; Mesquita, P.; Guarda, V.D. Stability of tiller population of continuously stocked marandu palisade grass fertilized with nitrogen. *Pesqui. Agropecu. Bras.* **2010**, *45*, 213–220. [CrossRef]
121. Paiva, A.J.; da Silva, S.C.; Pereira, L.E.T.; Caminha, F.O.; Pereira, P.M.; Guarda, V.D.A. Morphogenesis on age categories of tillers in marandu palisadegrass. *Sci. Agric.* **2011**, *68*, 626–631.
122. Sbrissia, A.F.; da Silva, S.C.; Sarmento, D.O.L.; Molan, L.K.; Andrade, F.M.E.; Gonçalves, A.C.; Lupinacci, A.V. Tillering dynamics in palisadegrass swards continuously stocked by cattle. *Plant Ecol.* **2010**, *206*, 349–359. [CrossRef]
123. Sousa, B.M.L.; Nascimento, D., Jr.; da Silva, S.C.; Freitas, H.C.M.; Rodrigues, C.S.; Fonseca, D.M.; Siveira, M.C.T.; Sbrissia, A.F. Structural and morphogenetic characteristics of andropogon grass submitted to different cutting heights. *Rev. Bras. Zootecn.* **2010**, *39*, 2141–2147. [CrossRef]
124. Pereira, L.E.T.; Paiva, A.J.; Geremia, E.V.; Da Silva, S.C. Regrowth patterns of elephant grass (*Pennisetum purpureum* Schum.) subjected to strategies of intermittent stocking management. *Grass Forage Sci.* **2015**, *70*, 195–204. [CrossRef]
125. Carvalho, P.C.F. Sward management as a generator of suitable grazing environments for animal production. In *Teoria e Prática da Produção Animal em Pastagens*; Pedreira, C.G.S., Moura, J.C., da Silva, S.C., Eds.; FEALQ: Piracicaba, São Paulo, Brazil, 2005; pp. 7–32.
126. Carvalho, P.C.F.; Trindade, J.K.; da Silva, S.C.; Bremm, C.; Mezzalira, J.C.; Nabinger, C.; Amaral, M.F.; Carassai, I.J.; Martins, R.S.; Genro, T.C.M.; *et al.* Consumo de forragem por animais em pastejo: Analogias e simulaçoes em pastoreio rotativo. In *Simpósio Sobre Manejo da Pastagem—Intensificação de Sistemas de Produção Animal em Pasto*; da Silva, S.C., Pedreira, C.G.S., Moura, J.C., Faria, V.P., Eds.; FEALQ: Piracicaba, Sa o Paulo, Brazil, 2009; pp. 61–93 . (In Portuguese)
127. Fonseca, L.; Mezzalira, J.C.; Bremm, C.; Filho, R.S.A.; Gonda, H.L.; Carvalho, P.C.F. Management targets for maximising the short-term herbage intake rate of cattle grazing in *Sorghum bicolor*. *Livest. Sci.* **2012**, *145*, 205–211. [CrossRef]
128. Mezzalira, J.C.; Carvalho, P.C.F.; Fonseca, L.; Bremm, C.; Cangiano, C.H.; Gonda, H.L.; Laca, E.A. Behavioural mechanisms of intake rate by heifers grazing swards of contrasting structures. *Appl. Anim. Behav. Sci.* **2014**, *153*, 1–9. [CrossRef]
129. Trindade, J.K.; da Silva, S.C.; Souza, S.J., Jr.; Giacomini, A.A.; Zeferino, C.V.; Guarda, V.D.A.; Carvalho, P.C.F. Composição morfológica da forragem consumida por bovinos de corte durante o rebaixamento do capim-marandu submetido a estratégias de pastejo rotativo. *Pesqui. Agropecu. Bras.* **2007**, *42*, 883–890. (In Portuguese) [CrossRef]
130. Souza, S.J., Jr. Modificações na estrutura do dossel, comportamento ingestivo e composição da dieta de bovinos durante o rebaixamento do capim-mulato submetido a estratégias de pastejo rotativo. Ph.D. Thesis, Universidade de São Paulo, Piracicaba, Brazil, 13 October 2011.
131. Zanini, G.D.; Santos, G.T.; Schmitt, D.; Padilha, D.A.; Sbrissia, A.F. Distribution of stem in the vertical structure of Aruana guineagrass and Annual ryegrass pastures subjected to rotational grazing by sheep. *Cienc. Rural* **2012**, *42*, 882–887.
132. Fonseca, L.; Carvalho, P.C.F.; Mezzalira, J.C.; Bremm, C.; Galli, J.R.; Gregorini, P. Effect of sward surface height and level of herbage depletion on bite features of cattle grazing *Sorghum bicolor* swards. *J. Anim. Sci.* **2013**, *91*, 4357–4365. [CrossRef] [PubMed]
133. Gimenes, F.M.A.; Da Silva, S.C.; Fialho, C.A.; Gomes, M.B.; Berndt, A.; Gerdes, L.; Colozza, M.T. Ganho de peso e produtividade animal em capim-marandu sob pastejo rotativo e adubação nitrogenada. *Pesqui. Agropecu. Bras.* **2011**, *46*, 751–759. (In Portuguese) [CrossRef]
134. Knoke, T.; Bendix, J.; Pohle, P.; Hamer, U.; Hildebrandt, P.; Roos, K.; Gerique, A.; Sandoval, M.L.; Breuer, L.; Tischer, A.; *et al.* Afforestation or intense pasturing improve ecological and economic value of abandoned tropical farmlands. *Nat. Commun.* **2014**, *5*, 5612. [CrossRef] [PubMed]
135. Carvalho, C.A.B.; da Silva, S.C.; Sbrissia, A.F.; Fagundes, J.L.; Carnevalli, R.A.; Pinto, L.F.M.; Pedreira, C.G.S. Carboidratos não estruturais e acúmulo de forragem em pastagens de *Cynodon* spp. sob lotação continua. *Sci. Agric.* **2001**, *58*, 667–674. (In Portuguese)

136. Da Silva, S.C.; Pereira, L.E.T.; Sbrissia, A.F.; Hernández-Garay, A. Carbon and nitrogen reserves in marandu palisade grass subjected to intensities of continuous stocking management. *J. Agric. Sci.* **2014**, *7*, 1–15. [CrossRef]
137. Corre, N.; Bouchart, V.; Ourry, A.; Boucaud, J. Mobilization of nitrogen reserves during regrowth of defoliated *Trifolium repens* L. and identification of potential vegetative storage proteins. *J. Exp. Bot.* **1996**, *47*, 1111–1118. [CrossRef]
138. White, L.M. Carbohydrate reserves of grasses: A review. *J. Range Manag.* **1973**, *26*, 13–18. [CrossRef]
139. Da Silva, S.C.; Gimenes, F.M.A.; Sarmento, D.O.L.; Sbrissia, A.F.; Oliveira, D.E.; Hernadez-Garay, A.; Pires, A.V. Grazing behaviour, herbage intake and animal performance of beef cattle heifers on marandu palisade grass subjected to intensities of continuous stocking management. *J. Agric. Sci.* **2013**, *151*, 727–739. [CrossRef]
140. Ward, V.Y.; Blaser, R.E. Carbohydrates feed reserves and leaf. *Crop Sci.* **1961**, *1*, 366–370. [CrossRef]
141. Schnyder, H.; de Visser, R. Fluxes of reserve-derived and currently assimilated carbon and nitrogen in perennial ryegrass recovering from defoliation. The regrowing tiller and its component functionally distinct zones. *Plant Physiol.* **1999**, *119*, 1423–1436. [CrossRef] [PubMed]
142. Thornton, B.; Millard, P.; Bausenwein, U. Reserve formation and recycling of carbon and N during regrowth of defoliated plants. In *Grassland Ecophysiology and Grazing Ecology*; Lemaire, G., Hodgson, J., Moraes, A., Nabinger, C., Carvalho, P.C.F., Eds.; CAB International: Wallingford, UK, 2000; pp. 85–99.
143. Fulkerson, W.J.; Donaghy, D.J. Plant soluble carbohydrate reserves and senescence—Key criteria for developing an effective grazing management system for ryegrass-based pastures: A review. *Aust. J. Exp. Agric.* **2001**, *41*, 261–275. [CrossRef]
144. Turner, L.R.; Donaghy, D.J.; Lane, P.A.; Rawnsley, R.P. Effect of defoliation interval on water-soluble carbohydrate and nitrogen energy reserves, regrowth of leaves and roots, and tiller number of cocksfoot (*Dactylis glomerata* L.) plants. *Aust. J. Agric. Res.* **2006**, *57*, 243–249. [CrossRef]
145. Lestienne, F.; Thornton, B.; Gastal, F. Impact of defoliation intensity and frequency on N uptake and mobilization in Lolium perenne. *J. Exp. Bot.* **2006**, *57*, 997–1006. [CrossRef] [PubMed]
146. Volaire, F.; Barkaoui, K.; Norton, M. Designing resilient and sustainable grasslands for a drier future: Adaptive strategies, functional traits and biotic interactions. *Eur. J. Agron.* **2014**, *52*, 81–89. [CrossRef]
147. Agnusdei, M.G.; Mazzanti, A. Frequency of defoliation of native and naturalized species of the Flooding Pampas (Argentina). *Grass Forage Sci.* **2001**, *56*, 344–351. [CrossRef]
148. Pillar, V.D.P.; Müller, S.C.; Castilhos, Z.M.D.S.; Jacques, A.V.A. Campos Sulinos-conservação e uso sustentável da biodiversidade. In *Secretaria de Biodiversidade e Florestas, Departamento de Conservação da Biodiversidade*, 1st ed.; Ministério do Meio Ambiente—MMA: Brasília, Distrito Federal, Brazil, 2009. (In Portuguese)
149. Parsons, A.J.; Rowart, J.S.; Newton, P.C.D. Managing pasture for animals and soil carbon. *Proc. N. Z. Grassl. Assoc.* **2009**, *71*, 77–84.

# Permissions

All chapters in this book were first published in MDPI; hereby published with permission under the Creative Commons Attribution License or equivalent. Every chapter published in this book has been scrutinized by our experts. Their significance has been extensively debated. The topics covered herein carry significant findings which will fuel the growth of the discipline. They may even be implemented as practical applications or may be referred to as a beginning point for another development.

The contributors of this book come from diverse backgrounds, making this book a truly international effort. This book will bring forth new frontiers with its revolutionizing research information and detailed analysis of the nascent developments around the world.

We would like to thank all the contributing authors for lending their expertise to make the book truly unique. They have played a crucial role in the development of this book. Without their invaluable contributions this book wouldn't have been possible. They have made vital efforts to compile up to date information on the varied aspects of this subject to make this book a valuable addition to the collection of many professionals and students.

This book was conceptualized with the vision of imparting up-to-date information and advanced data in this field. To ensure the same, a matchless editorial board was set up. Every individual on the board went through rigorous rounds of assessment to prove their worth. After which they invested a large part of their time researching and compiling the most relevant data for our readers.

The editorial board has been involved in producing this book since its inception. They have spent rigorous hours researching and exploring the diverse topics which have resulted in the successful publishing of this book. They have passed on their knowledge of decades through this book. To expedite this challenging task, the publisher supported the team at every step. A small team of assistant editors was also appointed to further simplify the editing procedure and attain best results for the readers.

Apart from the editorial board, the designing team has also invested a significant amount of their time in understanding the subject and creating the most relevant covers. They scrutinized every image to scout for the most suitable representation of the subject and create an appropriate cover for the book.

The publishing team has been an ardent support to the editorial, designing and production team. Their endless efforts to recruit the best for this project, has resulted in the accomplishment of this book. They are a veteran in the field of academics and their pool of knowledge is as vast as their experience in printing. Their expertise and guidance has proved useful at every step. Their uncompromising quality standards have made this book an exceptional effort. Their encouragement from time to time has been an inspiration for everyone.

The publisher and the editorial board hope that this book will prove to be a valuable piece of knowledge for researchers, students, practitioners and scholars across the globe.

# List of Contributors


**Christine H. Gelley**
Agriculture & Natural Resources Educator, Ohio State University Extension, Caldwell, OH 43724, USA

**Amanda J. Ashworth**
Poultry Production and Product Safety Research Unit, USDA-ARS, Fayetteville, AR 72701, USA

**Patrick D. Keyser**
Department of Forestry, Wildlife and Fisheries, University of Tennessee, Knoxville, TN 37996, USA

**Renata L. G. Nave**
Department of Plant Sciences, University of Tennessee, Knoxville, TN 37996, USA

**Justin D. Rhinehart**
Department of Animal Science, University of Tennessee, Knoxville, TN 37996, USA

**Celal Cevher**
Field Crops Central Research Institute, Şehit Cem Ersever 9-11, Yenimahalle, Ankara 06170, Turkey

**Bulent Altunkaynak**
Department of Statistics, Faculty of Science, Gazi University, Beşevler, Ankara 06500, Turkey

**Surendra Bhattarai, Dilip Biswas and Bill Biligetu**
Crop Development Center of Department of Plant Sciences, University of Saskatchewan, 51 Campus Drive, Saskatoon, SK S7N 5A8, Canada

**Yong-Bi Fu**
Plant Gene Resources of Canada, Saskatoon Research and Development Centre, Agriculture and Agri-Food Canada, 107 Science Place, Saskatoon, SK S9H 3X2, Canada

**Tim Vleugels and Gerda Cnops**
Plant Sciences Unit, Flanders Research Institute for Agriculture, Fisheries and Food (ILVO), Caritasstraat 39, 9090 Melle, Belgium

**Helga Amdahl**
Graminor Breeding Ltd., Hommelstadvegen 60, NO-2322 Ridabu, Norway

**Isabel Roldán-Ruiz**
Plant Sciences Unit, Flanders Research Institute for Agriculture, Fisheries and Food (ILVO), Caritasstraat 39, 9090 Melle, Belgium
Department of Plant Biotechnology and Bioinformatics, Ghent University, Technologiepark 71, 9052 Zwijnaarde, Belgium

**Alec D. Mackay**
AgResearch Grasslands Private Bag 11-008, Palmerston North 4442, New Zealand

**Arif Hasan Khan Robin**
Department of Genetics and Plant Breeding, Bangladesh Agricultural University, Mymensingh 02202, Bangladesh

**David F. Chapman**
DairyNZ, Lincoln University, Lincoln 7647, New Zealand

**François Gastal**
INRA, Unité Expérimentale Fourrages Environnement Ruminants, INRA-FERLUS, CS 80006, Lusignan 86600, France
INRA, Unité de Recherche Prairies et Plantes Fourragères, INRA-URP3F, CS 80006, Lusignan 86600, France

**Gilles Lemaire**
INRA, Unité de Recherche Prairies et Plantes Fourragères, INRA-URP3F, CS 80006, Lusignan 86600, France

**Jorge F. S. Ferreira, Monica V. Cornacchione, Xuan Liu and Donald L. Suarez**
US Salinity Laboratory, 450 W. Big Springs Rd., Riverside, CA 92507, USA

**Philippe Barre and Abraham J. Escobar-Gutiérrez**
INRA, UR4, Le Chêne RD 150, Lusignan 86600, France

**Lesley B. Turner**
IBERS, Aberystwyth University, Gogerddan Campus, Aberystwyth SY23 3EE, UK

**Jennifer W. MacAdam**
Department of Plants, Soils and Climate, Utah State University, 4820 Old Main Hill, Logan, UT 84322, USA

**Juan J. Villalba**
Department of Wildland Resources, Utah State University, 5230 Old Main Hill, Logan, UT 84322, USA

**Máximo Lorenzo**
INTA, Estación Experimental Balcarce, C.C. 276, Balcarce 7620, Argentina

**Silvia G. Assuero**
Laboratorio de Fisiología Vegetal, Facultad de Ciencias Agrarias, Universidad Nacional de Mar del Plata, C.C. 276, Balcarce 7620, Argentina

**Jorge A. Tognetti**
Laboratorio de Fisiología Vegetal, Facultad de Ciencias Agrarias, Universidad Nacional de Mar del Plata, C.C. 276, Balcarce 7620, Argentina
Comisión de Investigaciones Científicas de la Provincia de Buenos Aires, La Plata 1900, Argentina

**Racheal H. Bryant**
Faculty of Agriculture and Life Sciences, Lincoln University, Lincoln 7647, New Zealand

**Cory Matthew and John Hodgson**
Institute of Agriculture and Environment, Massey University, Private Bag 11-222, Palmerston North 4442, New Zealand

**Athole H. Marshall, Matthew Lowe and Rosemary P. Collins**
Institute of Biological, Environmental and Rural Sciences, Aberystwyth University, Gogerddan, Aberystwyth, Ceredigion SY233EE, UK

**Masahiko Hirata**
Department of Animal and Grassland Sciences, Faculty of Agriculture, University of Miyazaki, Miyazaki 889-2192, Japan

**Louis J. Irving**
Faculty of Life and Environmental Science, University of Tsukuba, 1-1-1 Tennodai, Tsukuba 305-8572, Japan

**Sila Carneiro da Silva**
Animal Science Department, University of São Paulo, E.S.A. "Luiz de Queiroz", Av. Pádua Dias, 11, CEP 13418-900, Piracicaba, SP, Brazil

**André Fischer Sbrissia**
Centro de Ciências Agroveterinárias, Santa Catarina State University, Avenida Camões, 2090, CEP 88520-000, Lages, SC, Brazil

**Lilian Elgalise Techio Pereira**
Animal Science Department, University of São Paulo, Faculdade de Engenharia de Alimentos e Zootecnia, Av. Duque de Caxias Norte, 225, CEP 13635-900 Pirassununga, SP, Brazil

# Index

www.ingramcontent.com/pod-product-compliance
Lightning Source LLC
Chambersburg PA
CBHW081215180926
51613CB00003B/249